MANUAL OF EXPERIMENTS
IN
APPLIED PHYSICS

E. V. SMITH
M.Sc., A.Inst. P.

*Department of Applied Sciences
Rugby College of Engineering
Technology*

LONDON
BUTTERWORTHS

ENGLAND:	BUTTERWORTH & CO. (PUBLISHERS) LTD.
	LONDON: 88 Kingsway, W.C.2
AUSTRALIA:	BUTTERWORTH & CO. (AUSTRALIA) LTD.
	SYDNEY: 20 Loftus Street
	MELBOURNE: 343 Little Collins Street
	BRISBANE: 240 Queen Street
CANADA:	BUTTERWORTH & CO. (CANADA) LTD.
	TORONTO: 14 Curity Avenue, 374
NEW ZEALAND:	BUTTERWORTH & CO. (NEW ZEALAND) LTD.
	WELLINGTON: 49/51 Ballance Street
	AUCKLAND: 35 High Street
SOUTH AFRICA:	BUTTERWORTH & CO. (SOUTH AFRICA) LTD.
	DURBAN: 33/35 Beach Grove

©
Butterworth & Co. (Publishers) Ltd.
1970

Suggested U.D.C. number: 53 (076·5)

SBN 408 57900 5 Standard
SBN 408 70009 2 Limp

Made and printed by offset in Great Britain by
William Clowes and Sons, Limited, London and Beccles

MANUAL OF EXPERIMENTS
IN
APPLIED PHYSICS

PREFACE

The core of this book is based upon experiments used by the author over a number of years for undergraduate teaching in Applied Physics and Engineering Science courses at both Honours and Pass Degree level.

About this core will be found other experiments, which it is hoped, are suitable on the one hand for courses of Higher National Certificate and Diploma level in Applied Physics and on the other for postgraduate level work, such as for example those in Nuclear Physics.

Any book of this nature will inevitably spark off the ever-ready arguments as to what constitutes Pure and what Applied Physics. Certainly a few of these experiments, in particular those in the Applied Optics section, will be familiar to many students undergoing Pure Physics courses but one might argue that they have an applied flavour, in their context here, which justifies their inclusion under this title. Of the bulk of the experiments it is hoped that little doubt exists of their applied nature.

The divisions of Physics of Engineering Materials and Mechanics are not bound by rigid frontiers and the exchange of an experiment from one section to the other could no doubt be justified according to the various points of view adopted in approaching the selection of experiments given.

Other selections will surely spring to the mind of the reader as perhaps more appropriate; the author trusts, that in spite of this, the volume will prove useful to students in both universities and technical colleges involved in Applied Physics and Engineering Science courses.

The experiments owe much to the courses in Applied Physics which ran and are running at the University of Salford (then the College of Advanced Technology) and the College of Engineering Technology, Rugby, as well as those of the Harwell Reactor School. The author is grateful of the facilities in these institutions which have made this work possible.

Thanks are due to colleagues both of the past and the present, for helpful discussions and advice upon the many problems that arose in the preparation of the manuscript. Particular thanks must go to the scientific advisers acting for the publishers who 'pruned' and

PREFACE

streamlined the work, as well as providing a number of ideas to supplement its presentation.

A large debt of gratitude must go to the authors of those books and papers quoted in the references given at the end of each experiment. These, in some cases, have provided background or additional information, but in others have formed the basis of the experiment itself; it is hoped that no acknowledgement has been omitted in this regard.

In the preparation of a manuscript of this sort a vast amount of typing is obviously necessary; I wish to thank my wife for her patience, her encouragement and her assistance with all stages of the script.

Finally the work is dedicated to the students without whose efforts in carrying out these experiments, this work could not have been written; their efforts are displayed in the experimental results presented.

E.V.S.

CONTENTS

Preface v

Section A. Electronics and Solid State Physics

A1. Electronics

A1.1   To Investigate the Characteristics of a Diode Valve ..   1
A1.2   Investigation of Triode Characteristics and the Use of the Triode in Amplification   17
A1.3   To Determine the Static Characteristic of a Pentode   24
A1.4   To Investigate the Frequency Response of a Resistance-Capacitance Coupled Amplifier..   28
A1.5   Use of the Cathode Ray Oscilloscope (*a*) To Determine the Sensitivity of the Instrument as a Voltage Measuring Device, and (*b*) as a Phase Comparator ..   33
A1.6   To Determine the Selectivity or Quality Factor Q of an *LC* Circuit   40
A1.7   To Calibrate the Tuning Capacitor of an Oscillator by Frequency Measurement   43
A1.8   To Verify that the Period of the Transition between the Two States of an Astable Multivibrator is Proportional to the Coupling Capacitances used in the Circuit   49
A1.9   To Investigate the Characteristics of Diode Rectifier Power Supplies   54
A1.10 To Determine the Characteristic Impedance and the Velocity Constant of a Transmission Line   59
A1.11 To Plot Equipotentials Using the Electrolytic Tank ..   69

A2. Solid State Physics

A2.1   To Determine the Electrical Conductivity of a *p*-type Semiconducting Silicon Specimen   78
A2.2   To Measure the Hall Coefficient, the Number of Charge Carriers per Unit Volume and the Carrier Mobility in a Doped Semiconductor   83
A2.3   To Determine the Energy Gap of a Specimen of Intrinsic Semiconducting n-type Germanium   89
A2.4   To Investigate the Characteristics of a Thermistor ..   95
A2.5   To Plot the Current/Voltage Characteristic of a Junction Diode   98

CONTENTS

A2.6	To Plot the Collector and Base Characteristics of a Junction Transistor	104
A2.7	To Measure the Current Gain of a Transistor Amplifier Connected in the Common Emitter Mode ..	108
A2.8	To Determine the Energy Gap of Germanium from Investigation of the Variation of the Saturation Current with Change in Temperature under the Conditions of Reverse Bias for a Junction Diode ..	113
A2.9	The Use of a Semiconducting Thermoelectric Material (Bismuth Telluride) to Determine the Dew Point by Observation of the Peltier Effect	118
A2.10	To Determine the Voltage-Current Characteristic of a Varistor	121
A2.11	To Investigate the Effect of Change in Temperature down to Liquid Nitrogen Temperature upon the Breakdown Potential of a Voltage Reference Diode..	124

Section B. Nuclear and Radiation Physics

B1. Nuclear Physics

B1.1	To Plot the Characteristic Curve of a Geiger Counter and To Determine the Resolving Time of the Counter	129
B1.2	The Statistics of Counting of the Particles Emitted from a Radioactive Source	136
B1.3	To Investigate the Absorption of Beta Rays from a Radioactive Substance by Aluminium..	140
B1.4	To Determine the Characteristics of a Scintillation Counter	144
B1.5	To Determine the Energy Resolution of a Scintillation Counter Using Gamma Rays from ^{137}Cs	149
B1.6	To Investigate the Absorption of Gamma Radiation by Lead and to Determine the Mass Absorption Coefficient and the Half Value Thickness of the Absorber	155
B1.7	To Determine the Slowing Down Length of Neutrons in Paraffin Wax	158
B1.8	To Measure the Diffusion Length of Neutrons in a Moderating Medium, e.g. Graphite or Paraffin Wax ..	168
B1.9	To Determine the Range of Alpha Particles Emitted from a Radioactive Source	175
B1.10	To Determine the Half-life of Neutron-irradiated Indium Foil	180
B1.11	To Investigate the Simultaneous Emission of Beta and Gamma Activity from Sodium-22 and to Verify the	

CONTENTS

Empirical Formula Relating the Mass Absorption Coefficient to the Maximum Energy in the Beta Spectrum 184

B2. Radiation Physics

B2.1 To Analyse an X-ray 'Powder' Photograph for a Substance having a Cubic Crystal Lattice 187
B2.2 The Use of the ASTM Index Cards to Identify Substances in a Mixture by the X-ray Powder Method .. 193
B2.3 Determination of the Linear Coefficient of Expansion of a Metal by Use of 'Back Reflection' X-rays .. 197
B2.4 To Determine the Lattice Spacing of a Single Crystal by the X-ray Single Crystal Rotation Method .. 200
B2.5 To Investigate the Absorption of X-radiation by Copper 205

Section C. Applied Optics, Spectroscopy, Photometry

C1. Applied Optics

C1.1 To Determine the Specific Rotation of a Solution of Sugar 213
C1.2 To Investigate the Relationship between Optical Activity and Wavelength 217
C1.3 To Determine the Linear Coefficient of Expansion of Brass by Fizeau's Interference Method 220
C1.4 To Verify Cauchy's Equation Using the Line Spectra of the Elements Neon and Mercury 223
C1.5 To Investigate the Faraday Effect and to Determine Verdet's Constant 229
C1.6 To Calibrate the Babinet Compensator 232
C1.7 The Use of the Rayleigh Refractometer to Determine the Refractive Index of Air at N.T.P. 236
C1.8 To Determine the Thickness of Thin Films Using Fraunhöfer Diffraction Fringes 240
C1.9 The Use of a Gas Laser and a Steel Rule to Measure the Wavelength of the Monochromatic Light from the Laser 247
C1.10 To Determine the Wavelength of the Light Emitted by the Neon Gas Laser and to Verify the Law Governing Interference from a Young's Double Slit 250

C2. Spectroscopy

C2.1 The Qualitative Analysis of a Sample of Brass Using the Hilger Medium Quartz Spectrograph 254

CONTENTS

C2.2	To Determine the Rydberg Constant	258
C2.3	Use of the Spectrophotometer to Verify Beer's Law Governing Absorption of Light by Coloured Media..	263
C2.4	To Determine the Magnitude of the Energy Gap in the Band Structure of Indium Antimonide from the Infrared Spectrum of the Semiconductor	267
C2.5	To Verify Lambert's Law by the Use of the Spekker Absorptiometer	271
C2.6	To Determine the Characteristic Density-Exposure Curve of a Photographic Film using the Hilger Microdensitometer	275

C3. Photometry

C3.1	To Plot the Polar Curve of a Filament Lamp and to Determine its Mean Spherical Intensity	279
C3.2	To Compare the Luminous Intensities of Two Lamps Emitting Light of Different Colours Using a Flicker Photometer	283
C3.3	To Investigate the Characteristic Exposure Curve of Photographic Paper	287

Section D. The Physics of Engineering Materials

D1. Metals

D1.1	To Observe the Plastic Deformation Processes in Lithium Fluoride Crystals	291
D1.2	To Investigate the Creep of Metals	294
D1.3	To Construct the Equilibrium Diagram Appropriate to a Series of Lead–Antimony Alloys of Varying Composition	299
D1.4	To Investigate the Effect of Variation in Carbon Content upon the Tensile Properties of Steel	305
D1.5	To Illustrate the Zone Refining Process using Naphthalene	311
D1.6	To Investigate the Characteristics of an Electrolytic Polishing Cell using a Copper or Brass Specimen ..	315
D1.7	Preparation of Magnetic Thin Films by Electrodeposition and Investigation of the Permeability of the Thin Film	318
D1.8	To Prepare a series of Nickel–Copper Alloys and to Verify that the Curie Point of the Alloys varies Linearly with Percentage Nickel Content	323
D1.9	To Grow Single Crystals of Cadmium from the Melt and to Confirm the Orientation of the Mode of Growth	327

CONTENTS

D1.10 To Investigate the Strain-Anneal Method of Single Crystal Growth using Aluminium 329

D2. Non-metals

D2.1 The Growth of Single Crystals of Hexamethylene Tetramine from the Vapour Phase 333
D2.2 To Determine the Difference Between the Principal Magnetic Susceptibilities of an Anisotropic Weakly Paramagnetic Crystal 336
D2.3 To Prepare a Ferrite Specimen and to Investigate the Variation of its Magnetic Properties with Composition 339
D2.4 To Investigate the Effect of Ferromagnetic Impurity in a Weakly Magnetic Specimen 346
D2.5 To Prepare Doped Cadmium Sulphide Photosensitive Devices and to Investigate their Performance .. 350
D2.6 The Electron Spin Resonance Method Used to Determine the Landé Splitting Factor 356
D2.7 Determination of the Gyromagnetic Ratio Using the Nuclear Magnetic Resonance Method 361
D2.8 To Observe Magnetic Domains in a Ferrimagnetic Garnet and to Use the Changes in the Domain Pattern to Interpret the Magnetization Hysteresis Cycle .. 364
D2.9 To Prepare a Thermistor and to Investigate the Variation of its Resistivity at Room Temperature with Varying Proportions of its Constituents 367
D2.10 To Investigate the Variation of Capacitance (and Power Factor) with Temperature Change in a Ferroelectric (Barium Titanate) 370

Section E. Applied Heat, Mechanics of Fluids and Solids

E1. Applied Heat

E1.1 To Investigate the Form of the Law Governing the Operation of a Thermocouple and to Determine the Neutral Temperature of the Couple 375
E1.2 To Calibrate a Chromel-Alumel Thermocouple by Means of a Cooling Method 379
E1.3 To Investigate the Variation in the Specific Heat of Graphite with Change in Temperature 386
E1.4 To Determine the Thermal Conductivity of a Bad Conductor in the Form of a Solid Cylindrical Shell .. 391
E1.5 The Radial Heat-flow Method Used to Determine the Thermal Conductivity of a Liquid 394

CONTENTS

E1.6 Ångström's Method Used to Measure the Thermal Diffusivity of a Copper Bar 400
E1.7 To Find the Temperature Distribution across the Cladding of a Nuclear Reactor Fuel Element by an Analogue Method, Using the Servomex Field Plotter 409
E1.8 To Determine the Latent Heat of Vaporization of Liquid Nitrogen from Vapour Pressure Measurement 416

E2. Mechanics of Fluids and Solids

E2.1 Cornu's Method to Determine the Elastic Moduli of a Perspex Specimen Using an Interference Technique 421
E2.2 To Investigate the Law Governing the Operation of a Strain Gauge and to Use the Gauge to Determine Young's Modulus for a Metal Beam 426
E2.3 To Analyse the Principal Stresses in a Loaded 'Araldite' Specimen Using the Photoelastic Technique .. 431
E2.4 To Determine the Porosity of Refractory Materials .. 437
E2.5 The Redwood No. 1 Viscometer Used to Investigate the Kinematic Viscosity of an Oil with Temperature Change 441
E2.6 The Use of the Servomex Field Plotter to Investigate the Variations of Pressure and Velocity over a Simulated Aerofoil 448
E2.7 To Determine the Speed of a Vacuum Pump and to Verify Gaede's Equation 456
E2.8 To Determine the Pumping Speed of a Vacuum System by the Steady State Method 460
E2.9 To Investigate the Variation of Pumping Speed of a Mercury-vapour Diffusion Pump with the Pressure in a Vacuum System 464
E2.10 To Investigate the Effect Upon the Pumping Speed of a Mercury-vapour Diffusion Pump, of Changes in the Supply of Power to the Heater of the Diffusion Pump 468
E2.11 To Measure a Low Pressure Using the McLeod Gauge 470
E2.12 To Investigate the Principles of Operation of the Pirani Gauge 474
E2.13 To Prepare a Thin Film of Silver by Evaporation in Vacuum 481

Section A. Electronics and Solid State Physics

Experiment No. A1.1

TO INVESTIGATE THE CHARACTERISTICS OF A DIODE VALVE

Introduction

The diode valve consists essentially of a filament and an anode sealed into an evacuated glass envelope. As the filament is heated electrons are emitted from its surface, and if the anode is maintained positive with respect to the filament a continual flow of electrons across the vacuum space will occur (the so-called 'anode current').

The aim of this experiment is to investigate the way in which this anode current varies with the voltage applied to the anode and further to verify the form of the laws governing the emission in two main portions of the resulting curve.

Theory

As the applied anode volts increase, more and more electrons are attracted to the anode until a condition of saturation is achieved [corresponding to the region BC on the curve (*Figure A1*) where the anode current is at its maximum].

This portion of the curve is governed by Richardson's equation that

$$I_{\text{saturation}} = AT^2 \exp\left(-\phi_0 e / kT\right)$$

where A = a constant
  e = fundamental charge on the electron
  exp = exponential constant = 2·718
  ϕ_0 = the work function of the material of the filament
  k = Boltzmann's constant
  I_{sat} = filament current density at saturation, and
  T = absolute temperature of the filament or cathode.

[The derivation of this law (see Appendix 1) is of value since it shows that the electron emission can be treated as analogous to the evaporation of molecules from the surface of a liquid.]

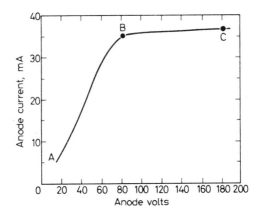

Figure A1. AB = Region of curve governed by 'three halves power law'. BC = Region governed by Richardson's law

Taking logarithms

$$\log_e I_a/T^2 = \log_e A - (\phi_0 e/kT).$$

Hence plotting $\log_e(I/T^2)$ against $1/T$ a straight line is obtained (*Figure A2*) whose slope gives $(\phi_0 e/k)$ whilst the intercept measures A.

In the region OA of the curve, however, we have a state of affairs where not all electrons are being drawn to the anode and a 'cloud' of electrons exists in the space between the filament and the anode. This 'space charge' of electrons tends to repel the electrons at the filament surface back into the emitter and thus limits the flow of current in the valve. Hence OA is known as the space charge limited region of the curve, and is governed by the 'three halves power law'

$$I_a = kV_a^{\frac{3}{2}}$$

where I_a = anode current
V_a = anode voltage
k = constant.

In order to verify the power 3/2, log *I* should be plotted against log *V* corresponding to readings taken in this region, when the slope

CHARACTERISTICS OF A DIODE VALVE

of the graph gives the power involved in the equation and the intercept gives k.

Derivation of the three halves power law (due to Child and Langmuir) is given in Appendix 2.

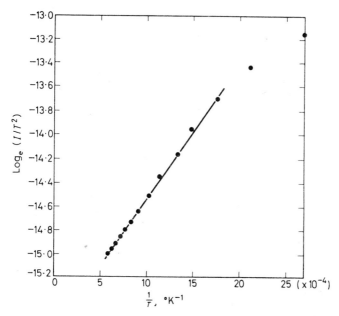

Figure A2. Graph of log (I/T^2) against $1/T$, the slope of which gives b and whose intercept gives A

Method

Set up the diode as shown in *Figure A3*. Apply an anode voltage of say 50 V and adjust the filament current to about 2·5 A, then allow the whole apparatus to 'warm up' for 10 minutes or so.

Using a fixed value of filament current (2·4 A in *Figure A1*) increase the anode voltage in 10 V steps from 0 to 100 V and tabulate the anode current at each step.

Repeat the procedure with the filament current set at lower values, e.g. 2·3 and 2·2 A in turn. Plot the characteristic as shown in *Figure A4*.

Compute log I and log V values corresponding to the region AB of the curve, and hence verify the general form of the 'three halves power law' (*Figure A8*)

$$I_a = k V_a^{\frac{3}{2}}$$

To verify Richardson's equation the temperature of the diode filament must be known in order to plot $\log (I/T^2)$ against $1/T$.

This may be obtained from data sheets provided by the valve manufacturers, but it is most instructive to estimate the absolute

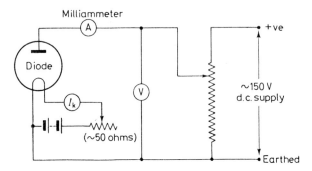

Figure A3. Circuit used to obtain diode characteristics

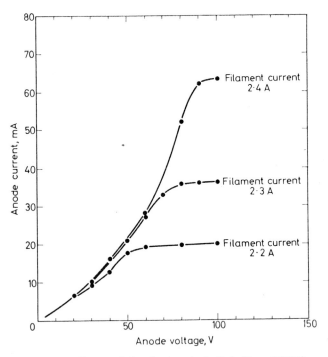

Figure A4. Characteristics of a thermionic diode (Type GRD 7)

temperature of the filament from measurements of its resistance. This may be done as a separate experiment by placing the filament in a bridge circuit as shown in *Figure A5*, where it forms one arm of

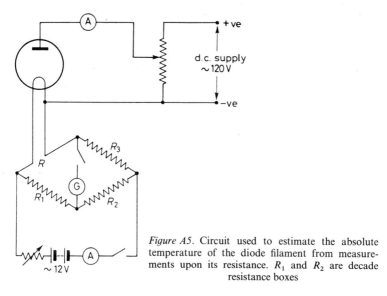

Figure A5. Circuit used to estimate the absolute temperature of the diode filament from measurements upon its resistance. R_1 and R_2 are decade resistance boxes

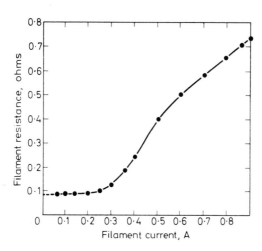

Figure A6. Graph showing the change in the resistance of the filament as the filament current varies

the Wheatstone network having resistances R_1, R_2, R (the filament resistance) and R_3.

Select a convenient value of R_2, pass a small current through the heater and adjust R_1 so that no deflection occurs on the galvanometer.

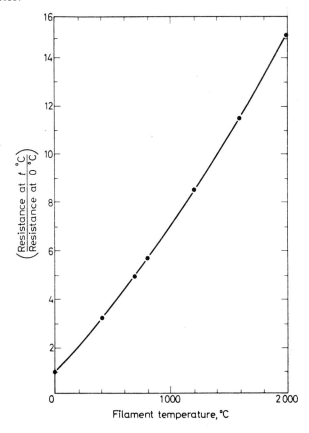

Figure A7. Graph of (R_t/R_0) against filament temperature

Tabulate heater current against the ratio (R_1/R_2). Increase the current in small steps and repeat the 'null' measuring procedure. Plot the results when a graph such as *Figure A6* should be obtained, from which by extrapolation the resistance of the filament at room temperature should be found. Then using

$$R_t = R_0\{1 + (5.238 \times 10^{-3})t + (0.7 \times 10^{-6})t^2 + (0.062 \times 10^{-9})t^3\}$$

(see third reference). R_0 the resistance at 0° C can be calculated and hence the ratio (R_t/R_0).

The resulting plot of R_t/R_0 against filament temperature in °K is shown in *Figure A7*.

A number of convenient values of temperature are chosen and the R_t/R_0 ratios taken to find the corresponding filament currents from the first graph (*Figure A6*). Hence log (I/T^2) and $1/T$ can be computed.

The results obtained are plotted as in *Figure A2* when a straight line results, verifying Richardson's equation. The slope of the graph is $-\phi_0 e/k$ from which the work function of the tungsten filament may be determined.

Discussion of Results

Three Halves Power Law. The slope of the log I_a against log V_a graph will give a result that may differ significantly from 1·5. This is due among other factors to:

(a) The theory assuming a plane emitter and a plane anode—both of infinite dimensions (Appendix 2)—when of course in practice we do not have this ideal geometry.

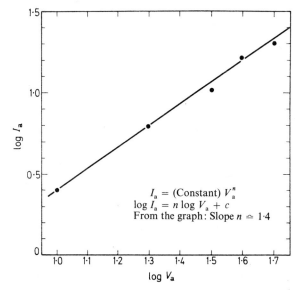

Figure A8. Graph to verify Langmuir's 'three halves power law'

(b) The cathode is assumed to be an equipotential surface and this is not necessarily so in this case.

Remember (1) that although the derivations of the two laws involved current density, actual currents are used in plotting the graph

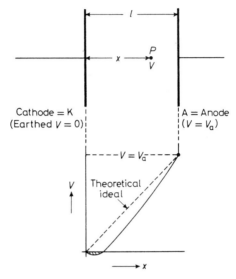

Figure A9. The idealized case of parallel plate electrodes and Langmuir's 'three halves power law'. Graph showing mode of variation of voltage between plates with distance of the point under consideration from the cathode. The shaded region represents the 'dip' that occurs due to the presence of a negative space charge near the cathode

and hence the constant A is a function of the emitting area of the filament, and (2) that I is in amps so that care must be taken to choose the correct system of units. The M.K.S. system recommends itself here and the values of Boltzmann's constant $k = 1.38 \times 10^{-23}$ joules $(°K)^{-1}$ and $e = 1.6 \times 10^{-19}$ coulombs are appropriate.

Further Work

The use of the thermionic diode in this way recommends itself for the verification of Stefan's fourth power law of radiation.

An ammeter is placed in the cathode circuit of the diode valve to measure the filament current and the Wheatstone bridge arrangement enables the resistance of the filament to be computed (*Figures A6* and *A7*).

CHARACTERISTICS OF A DIODE VALVE A1.1

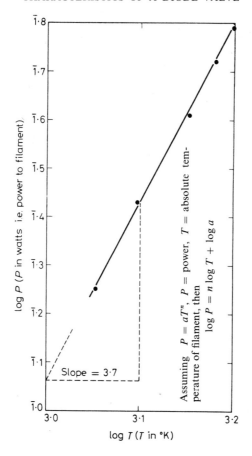

Power P to filament, Watts	Temperature T, °K	log P	log T
0·177	1 130	$\bar{1}$·2480	3·053
0·27	1 250	$\bar{1}$·4314	3·097
0·41	1 420	$\bar{1}$·6128	3·152
0·53	1 520	$\bar{1}$·7202	3·1818
0·6	1 580	$\bar{1}$·7782	3·199

Figure A10. Graph to verify Stefan's law

The accuracy obtained depends very strongly upon the graphs of resistance R_t against filament current and R_t/R_0 against filament temperature t, computed using Langmuir's expression. Whilst to a first approximation this latter graph appears to be a straight line an accurate computer calculation shows that the true graph deviates markedly from linearity (*Figure A7*). Small errors in R_t/R_0 give quite sizable inaccuracies in the estimate of the filament temperature and then the resulting $\log P$ against $\log T$ graph (*Figure A10*) may display points well away from the 'best fit' line.

An alternative method of presentation may be advisable in this case and plotting power P against T^4 is recommended (*Figure A11*).

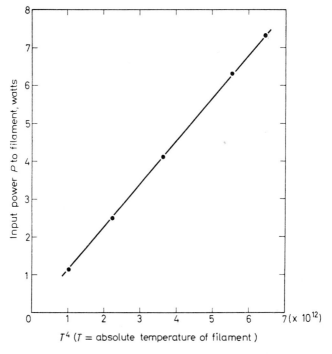

Figure A11. Verification of Stefan's law of radiation using a thermionic valve

The result shown does not display the accuracy that it at first suggests, since the valve emitter is not a true black-body radiator nor is all the input energy utilized in emission of radiation and therefore one would expect the index to be greater than 4 and not below it as the results imply.

The brief theory of this experiment to verify the Stefan–Boltzmann law is given in Appendix 3.

Using a retarding technique it is possible, also, to verify the law governing the distribution of velocities of the electrons emitted thermionically.

To do this a triode may be wired in such a way that the anode is used as the retarding electrode and a *negative* voltage V_R applied to it (*Figure A12*). The grid is used as the positive electrode (i.e. the

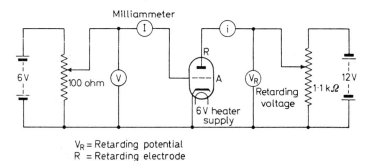

V_R = Retarding potential
R = Retarding electrode
A = Effective anode

Figure A12. Circuit used to investigate the law concerning the velocity distribution of thermionically emitted electrons. N.B. A triode is used but the anode is in fact held *negative* to provide the retarding potential, and the grid *positive*, so that I is effectively the anode current of a diode arrangement

anode of a diode arrangement) and the current I plotted against V_R for several different retarding voltages.

Schottky showed that the appropriate law was of the approximate form

$$I = (\text{constant}) \exp(-eV_R/kT)$$

whence log I against V_R should be a straight line. A typical graph verifying this is shown in *Figure A13*.

As the values of V_R increase the effects of space charge become significant and the graph is no longer linear as one enters the 'three halves power law' region.

Remembering that the current

$$I = Nev \quad \text{and} \quad \frac{mv^2}{2} = eV_R$$

where N = the number of electrons having velocity v
  e = the fundamental charge on the electron
  m = the mass of the electron
it is seen that this provides a method of investigating the electron velocity distribution.

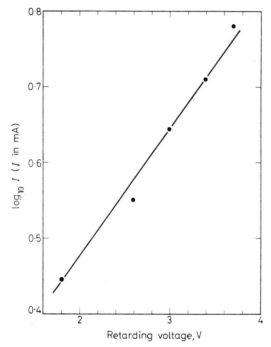

Figure A13. Graph of the logarithm of current against retarding voltage used to investigate the velocity distribution of thermionically emitted electrons

Note that the slope of the graph of *Figure A13* ($= 0.4343e/kT$) enables the absolute temperature T of the filament to be found.

Reading and References

More detailed treatment of thermionic emission, Richardson's equation and the three halves power law may be found in *Electricity and Magnetism*, vol. 1, by J. H. Fewkes and T. M. Yarwood (University Tutorial Press, 2nd edition, 1965).

CHARACTERISTICS OF A DIODE VALVE   A1.1

The diode itself is dealt with in *The Services Textbook of Radio*, vol. 3, *Electronics* (H.M.S.O. 1955).

The law quoted here governing variation of resistance is taken from *Advanced Practical Physics for Students* by B. L. Worsnop and H. T. Flint (Methuen, 1951).

Background reading upon Stefan's law may be found in *Thermal Physics* by P. M. Morse (Benjamin, 1965) or in *Heat and Thermodynamics* by J. K. Roberts and A. R. Miller (Blackie, 1958).

An alternative method of verifying Stefan's law using a filament lamp is given in *Intermediate Practical Physics* by T. M. Yarwood (Macmillan, 1955).

The law relating current I to retarding voltage V_R in the method of investigation of electron-velocity-distribution is due to Schottky, W. *Annln. Phys.* 1914, vol. 44, p. 1011.

APPENDIX 1

Derivation of Richardson's Law

The cloud of electrons may be treated as analogous to a gas and the laws of thermodynamics, in particular the Clausius–Clapeyron equation, applied to it.

Let the volume of the electrons in free space $= V_s$, let the volume of the electrons in the metal $= V_m$, let the energy absorbed in 'evaporation' of the electron from the surface into the space charge region $= W$, then by the Clausius–Clapeyron equation

$$W = T\frac{dP}{dT}(V_s - V_m)$$

where P represents pressure and T the absolute temperature.

In this case $\qquad V_s \gg V_m$

hence neglecting V_m and putting $V_s = V$,

$$W = VT\frac{dP}{dT}.$$

Now let $\phi =$ work function of the electron. Then energy required to remove N electrons from the surface of the emitter $= N\phi$.

Energy is also required to add electrons to the cloud of electrons in the space charge region at pressure P.

Therefore $\qquad W = (N\phi + PV) = VT\dfrac{dP}{dT}$

ELECTRONICS AND SOLID STATE PHYSICS

but $\quad PV = NkT$ where $k =$ Boltzmann's constant,

or $\quad V = \dfrac{NkT}{P}$

hence $\quad W = N(\phi + kT) = \dfrac{NkT^2}{P}\dfrac{dP}{dT}.$

Rearranging

$$\dfrac{dP}{dT} = \dfrac{P(\phi + kT)}{kT^2} \quad \text{or} \quad \int\dfrac{dP}{P} = \int\left(\dfrac{\phi}{kT^2} + \dfrac{1}{T}\right)dT + C$$

whence

$$\log_e p = \int\left(\dfrac{\phi}{kT^2}\right)dT + \log T + C \quad \text{or} \quad P = C_1 T \exp\left(\int \phi\, dT/kT^2\right)$$

assuming n electrons cm^{-3}

$$P = nkT$$

hence $\quad n = \dfrac{P}{kT} = \left(\dfrac{C_1}{k}\right)\exp\left(\int \phi\, dT/kT^2\right).$

Considering the number of electrons in the cloud crossing an arbitrary unit area per second (n_1) then n_1 may be assumed (1) proportional to n also (2) proportional to the r.m.s. velocity $\bar{c}$

hence $\quad n_1 = bn\bar{c}\quad$ where b is a constant

but from the kinetic theory the kinetic energy of the electrons

$$m\bar{c}^2 = \tfrac{3}{2}kT \quad \text{i.e.} \quad \bar{c} = \left(\dfrac{3kT}{2m}\right)^{\tfrac{1}{2}}$$

hence $\quad n_1 = BnT^{\tfrac{1}{2}}\quad$ where B is constant

$$\therefore n_1 = B\left[\left(\dfrac{C_1}{k}\right)\exp\left(\int \phi\, dT/kT^2\right)\right]T_1 = AT^{\tfrac{1}{2}}\exp\left(\int \phi\, dT/kT^2\right)$$

where A is a constant. Assuming that ϕ is a function of the absolute temperature, in particular that

$$\phi = \phi_0 + \tfrac{3}{2}kT$$

CHARACTERISTICS OF A DIODE VALVE   A1.1

where ϕ_0 represents the work function at $0°$ K

$$\int \frac{\phi \, dT}{kT^2} = \phi_0 \int \frac{dT}{kT^2} + \frac{3}{2} \int \frac{dT}{T} = \frac{-\phi_0}{kT} + \frac{3}{2} \log_e T.$$

hence $\quad n_1 = AT^{\frac{3}{2}} \exp(\phi_0/kT) \exp(\frac{3}{2} \log_e T)$

but $\quad\quad\quad\quad \exp(\frac{3}{2} \log_e T) = T^{\frac{3}{2}}$

$$\therefore n_1 = A_1 T^2 \exp(-\phi_0/kT)$$

hence $\quad\quad\quad I = AT^2 \exp(\phi_0/kT).$

APPENDIX 2

Although the rate of emission of electrons in the thermionic diode depends on Richardson's law, the rate at which they are collected by the anode depends on the anode potential. At voltages less than that producing saturation a negative cloud of electrons forms near the cathode, which as we have seen 'limits' the anode current.

To determine the three halves power law which governs this state of affairs in the valve, consider a voltage V between cathode and anode. Theoretically the graph of variation of voltage with the position x of the electron between the electrodes is a straight line (shown dotted in *Figure A9*) but in fact a dip occurs due to the effect of the negative space charge and at the turning point A

$$\frac{dV}{dx} = 0.$$

Since this region is very near the cathode we may assume approximately also that $V = 0$ at $x = 0$. By Poisson's equation

$$\frac{d^2 V}{dx^2} = 4\pi\rho \quad \text{where } \rho \text{ is the charge per unit volume}$$

hence if i = current per unit area

$\quad\quad i = \rho v \quad$ where $v =$ velocity of the electrons,

then $\quad\quad \dfrac{d^2 V}{dx^2} = \dfrac{4\pi i}{v} \quad$ but $\quad v = \left(\dfrac{2V_e}{m_0}\right)^{\frac{1}{2}}$

from consideration of kinetic and potential energy hence

$$\frac{d^2 V}{dx^2} = 4\pi i \left(\frac{m_0}{2V_e}\right)^{\frac{1}{2}}$$

15

ELECTRONICS AND SOLID STATE PHYSICS

We wish to convert this to terms of dV/dx and V

$$\frac{d^2V}{dx^2} = \frac{d}{dx}\left(\frac{dV}{dx}\right) = \frac{d}{dx}(U) \quad \text{where } U = \frac{dV}{dx}$$

but

$$\frac{dU}{dx} = \frac{dU}{dV}\frac{dV}{dx} = U\frac{dU}{dV} = \frac{1}{2}\frac{d}{dV}(U^2)$$

hence

$$\frac{d^2V}{dx^2} = \frac{1}{2}\frac{d}{dV}\left(\frac{dV}{dx}\right)^2$$

and

$$\therefore \frac{d}{dV}\left(\frac{dV}{dx}\right)^2 = 8\pi i \left(\frac{m_0}{2V_e}\right)^{\frac{1}{2}} = \left\{8\pi i \left(\frac{m}{2e}\right)^{\frac{1}{2}}\right\} V^{-\frac{1}{2}}.$$

Then by integrating

$$\left(\frac{dV}{dx}\right)^2 = 2\left\{8\pi i \left(\frac{m}{2e}\right)^{\frac{1}{2}}\right\} V^{\frac{1}{2}} + \text{constant } c$$

but at $x = 0$, $V = 0$ and $dV/dx = 0$, i.e. $c = 0$, hence

$$\left(\frac{dV}{dx}\right) = 4(\pi i)^{\frac{1}{2}}\left(\frac{m}{2e}\right)^{\frac{1}{4}} V^{\frac{1}{4}}.$$

Separating variables

$$\int_0^{V_{max}} \frac{dV}{V^{\frac{1}{4}}} = \left\{4(\pi i)^{\frac{1}{2}}\left(\frac{m}{2e}\right)^{\frac{1}{4}}\right\} \int_0^l dx$$

$$\therefore \frac{4}{3}\{V_{max}^{\frac{3}{4}}\} = \left\{4(\pi i)^{\frac{1}{2}}\left(\frac{m}{2e}\right)^{\frac{1}{4}}\right\} l$$

and squaring

$$V^{\frac{3}{2}} = \frac{9}{16}\left\{16\pi i \left(\frac{m}{2e}\right)^{\frac{1}{2}}\right\} l^2 \quad \text{or} \quad i = \left(\frac{\text{constant}}{l^2}\right) V^{\frac{3}{2}}.$$

APPENDIX 3

The Stefan–Boltzmann law governing radiation from a black body may be written

$$P_1 = \sigma(T^4 - T_0^4)$$

where P_1 = power radiated per unit area of emissive surface (i.e. the energy radiated per unit area per second)
  σ = Stefan's constant
  T = temperature of the filament in °K
  T_0 = temperature of the surroundings.

If it is assumed that all power supplied to the valve is utilized and that T_0 the absolute temperature of the surroundings may be neglected in comparison with T,

$$P_1 = \frac{I^2 R}{(\text{Area})} = \sigma T^4 \quad \text{or} \quad I^2 R = (\text{constant})T^4$$

where I = current through the filament and R = resistance of the filament.

R may be determined from the measurements as indicated, for the graph of *Figure A6* enables R_t/R_0 and R_0 to be found corresponding to a particular filament current and the temperature of the filament corresponding to chosen values of R_t/R_0 can be found from the second graph (*Figure A7*). Hence the power $P = I^2 R$ may be computed. Then assuming $P = A\sigma T^4 = aT^4$ where A = effective emitting area of filament, and a is a constant

$$\log P = n \log T + \log a$$

and plotting log P against log T as abscissa yields a straight line whose slope $n \simeq 4$ (*Figure A10*).

Experiment No. A1.2

INVESTIGATION OF TRIODE CHARACTERISTICS AND THE USE OF THE TRIODE IN AMPLIFICATION

Introduction

Essentially the triode has a cathode with a control grid close to it, and an anode at considerable potential above that of the cathode.

A small negative potential on the grid with respect to the cathode can counteract the effect of the high positive potential on the (more distant) anode and the relationships between anode current, grid voltage and anode voltage are known as the triode characteristics.

Theory

The graph of the anode current I_a against anode voltage V_a is known as the anode characteristic and that of anode current against grid voltage V_g as the grid (or mutual) characteristic. From these characteristics may be found the 'anode slope resistance' r_a, the amplification factor μ and the mutual conductance g_m of the valve.

$r_a = \partial V_a / \partial I_a$ measured at the straight line region of the graph with V_g constant

$$\mu = \frac{\delta V_a \text{ to produce given change in } I_a}{\delta V_g \text{ to produce same change in } I_a}$$

and $g_m = \delta I_a / \delta V_g$ for a constant value of anode voltage.

When the valve has no resistive load in the anode lead the resulting graphs are known as 'static' characteristics but when the anode current flows through a sizable resistance the graphs are significantly altered and are known as the dynamic characteristics of the valve.

It should be noted that even when no input signal is applied to the grid there are still significant values of anode current, anode voltage and grid voltage and this particular point on the characteristic curve is known as the quiescent point. The determination of this point is essential to the proper analysis of an amplifier circuit and to do so a 'load' line must first be drawn.

If it is assumed that the grid bias is so strongly negative that no anode current flows through the load resistance R, the voltage across the resistance is zero and hence the voltage across the valve is equal to the anode volts.

This point A is shown on the anode characteristics of *Figure A14*. If now it is assumed that the grid is so positive that there is perfect conduction across the valve, i.e. effectively the cathode and anode are 'electrically tied', then the voltage across the valve is zero and the anode current is given by the ratio (anode voltage/load resistance).

Thus for *known chosen* values of anode voltage and load resistance the point B (*Figure A14*) may be plotted. Joining points A and B gives the 'load' line and for a selected standing bias on the grid, the intersection of this line with the (I_a/V_a) characteristic gives the quiescent point Q; the corresponding values of anode voltage and anode current are known as the 'operational' or operating-point values.

In addition to the quiescent point, the 'cut-off' and saturation points, which define the limits of amplification, are important. With a strong negative potential on the grid the anode current is zero, i.e. 'cut-off' occurs and the minimum grid voltage to do this lies on the characteristic passing through A and shown dotted in *Figure A14*.

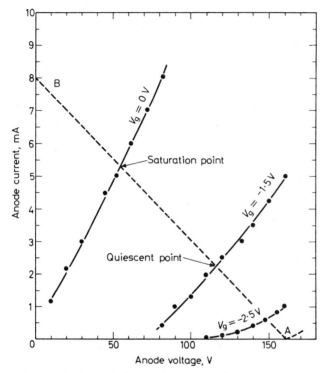

Figure A14. The anode characteristics of a triode. AB represents the 'load' line

When the input voltage on the grid is zero the valve is in a condition corresponding to optimum conduction and is said to be 'saturated'. Thus where the load line crosses the anode characteristic fixed by $V_g = 0$ we have the saturation point.

Method

Set up the circuit as shown in *Figure A15*. Pass the recommended current through the filament of the valve and adjust the rheostat to give about 2 V (negative) on the grid.

Keeping this value constant increase the anode voltage in convenient steps of say 10 V and note the anode current flowing in the milliammeter.

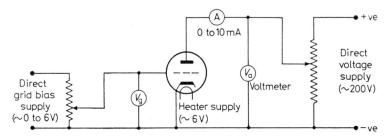

Figure A15. Circuit used in the determination of the static characteristics of a thermionic triode

Repeat the procedure for grid voltages $V_g = 0$ and $V_g = 3$ V. Plot the static anode characteristics as shown in *Figure A16* and from them calculate the valve parameters, μ, r_a and g_m.

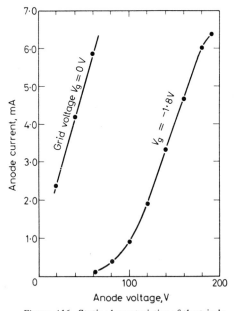

Figure A16. Static characteristics of the triode

Now choose a fixed convenient value of anode volts V_a and measure the anode current as the grid voltage is varied in steps of about 0·1 V from −2 V towards zero. Repeat the procedure for different values of anode voltage to obtain the mutual characteristics of the valve (*Figure A17*). From these characteristics, values of the

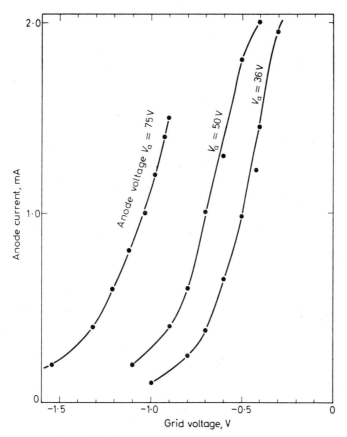

Figure A17. The mutual characteristics of a triode valve

valve parameters can again be calculated and compared with those obtained from anode characteristics.

Now introduce a large resistance (2–10 kΩ) into the anode lead (*Figure A18*) and in accordance with the theory the operating point,

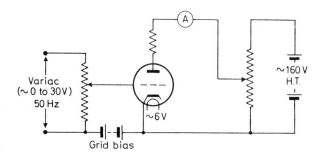

Figure A18. Circuit used to obtain the dynamic characteristics of the triode

the saturation point, the load line and the quiescent point for the chosen values of anode voltage, grid voltage, and load should be plotted on the anode characteristics.

Discussion of Results

Since two estimates of μ, r_a and g_m are made from separate graphs a good indication of the accuracy of the experiment is given.

Further Work

Using the valve to amplify alternating current applied to the grid, the amplification may be measured directly using the cathode ray oscilloscope and a circuit such as shown in *Figure A19*.

With the lead E connected to the earth point of the oscilloscope and A to the Y plates, a vertical trace is obtained on the screen, whose length is proportional to the alternating voltage across AB.

If the Y terminal is now disconnected and connected to F, a line of different length is obtained, but by adjustment of the rheostat the alternating potential upon the grid may be brought to such a value that the resulting output trace on the screen is of the same length as in the first case, so that the amplification M of the stage can be found since with point A connected to the Y plates and E to the earth

Length of trace $l \propto$ alternating voltage across AB.

TRIODE CHARACTERISTICS A1.2

With F connected to the Y plates

$l_1 \propto$ voltage across FE ($= M \times$ voltage across AB).

Readjust the rheostat to give trace of length as before.

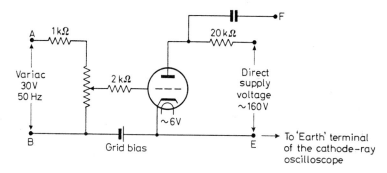

Figure A19. To display the dynamic characteristics of a triode valve upon the cathode ray oscilloscope. (1) First connect point A to the Y plates of the c.r.o. (2) Then connect point F to the Y plates

Then l is now $\propto$ output voltage V_{FE} ($\propto R_{AB}$ above)

but $V_{FE} = M \times$ (voltage across BD)

and Voltage across BD $\propto$ Resistance BD

hence $$M = \frac{\text{Resistance across AB}}{\text{Resistance BD}}.$$

Then using the formula

$$M = \frac{\mu R_L}{r_a + R_L} = \frac{\text{Resistance across AB}}{\text{Resistance across BD}}$$

$$\mu = \frac{(r_a + R_L) \times \text{Resistance across AB}}{R_L \times \text{Resistance across BD}}.$$

The result of this calculation should confirm the value of the amplification factor already obtained and enable strengthened conclusions to be drawn concerning the accuracy of the method.

Remember that the accuracy is limited by the sensitivity of the cathode ray oscilloscope (see Experiment No. A1.5) and by the care taken in adjusting the trace length in the second part to be the same as that in the first measurement made in the experiment.

Reading and References

Background reading upon the triode may be found in *Electronics* by P. Parker (Arnold, 1950) and in *Electronics for Scientists* by H. V. Malmstadt, C. G. Enke and E. C. Toren, jnr. (Benjamin, 1963).

Experiment No. A1.3

TO DETERMINE THE STATIC CHARACTERISTIC OF A PENTODE

Introduction

At frequencies of the order of 1 MHz the amplification obtained from a triode falls off (see Experiment No. A1.2) and instability results due to 'feedback' of energy from the anode to the grid via

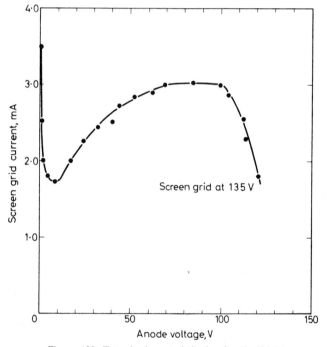

Figure A20. Tetrode characteristic showing the 'kink'

STATIC CHARACTERISTIC OF A PENTODE   A1.3

the comparatively low reactance path provided by the high interelectrode capacitance existing between them at high frequencies.

To overcome this effect a 'screen grid' is placed between the main control grid and the anode, this arrangement constituting the conventional tetrode, but this introduces a further complication, namely that secondary electrons emitted from the anode are attracted to the screen grid and produce a 'kink' in the resulting valve characteristic (*Figure A20*) giving rise to anomalous variation of anode current and a distorted output.

Theory

To overcome this last defect a third grid, the suppressor grid, is introduced. It is placed between the screen grid and the anode and is maintained strongly negative with respect to the anode by being connected to the cathode of the valve. Thus all the secondary electrons are attracted back to the anode and no 'kink' will appear in the curves. Electrons are slowed down between the screen grid and suppressor but accelerate again between suppressor and anode.

Method

Set up the circuit as shown in *Figure A21* with the suppressor grid connected to the cathode. The voltage on the screen grid should

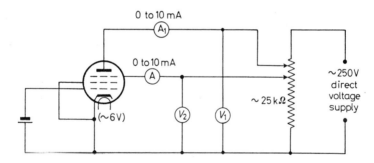

Figure A21. Circuit used to determine pentode characteristics

be set to about 80 V and the anode voltage increased in suitable steps from 0 to say 300 V, noting the anode current and screen grid current in each case.

Repeat the procedure with the control grid voltage at 1, 2 and 3 V in turn.

Finally reset the screen grid voltage first at 100 V then at 60 V and carry out the same sequence of readings in each case. Plot the anode and mutual characteristics as for the triode and also anode current variation with anode voltage for the various values of screen grid voltage (typical results for a 6J7 valve are shown in *Figures A22* and *A23*).

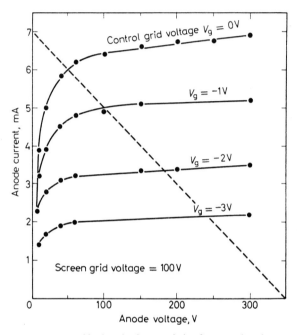

Figure A22. Anode characteristic of a pentode valve

Calculate the anode slope resistance, the amplification factor and the mutual conductance in the same way as for the triode (Experiment A1.2).

Discussion of Results

Note the anode resistance of a pentode is so large that generally the approximation

$$\text{Gain } M = g_m r_a$$

may be used, where g_m = mutual conductance and r_a = anode slope resistance.

The results obtained may be compared with the published values (see references).

A Word of Caution. The anode and screen grid are both at high positive potential—both collect electron current from the cathode. If the potential from the anode is removed this current will flow

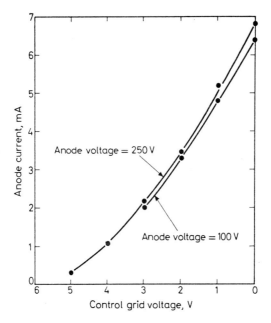

Figure A23. The mutual characteristics of a pentode valve

entirely to the screen which may as a result become overheated. For this reason care should be taken to remove the potential from BOTH screen grid and anode simultaneously (it is for this reason that comparatively low values of screen grid voltage are used when anode voltage is raised from zero to 300 V in the first part of this experiment).

If the screen grid potential is fixed, as is often done by connecting to the H.T. supply through a suitable resistance, the screen current at the quiescent point is required in order to determine the value of the resistance necessary. This may be obtained from the screen grid characteristics.

As was the case with the triode the valve parameters may be obtained from both anode and mutual characteristics and thus a reliable estimate of the accuracy of the results made. In considering any discrepancy between the values obtained and published values, remember that somewhat artificial operating conditions have been employed here.

It should be noted that the characteristics plotted (*Figure A22*) show unequal 'spacing' corresponding to equal increments of change in the control grid voltage. This means that some distortion must be present in the output as can be verified by constructing the appropriate load line.

Further Work

Whilst the tetrode is now virtually obsolete, for the reasons stated above, it is instructive to 'strap' the screen and suppressor grids of the pentode together and use the valve as a tetrode. The 'kink' obtained in the resulting characteristics may be clearly observed (*Figure A20*).

Reading and References

Background reading upon the pentode, its operation and application, may be found in *Electronics from Theory to Practice* by J. E. Fisher and H. B. Gatland (Pergamon, 1966).

Published values of valve parameters may be found in the valve data books of the respective manufacturers or in the *International Radio Tube Encyclopaedia* B. B. Babani (Bernards, 3rd edition, 1958).

Experiment No. A1.4

TO INVESTIGATE THE FREQUENCY RESPONSE OF A RESISTANCE-CAPACITANCE COUPLED AMPLIFIER

Introduction

It is often found that a small signal after single stage amplification is still insufficient to be of practical use and it is then necessary to amplify further by passing the signal through a second stage of

FREQUENCY RESPONSE OF RC AMPLIFIER   A1.4

amplification. The overall gain is then the product of the two individual stage amplifications but the process poses the problem of eliminating the direct voltage from the first stage output as this would interfere with the operation of the second stage.

The usual method of doing this is by resistance-capacitance coupling where a 'coupling' capacitor C_c (*Figure A24*) is used to eliminate any direct voltage and only alternating voltage appears across the grid resistor R_g.

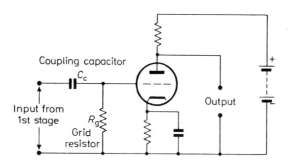

Figure A24. The coupling capacitor arrangement in a resistance-capacitance coupled amplifier. The coupling capacitor eliminates the direct voltage component so that only alternating voltage crosses the grid resistor

The aim of this experiment is to plot the frequency response curve for the amplifier and to note what changes take place when the values of the coupling capacitor and the grid resistor are varied.

Theory

In practice the gain of an amplifier cannot be constant for all frequencies but with any one design there is a 'bandwidth' over which the gain is reasonably constant in the mid-frequency range and is given by

$$M = \frac{-\mu R_L}{(r_a + R_L)}$$

where M = gain of the amplifier
 μ = amplification factor of the valve used
 r_a = anode slope resistance of the valve, and
 R_L = load resistance in the anode lead.

At low frequencies the gain decreases because the coupling capacitor and the grid resistor act as a voltage divider and as the frequency decreases the reactance of the capacitor increases in accordance with

$$X_c = \frac{1}{2\pi f C_c}$$

so that an increasing fraction of voltage appears across the coupling capacitor. The gain also falls at high frequencies due to the combined effects of the stray capacitances which arise (unintentionally) between the grid wire and the chassis, the grid and the cathode, and the grid and the anode. These capacitances act as a parallel combination and give a total 'shunt' capacitance C_s.

Hence the input voltage across C_s acts through a total shunt resistance consisting of the equivalent resistance of the first stage and the grid resistor R_g in parallel.

In this experiment the gain is given by

$$M = \frac{\text{Output voltage}}{\text{Input voltage}}$$

and the output voltage is measured on a cathode ray oscilloscope, the oscillator input voltage being kept constant.

Method

A suitable circuit is shown in *Figure A25* and with the amplifier operating under normal working conditions a constant input voltage from a beat frequency oscillator is fed to the amplifier.

Figure A25 shows a typical set of component values to give such a working arrangement. The frequency is set at say 20 Hz and the output voltage is read off as a trace-height upon an oscilloscope. If trace heights are used rather than direct voltage readings then the gain under normal working conditions at middle frequencies is given by

$$G = \frac{S \times h}{k V_{\text{input}}}$$

where S = sensitivity of c.r.o.
h = trace height
k = reduction factor of potentiometer

i.e. in diagram

$$\left(\frac{\text{1st resistance value}}{\text{sum of both resistances}}\right)$$

(or more conveniently directly as output voltage on a suitable voltmeter).

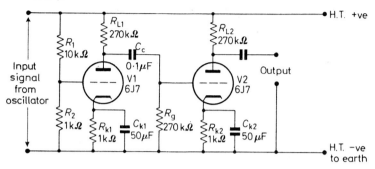

Figure A25. Circuit for investigating the response of a resistance-capacitance coupled amplifier, with typical component values for use with 6J7 valves

R_{L1}, Load resistor of 1st valve
R_{L2}, Load resistor of 2nd valve
C_c, Coupling capacitor
R_g, Grid resistor for 2nd valve

R_{k1}, Cathode resistor for 1st valve
R_{k2}, Cathode resistor for 2nd valve
R_1, R_2, Input potential divider network

The process is then repeated varying the frequency in convenient steps up to 10 kHz and a graph is plotted of output voltage (or trace height) against frequency. A result such as *Figure A26* curve (*a*) is obtained.

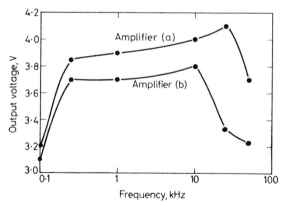

Figure A26. Graphs of the frequency response of a resistance-capacitance coupled amplifier

The value of the coupling capacitance is now reduced when according to the equation given in the theory the frequency at which the gain falls off on the low frequency side is higher than in the previous case (the product fC remaining constant).

This is evident in *Figure A26* curve (b).

To investigate high frequency response a capacitor is deliberately added (between output and earth) to the original circuit (*Figure A27*).

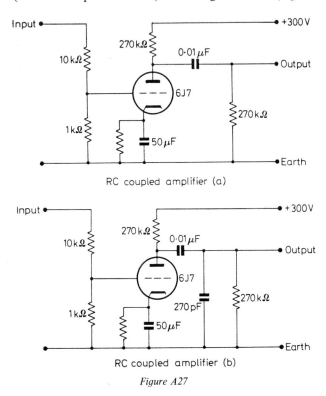

RC coupled amplifier (a)

RC coupled amplifier (b)

Figure A27

This represents extra shunt capacitance and again this increase causes the high frequency limit of linear response to be reduced in accordance with the theory (again, fC remaining constant).

Discussion of Results

The low and high frequency limits f_1 and f_2 should be calculated from

$$f_1 = \frac{1}{2\pi C_c R_g} \quad \text{and} \quad f_2 = \frac{1}{2\pi C_s R_L}$$

where C_c = coupling capacitance
       R_L = load resistance
       R_g = grid resistance
       C_s = shunt capacitance

and compared with those obtained from the graph. The discrepancies will enable typical design tolerances in fabricating amplifier circuits to be estimated.

Further Work

Not only should the effect of reduced capacitance at the low frequency end of the operating range of the amplifier be noted but observation made of its effect at high frequencies.

Reading and References

More detailed theory of amplifier circuits is given in *Electronics for Scientists* by H. V. Malmstadt, C. G. Enke and E. C. Toren, jnr. (Benjamin, 1963), p. 161, including a section on low and high frequency response.

Design data may be found in such texts as *Electronic and Radio Engineering* by F. E. Terman (McGraw-Hill, 4th edition, 1955).

Experiment No. A1.5

USE OF THE CATHODE RAY OSCILLOSCOPE (*a*) TO DETERMINE THE SENSITIVITY OF THE INSTRUMENT AS A VOLTAGE MEASURING DEVICE, AND (*b*) AS A PHASE COMPARATOR

Introduction

The cathode ray tube is familiar to all from its use in television. The theory below will show that the deflection of the electron beam on the screen is proportional to the voltage applied to the deflector plates.

ELECTRONICS AND SOLID STATE PHYSICS

Theory

(a) *Deflection Sensitivity*

Referring to *Figure A28* the electric field E between the deflector plates, whose separation is d, is given by

$$E = V/d$$

where V = potential difference across the plates. Hence

$$\text{force on the electron beam} = Ee = \frac{Ve}{d}$$

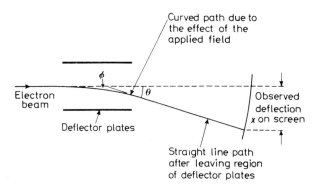

Figure A28. Deflection of beam of the cathode ray tube

where e = the charge on the electron and the acceleration of the electrons

$$a = \frac{Ve}{dm}$$

where m = mass of the electron, but if the initial velocity of the electron upon entering the region of the deflector plates is u then

$$u = \frac{l}{t}$$

or the time t taken to cross the region between the plates

$$t = \frac{l}{u}$$

34

USE OF THE C.R.O.

In this time the electron acquires a velocity v at right angles to its original direction of motion such that

$$v = at = \left(\frac{eV}{md}\right)\frac{l}{u}$$

Since from the diagram

$$\tan\theta = \frac{v}{u} = \frac{y}{L}$$

then

$$y = \frac{eVl}{md}\frac{L}{u^2}$$

where y represents the deflection of the spot on the screen of the tube (along the abscissa).

Again for a given accelerating voltage, u is constant and thus

$$y = \text{(constant)}\, V.$$

(b) *Phase Comparison*

If the basic alternating voltage to the applied circuit of *Figure A29*

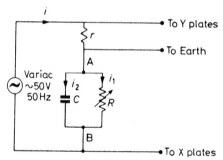

Figure A29. Circuit used to make phase comparison

is considered to be truly sinusoidal then the instantaneous charge on the capacitor C may be written

$$Q = Q_0 \sin\omega t$$

where $\omega = 2\pi \times$ frequency at the alternating source
t = time
Q = instantaneous value of charge
Q_0 = maximum charge

ELECTRONICS AND SOLID STATE PHYSICS

and since $V = Q/C$ the instantaneous voltage across AB becomes

$$V = \frac{Q_0 \sin \omega t}{C}$$

but the voltage across AB is given by $V = i_1 R$ hence

$$i_1 = \frac{Q_0 \sin \omega t}{RC} \quad \text{and} \quad i_2 = \frac{dQ}{dt} = \omega Q_0 \cos \omega t$$

hence

$$i = i_1 + i_2 = Q_0 \left\{ \frac{\sin \omega t}{RC} + \omega \cos \omega t \right\}$$

$$= Q_0 \left\{ \left(\frac{1}{CR} \right) \sin \omega t + \omega \cos \omega t \right\}$$

This may be rewritten as

$$i = Q_0 A \{ \sin \omega t \cos \delta + \cos \omega t \sin \delta \}$$

where

$$\cos \delta = \frac{1}{A(RC)}, \quad \sin \delta = \frac{\omega}{A} \quad \text{and} \quad \tan \delta = \omega CR$$

whence

$$i = AQ_0 \sin (\omega t + \delta)$$

and the voltage across the resistance R becomes

$$V_1 = AQ_0 R \sin (\omega t + \delta),$$

δ now representing the phase difference between the voltage across R and that across the resistance-capacitance combination (RC).

Using the circuit of *Figure A29* these voltages are now applied to the Y and X plates respectively of the cathode ray oscilloscope so that in general an ellipse is produced, from the shape of which the phase difference may be found in the following manner.

The trace is similar to that shown in *Figure A30* where θ, $2a$ and $2b$ have the connotations indicated.

At P the x deflection is given by

$$x = a \quad (\text{i.e. } \sin \omega t = 1)$$

and the y deflection by

$$y = b \sin (\omega t + \delta)$$

USE OF THE C.R.O. A1.5

since the voltage across the Y plates is $\hat{\delta}$ out of phase. Again
$$b \sin(\omega t + \delta) = b\{\sin \omega t \cos \delta + \cos \omega t \sin \delta\}$$
and it is seen that
$$\sin \omega t = 1 \quad \text{so that} \quad \cos \omega t = 0.$$

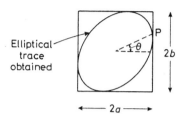

Figure A30. C.R.O. used as a phase comparator

Hence y in this case becomes
$$y\{b \sin(\omega t + \delta)\} = b \cos \delta + 0$$
so that
$$\frac{y}{x} = \frac{b \cos \delta}{a}$$
but
$$\frac{y}{x} = \tan \theta \quad \text{from the diagram}$$
hence
$$\cos \delta = \frac{a}{b} \tan \theta$$

and by carefully measuring a, b and the ratio $\tan \theta = y/x$, δ the phase angle between the two applied voltages may be calculated.

Method

(a) To determine the voltage sensitivity of the cathode ray oscilloscope. Set up the circuit as shown in *Figure A31* with a 200 kΩ potentiometer across an H.T. supply of about 120 V d.c.

Attach the variable tapping of the potentiometer to the X plates of the oscilloscope and the negative side of the supply to the 'earth' terminal of the instrument. Choose various voltages from 10 V upwards in steps of 10 V and note the deflection of the spot along the x axis. Plot a graph of x against the applied voltage as shown (*Figure A32*).

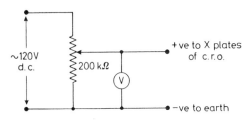

Figure A31. To use the cathode ray oscilloscope as a voltmeter

Repeat the observations for displacements along the y axis when the potentiometer tapping is removed from the X-plates terminal and connected to the Y plates.

(b) To calculate the phase difference between the applied voltages. Set up the circuit as shown in *Figure A29* using the 50 Hz supply through a variac and resistance r of the order of 5 kΩ.

Select a value of R, say, 15 kΩ and obtain the pertinent parameters of the ellipse in the way suggested above. Vary R and repeat the procedure for each chosen value in turn.

Discussion of Results

In measurement of voltage sensitivity there will in general be significant difference between that of the Y plates and that of the X plates although the graphs of applied voltage against deflection should be linear (*Figure A32*).

The results obtained for δ should be checked by direct calculation from the formula

$$\tan \delta = \omega CR$$

relating the phase angle to the chosen values of the components.

The resistance values should be accurately checked using a Post Office box and the capacitance by use of a Schering (or other suitable) capacitance bridge.

From the results obtained by the two independent methods, conclusions as to the accuracy of the method may be drawn.

Perhaps the weakest point in the procedure is introduced by the estimations of a, b, x and y. To trace the ellipse onto paper from which the measurements may be taken can be more satisfactory

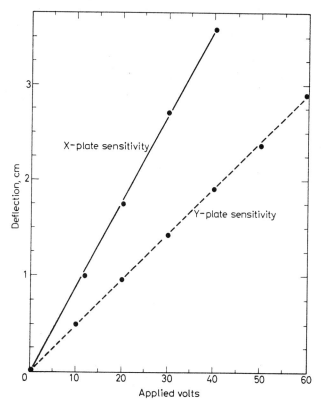

Figure A32. Graph showing sensitivity of c.r.o. used as a voltmeter

than reading off measurements from the screen, if great care is exercised.

Further Work

The use of the oscilloscope as a null indicator and in frequency measurement, as well as for many other purposes, is dealt with in *Oscilloscope Measuring Techniques* by J. Czech (Phillips Technical Library, 1965). This will provide ample suggestions for further work employing the instrument.

ELECTRONICS AND SOLID STATE PHYSICS

Reading and References

P. Parker's *Electronics* (Arnold, 1958) p. 60 discusses deflection sensitivity in more detail and the effect of various shapes of deflector plates upon it.

The operation of the cathode-ray oscilloscope controls should be given in the manufacturer's instruction booklet, but general notes upon this will be found in *Laboratory Physics* by J. H. Avery and A. W. K. Ingram (Heinemann, 1961), as well as advice upon building an oscilloscope in an article on 'The Oscilloscope and its Applications' by P. Cairns, which may be found in *Practical Electronics* for April 1967.

Experiment No. A1.6

TO DETERMINE THE SELECTIVITY OR QUALITY FACTOR Q OF AN LC CIRCUIT

Introduction

A circuit comprising inductance L and capacitance C has a natural frequency of oscillation determined by the equation

$$f_0 = \frac{1}{2\pi(LC)^{\frac{1}{2}}}$$

where f_0 represents the resonant frequency.

Forced oscillations can be imposed however, using a suitable oscillator but the resulting amplitude of the oscillations will depend upon the difference between the 'forcing' frequency of the oscillator and the natural frequency of the resonant circuit. When the two frequencies are equal the amplitude tends to infinity.

Theory

The tendency to infinity is quenched by the finite resistance of the circuit and the resulting graph of voltage across the LC circuit against the frequency is as shown in *Figure A33*.

The quality factor may then be defined as

$$Q = \frac{f_0}{f_2 - f_1}$$

SELECTIVITY OF AN LC CIRCUIT   A1.6

where f_1 and f_2 represent the frequencies at the points where the voltage = 0·707 times the maximum voltage (known as the 'half power points').

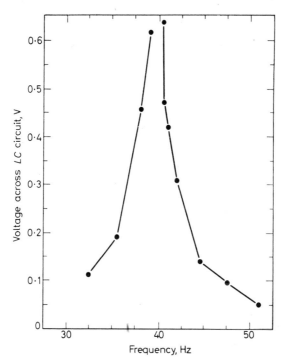

Figure A33. Frequency response curve of an *LC* combination

Method

An inductance L_1 (*Figure A34*) is connected across the oscillator and 'coupled' to the inductance L_2 which is itself in series with the variable capacitor C forming a series resonant circuit.

The voltage across the *LC* combination is then measured by the valve-voltmeter.

After fixing the generator voltage at about 5 V and the frequency at some convenient value, the capacitance is varied from say 100 pF to 1 200 pF in convenient steps and the corresponding valve-voltmeter readings of voltage across the *LC* circuit are noted. A graph of this voltage against capacitance is then plotted resulting in a curve such as shown in *Figure A35*.

Keeping the oscillator voltage the same as before, the capacitance is now fixed in the region of the value giving maximum voltage in

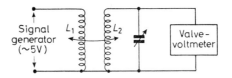

Figure A34. Circuit used to determine the selectivity Q of an *LC* combination. L_1 and L_2 represent coupled inductances

the first graph and the frequency of the oscillator output is varied from, for example, 5 to 100 Hz in convenient steps such as are indicated in *Figure A33*.

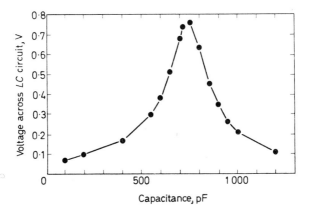

Figure A35. Determination of the selectivity of an *LC* circuit

The resulting resonance curve enables the selectivity of the series *LC* circuit to be determined.

Discussion of Results

The accuracy of the method depends to a large extent upon the closeness of the graphical points (i.e. the smallness of the chosen frequency intervals) taken, about the peak of the resonance curve and the 'half power points'.

Further Work

The inductance L_2 of the above circuit has of course inherent resistance associated with it but the experiment may be repeated placing a small resistance of the order of 5 Ω in series with the inductance to note the effect it has upon the curve by damping the oscillations.

It is also possible to repeat the experiment for a parallel *LCR* circuit and to compare the results with those obtained above.

The experiment may be extended still further by finding the resonant frequency corresponding to a number of different values of *C*. Plotting the capacitance against this resonant frequency value results in a straight line of slope $1/(4\pi^2 L)$, verifying the fundamental law governing resonance and providing a measure of the inductance used in the circuit.

Reading and References

The background theory of the quality factor of a resonant circuit is treated in *Degree Physics* (Part V) *Electricity and Magnetism* by C. J. Smith (Arnold, 3rd edition, 1963) or in *Electricity and Magnetism*, vol. 1, by J. H. Fewkes and T. M. Yarwood (University Tutorial Press, 2nd edition, 1965).

Experiment No. A1.7

TO CALIBRATE THE TUNING CAPACITOR OF AN OSCILLATOR BY FREQUENCY MEASUREMENT

Introduction

A variable capacitor with its dial reading simply in degrees is used as the tuning capacitor of a Hartley type series-fed oscillator (*Figure A36*). The aim here is to draw a calibration curve for the capacitor enabling the true capacitance to be determined from the dial reading.

Theory

When oscillations are produced in the circuit we know that

$$\omega_0^2 LC = 1$$

where ω_0 = resonant pulsatance
 L = inductance in the circuit
 C = capacity in the circuit

also

$$\omega = 2\pi f \quad (f = \text{frequency})$$

i.e.

$$4\pi f_0^2 LC = 1 \quad \text{or} \quad C = \frac{1}{4\pi f_0^2 L}.$$

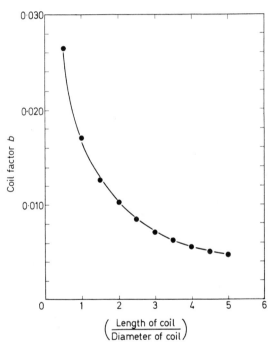

Figure A36. Graph showing the variation of the factor b (used in the determination of the inductance of a solenoid) with the ratio of (length of coil/diameter of coil) based on values given in *Radio Engineers Handbook* by F. E. Terman (McGraw-Hill, 4th edition, 1955)

Now the inductance L is a function of the geometry of the coil and in fact

$$L = bN^2 d \quad \text{(in micro-henries)}$$

where N = number of turns on the coil
 d = diameter of the coil
 l = the length of the coil
and b = a factor depending upon the ratio of the length to the diameter of the coil.

(A graph from which b may be determined (*Figure A36*) is shown, based on data given in the last reference.)

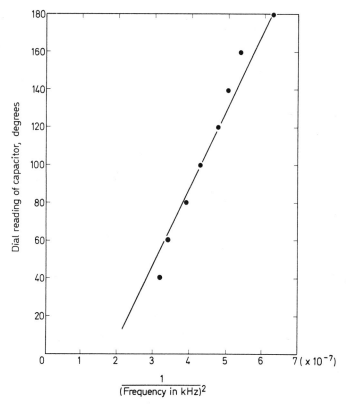

Figure A37. Capacitor calibration—graph of capacitor dial reading against $1/(\text{frequency})^2$

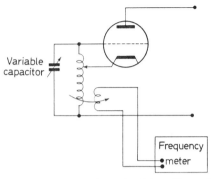

Figure A38. Hartley series-fed oscillator circuit

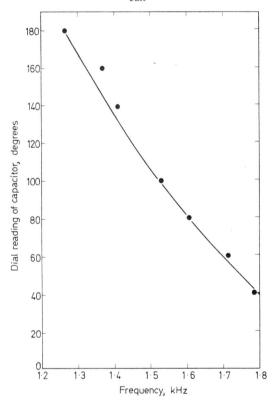

Figure A39. Calibration of a capacitor by frequency measurement

Thus

$$C = \frac{1}{4\pi^2 b N^2 d}\left(\frac{1}{f^2}\right) \quad \text{(see Figure A37)}.$$

The constant value of b for the coil used is given and assuming it is air cored ($\mu = 1$ on the c.g.s. system or $\mu = \mu_0 = 10^7/4\pi$ in m.k.s. units) the capacitance C can be calculated if the frequency is known.

Method

To measure the frequency the series-fed oscillator may be connected to a frequency meter as shown in *Figure A38*.

The capacitor dial is set to say 10° and the frequency meter is adjusted to give a null reading on the detector (i.e. a minimum note if headphones are used). The frequency-meter reading is then noted.

The procedure is repeated for every 10 or 20° and readings of the

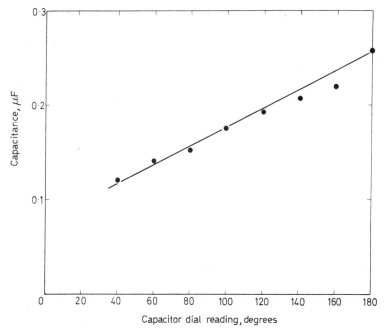

Figure A40. Calibration of the tuning capacitor of an oscillator by measurement of frequency

capacitor dial and frequency-meter noted as before. The graph of *Figure A39* shows the type of results obtained.

The true capacitance is calculated from the theoretical formula given and this is plotted against the capacitor reading to obtain the calibration graph for the capacitor used in the oscillator circuit.

Discussion of Results

Perhaps the most difficult part of the experiment is the detection of the balance point. In general there will be harmonic frequencies present which can make it difficult to obtain a sharp 'null' point.

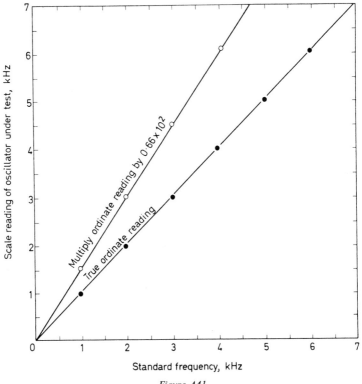

Figure A41

The graph of capacitor dial reading against calculated capacitance shows a typical 'scatter' of points about the best-fit line and gives a good indication of the order of accuracy involved in the experiment (*Figure A40*).

Further Work

The oscillator might be calibrated against a standard quartz-crystal oscillator. *Figure A41* shows a typical graph of the frequency reading of a beat frequency oscillator against the standard frequency of the quartz crystal.

Reading and References

Brief background reading upon oscillators and the conditions required for oscillation appropriate to each type of oscillator may be found in *Handbook for Science Masters and Lecturers* by I. L. Muler and K. E. J. Bowden (Advance Components Ltd., 1963). Typical component values are given in this reference.

A chapter on oscillators (including transistor types) may be found in *Principles of Electronics* by M. R. Gavin and J. E. Houldin, (English Universities Press, 5th impression, 1965).

Data concerning the variation of the inductance factor b with the length to diameter ratio of the coil is to be found in *Electronic and Radio Engineering* by F. E. Terman (McGraw-Hill, 4th edition, 1955).

Experiment No. A1.8

TO VERIFY THAT THE PERIOD OF THE TRANSITION BETWEEN THE TWO STATES OF AN ASTABLE MULTIVIBRATOR IS PROPORTIONAL TO THE COUPLING CAPACITANCES USED IN THE CIRCUIT

Introduction

The multivibrator is effectively a resistance-capacitance-coupled amplifier with the output of its second stage coupled back to the input of the first stage.

Three types are possible:

(1) The astable multivibrator which has capacitor couplings C_1 and C_2.

(2) The monostable multivibrator where one coupling is resistive and the other capacitive.

(3) The bistable multivibrator where both couplings are purely resistive.

In this experiment the first type is used and it is found to have no stable state but to reverse periodically from the condition where the valve of the first stage conducts (with the second valve 'cut off') to the opposite condition where the second stage conducts and the first valve is cut off.

The time period between these two states is governed primarily by the time-constant of the circuit and the aim here is to show that the period is linearly proportional to the coupling capacitance.

Theory

Figure A42 shows the waveforms which are obtained using the astable multivibrator. At the outset although both valves will operate, one current flow will be greater than the other.

Assuming that the first valve conducts more readily than the second, then its anode voltage V_{a1} will drop slightly relative to that of the second. This sends a negative pulse through the first coupling capacitor resulting in a reduction of electron flow through the second valve. The effect is to give a further increase in anode voltage V_{a2} of the second valve and a positive pulse is driven through the second coupling capacitor (C_2) to the grid of the first valve.

Thus the system grows increasingly unstable until the first valve is fully conducting and the second valve is fully 'cut off'.

However, this action charges the coupling capacitors and upon reaching a condition of maximum anode potential on the one stage the first capacitor begins to discharge, the time of discharge being determined by the time constant $R_{g1}C_1$ where R_{g1} represents the appropriate grid resistance. This discharge causes change in anode current resulting in a reversal of the above sequence of operations and the second valve now conducts whilst the first stage is 'cut off'.

If the time constants of the two couplings are the same, symmetrical pulses each of the same duration are obtained, but with different values of C_1 and C_2 an asymmetrical system of pulses is seen.

Method

A typical circuit is shown in *Figure A43* and a double beam cathode ray oscilloscope is employed so that pairs of waveforms may be observed directly (and simultaneously) on the screen.

First using appropriate voltage scales, the anode voltage-waveforms V_{a1} and V_{a2} are recorded on the c.r.o. screen. The patterns should be photographed or sketched, as in *Figure A42 a* and *b*.

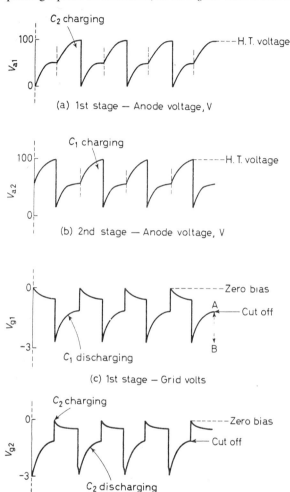

Figure A42. The multivibrator: typical waveforms as seen on the oscilloscope screen

The oscilloscope amplifier leads are removed from the anode leads and now used to measure V_{g1} and V_{g2}, the grid voltages, the waveforms again being recorded. The voltage rise between the points A and B on *Figure A42c* should be noted.

Making use of the variable frequency timebase on the c.r.o., adjustment is made until the timebase frequency is synchronized with that of the transition between the two saturation states of the multivibrator. The period and the coupling capacitor values are then noted.

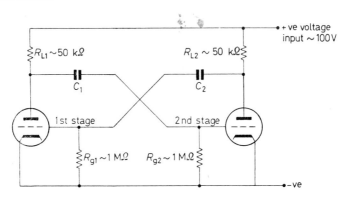

Figure A43. Astable multivibrator circuit

The values of the capacitors are now changed and a series of different discharge times measured by varying the timebase frequency of the c.r.o. until synchronization with the frequency of the waveform occurs in each case.

A graph of the capacitor values against the period of the circuit should then be plotted (*Figure A44*).

Discussion of Results

The action of the circuit is most clearly seen by reference to the waveforms and it is important to note that the anode voltage waveforms are similar but 180° out of phase, as are those of the two grids.

The regions of the graphs at which the capacitors are charging and discharging in turn, as well as the grid cut-off points, should be noted.

Whilst the result of the graph of *Figure A44* is a typical one, it will depend to a large extent upon the accuracy of the chosen capacitors, and remember that for typical radio-type capacitors the tolerance may be $\pm 20\%$.

Further Work

The graphical results quoted are those for a multivibrator using thermionic valves but the transistor is more versatile and convenient.

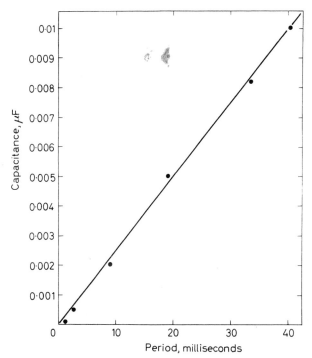

Figure A44. The multivibrator: graph of coupling capacitance against the period of oscillation associated with the circuit

For this reason both types of circuit are shown here (*Figures A43* and *A45*. If time allows it is worthwhile to compare the action of the three types of multivibrator mentioned in the introduction and to note the uses to which each may be put.

Reading and References

The classification of relaxation oscillators, including the multivibrators, is dealt with in *Electronics from Theory into Practice* by J. E. Fisher and H. B. Gatland (Pergamon, 1966), under the chapter on 'Waveform Generators'.

ELECTRONICS AND SOLID STATE PHYSICS

The Handbook for Science Masters and Lecturers by I. L. Muler and H. E. Bowden (Advance Components Ltd., 1963) gives an outline of this experiment and quotes typical components and equip-

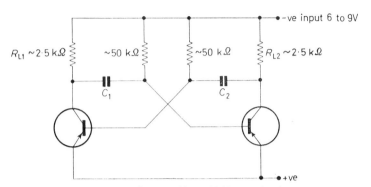

Figure A45. Transistor astable multivibrator circuit

ment which may be used. It suggests an alternative method of measuring the multivibrator frequency using a signal generator and Lissajous' patterns rather than synchronization with the timebase.

Electronics for Scientists by H. V. Malmstadt, C. G. Enke and E. C. Toren (Benjamin, 1963) gives worked examples of the theoretical frequency for a number of different component values typical for the multivibrator circuit.

Experiment No. A1.9

TO INVESTIGATE THE CHARACTERISTICS OF DIODE RECTIFIER POWER SUPPLIES

Introduction

It is convenient to use rectified a.c. supplies rather than batteries for operating electronic equipment and the aim of this experiment is to investigate the inherent characteristics of half-wave and full-wave rectifiers with (*a*) choke and (*b*) capacitor input from the rectifier to the filter of the supply.

CHARACTERISTICS OF DIODE RECTIFIER POWER SUPPLIES A1.9

Theory

The power supply is arranged so that it may be used in one of the following two ways:

(*a*) In conjunction with a full-wave (double-diode) rectifier system having two anodes. In this case both anodes are connected to the H.T. winding of a transformer so that one diode unit conducts on the positive half-cycle of the alternating current and the other on the negative half-cycle (*Figures A46* and *A51*).

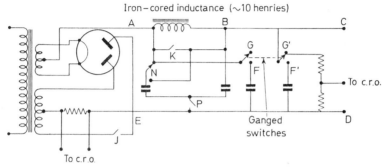

Figure A46. Circuit used to investigate power supplies and their characteristics. N.B. The circuit may be conveniently split into three parts. (1) The rectifying unit (to the left of AE). (2) The filter unit (between A and B). (3) The load circuit (to the right of B)

(*b*) With a half-wave (single-diode) rectifier which conducts only on alternative half-cycles.

The diode arrangements may be conveniently mounted such that suitable tappings are provided on the panel to enable choke input (*Figure A47*) or capacitor input (*Figure A48*) to be employed.

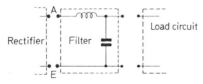

Figure A47. Choke input: arrangement given by opening K, switching N down and closing P

In designing a power supply three problems present themselves:
(1) The initial rectification of the a.c. input.
(2) The conversion of the resulting fluctuating direct voltage to

'smoothed' uniform voltage. This is done by a suitable filter circuit which may be of the form shown in *Figure A47* or *Figure A48*.

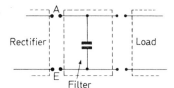

Figure A48. Capacitor input: given by closing K, switching N upwards and closing P

(3) That of internal resistance which by the very nature of the circuit is affected by the resistance of the secondary winding of the transformer, the resistance across the rectifying diode and the reactance of the iron cored inductance (choke) in the filter circuit (which varies approximately inversely with the current).

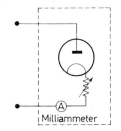

Figure A49. Valve arrangement to act as variable load for supply

The circuit as a whole may thus be conveniently considered in three units: (*a*) the rectifier, (*b*) the filter circuit and (*c*) the load resistance circuit. This latter should be made up of two parts, a fixed resistance section which is chosen to limit current to a safe maximum and a variable resistance which may conveniently be supplied by a valve circuit (*Figure A49*).

Method

A circuit of the form shown in *Figure A46* is used. The key J is opened so that the diode acts as a half-wave rectifier. The key K is opened, key N is switched down and key P is closed to give a choke input filter.

By means of a suitable d.c. voltmeter (300 V range) the voltages between DA and DB are measured as the valve current in the load circuit (*Figure A49*) is increased slowly from zero to maximum, in steps of about 5 mA.

The key J is then closed so that the diode now operates on full-wave rectification and the voltage is again measured as the valve current increases.

Finally key K is closed, key N is switched upwards (*Figure A46*) and key P is kept closed to give a capacitor input filter. The voltages across DA and DB are taken as before, starting with minimum current through the valve acting as 'load'.

Graphs of output volts against the load current in the valve are plotted as shown in *Figure A50*.

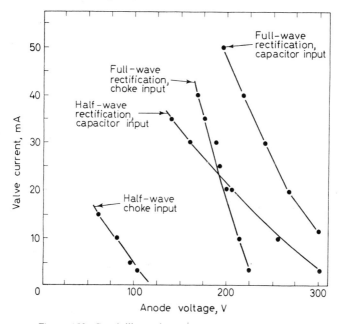

Figure A50. Graph illustrating power supply characteristics

Using the ganged switch in position G (*Figure A46*) the current and voltage waveforms for both half- and full-wave rectification with each type of filter circuit may be displayed on the cathode ray oscilloscope and sketches or photographs of them obtained (*Figure A51*).

Discussion of Results

The graphs show (*Figure A50*) that: (1) the output voltage is never truly constant but varies with the load on the supply circuit; (2) the full-wave rectifier with a capacitor input to the filter produces the highest output voltage but that this decreases more rapidly with increased load than it does for choke-input to the filter; (3) the half-wave rectifier, as anticipated, is not as effective as the full-wave system.

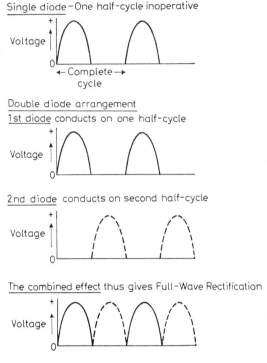

Figure A51. Half-wave and full-wave rectification by single and double-diode systems

The cathode ray oscilloscope waveforms show that the alternating ripple voltage increases as the load increases and that, as anticipated, the ripple is much greater for half-wave rectification than for the full-wave arrangement.

Further Work

As stated above the internal resistance of the supply varies under the influence of many factors and it is instructive to calculate the differential internal resistance for each type of circuit used, by taking the ratio of the maximum slope to the minimum slope of graphs of the output voltage against load current.

Reading and References

A useful chapter on power supplies may be found in *Electronics from Theory to Practice* by J. E. Fisher and H. B. Gatland (Pergamon, 1966). This includes a number of design examples and also mentions the use of semiconductor diodes in power supply circuits. An alternative treatment of the theory of rectification may be found in *Electronics* by R. H. Mattson (Wiley, 1966).

Experiment No. A1.10

TO DETERMINE THE CHARACTERISTIC IMPEDANCE AND THE VELOCITY CONSTANT OF A TRANSMISSION LINE

Introduction

In general any system of two (or more) separate current-carrying conductors is able to propagate electromagnetic waves in the dielectric medium between the conductors. In particular one form of the waves, known as the principal wave, may be propagated at any frequency and a system employing this principal wave is referred to as a transmission line.

Theory

In long cables or transmission lines used to convey electrical power over great distances the capacitance and self-inductance may be considered as being distributed along the line. (Recall that in considering the capacitance of a cylindrical capacitor it was necessary to postulate an infinite length and first to consider capacity per unit length of the system.)

Hence if C, L, R and G are the capacitance, inductance, resistance and conductance per unit length of the line in each case, then the impedance Z per unit length is given by

$$Z = (R + j\omega L)$$

and the admittance Y per unit length by

$$Y = (G + j\omega C)$$

hence the fall in potential $-\delta V$ along a length δx of the line is given by

$$-\delta V = ZI\,\delta x$$

or

$$\frac{\partial V}{\partial x} = -ZI \qquad (1)$$

Note the use of the partial derivative here since V depends upon time as well as distance x.
Similarly

$$-\delta I = VY\,\delta x$$

or

$$-\frac{\partial I}{\partial x} = YV. \qquad (2)$$

We can now link the results of (1) and (2) by considering second derivatives

$$\frac{\partial^2 V}{\partial x^2} = -\frac{\partial}{\partial x}(ZI) = -Z\left(\frac{\partial I}{\partial x}\right) = ZYV \qquad (3)$$

and

$$\frac{\partial^2 I}{\partial x^2} = \frac{\partial}{\partial x}(YV) = Y\frac{\partial V}{\partial x} = YZI \qquad (4)$$

Solutions of these equations are of the form

$$I = I_1 \exp(\gamma x) + I_2 \exp(-\gamma x)$$

and

$$V = V_1 \exp(\gamma x) + V_2 \exp(-\gamma x) \qquad (5)$$

where γ is known as the propagation constant and is equal to $(YZ)^{\frac{1}{2}}$, i.e.

$$\gamma = \{(R + j\omega L)(G + j\omega C)\}^{\frac{1}{2}} \qquad (6)$$

Note that since Z and Y are complex γ is complex

$$\gamma = \alpha + j\beta$$

CHARACTERISTICS OF A TRANSMISSION LINE   A1.10

where α is found from physical considerations (below) to be an attenuation constant and β a phase constant determining the electromagnetic wave velocity associated with the line.

In equation (5), I and V are instantaneous values of current and voltage given by the real and imaginary parts of $I\exp(j\omega t)$ and $V\exp(j\omega t)$ hence

$$I\exp(j\omega t) = I_1 \exp(\alpha x)\exp\{j(\omega t + \beta x)\}$$
$$+ I_2 \exp(-\alpha x)\exp\{j(\omega t - \beta x)\} \quad (7)$$

and

$$V\exp(j\omega t) = V_1 \exp(\alpha x)\exp\{j(\omega t + \beta x)\}$$
$$+ V_2 \exp(-\alpha x)\exp\{j(\omega t - \beta x)\} \quad (8)$$

The second (I_2 or V_2) term represents a wave moving left to right and suffering attenuation. The first (I_1 or V_1) term represents a similar wave travelling in the opposite direction.

By substitution of results of equation (5) in equation (2) we obtain

$$-\frac{\partial I}{\partial x} = -\frac{\partial}{\partial x}\{I_1 \exp(\gamma x) + I_2 \exp(-\gamma x)\}$$
$$- \gamma\{I_1 \exp(\gamma x) - I_2 \exp(-\gamma x)\} \quad (9)$$
$$= YV_1 \exp(\gamma x) + YV_2 \exp(-\gamma x). \quad (10)$$

This is true for all values of x and equating coefficients

$$-\gamma I_1 = YV_1 \quad \text{and} \quad \gamma I_2 = YV_2$$

hence

$$\frac{V_2}{I_2} = -\frac{V_1}{I_1} = \left(\frac{Z}{Y}\right)^{\frac{1}{2}} = \left(\frac{R + j\omega L}{G + j\omega C}\right)^{\frac{1}{2}} = Z_0 \quad (11)$$

the characteristic impedance.

At frequencies of 1 MHz and above, the resistance R and conductance G are negligible compared with ωL and ωC so that

$$\beta = \omega(LC)^{\frac{1}{2}} \quad (12)$$

and

$$Z_0 = \left(\frac{L}{C}\right)^{\frac{1}{2}} \quad (13)$$

also the velocity of propagation of the wave in the line

$$v_L = \frac{\omega}{\beta} = \frac{1}{(LC)^{\frac{1}{2}}} \quad (14)$$

The ratio of the velocity in the line to the velocity of the electromagnetic wave in free space is known as the velocity constant p of the line

$$\left(\frac{C_s}{C}\right)^{\frac{1}{2}} = \frac{1}{\varepsilon^{\frac{1}{2}}} \tag{15}$$

For a loss-free line $\gamma = j\beta$ and we have

$$I = I_1 \exp(j\beta x) + I_2 \exp(-j\beta x) \tag{16}$$

and

$$V = V_1 \exp(j\beta x) + V_2 \exp(-j\beta x) \tag{17}$$

but by equation (11)

$$V_1 = Z_0 I_1 \quad \text{and} \quad V_2 = -Z_0 I_2.$$

Unless $Z_L = Z_0$

$$Z_L = \frac{V_1 + V_2}{I_1 + I_2} = \frac{V_2}{I_2}\left\{\frac{1 + \rho}{1 - \rho}\right\}$$

where V_2 = maximum voltage amplitude associated with the reflected wave in the transmission line
V_1 = maximum voltage amplitude associated with the incident wave in the transmission line
I_1 = maximum current amplitude associated with the incident wave in the transmission line
I_2 = maximum current amplitude associated with the reflected wave in the transmission line

and ρ = reflection coefficient = $V_1/V_2 = -I_1/I_2$ at $x = 0$ (see *Figure A52*),

but

$$\frac{V_2}{I_2} = Z_0 \quad \therefore \quad Z_L = Z_0\left\{\frac{1 + \rho}{1 - \rho}\right\}.$$

Hence current is given by

$$I = I_2 \{\exp(-j\beta x) - \rho \exp(+j\beta x)\}$$

and voltage

$$V = V_2 \{\exp(-j\beta x) + \exp(j\beta x)\}$$
$$= I_2 Z_0 \{\exp(-j\beta x) + \exp(+j\beta x)\}$$

therefore the input impedance

$$Z_{in} = Z_0 \left\{ \frac{\exp(j\beta l) + \rho \exp(-j\beta l)}{\exp(j\beta l) - \rho \exp(-j\beta l)} \right\}$$

which substituting for

$$\rho = \frac{Z_L - Z_0}{Z_L + Z_0}$$

gives

$$Z_{in} = Z_0 \frac{Z_L + jZ_0 \tan \beta l}{Z_0 + jZ_L \tan \beta l}$$

and when the line is *short circuited*

$$Z_L = 0 \quad \text{then} \quad Z_{in} = +jZ_0 \tan \beta l$$

and when

$$Z_L \to \infty \quad Z_{in} = -jZ_0 \cot \beta l$$

also when

$$Z_L = Z_0 \quad Z_{in} = Z_0$$

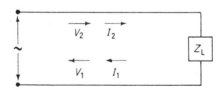

Figure A52. Diagram showing incident and reflected voltages and currents in a transmission line terminated by an impedance not equal to its characteristic impedance. Z_L represents the load impedance ($\neq Z_0$). Arrows indicate incident and reflected waves.

V_2 = the maximum voltage amplitude associated with the reflected wave in the transmission line
I_2 = the maximum current amplitude associated with the reflected wave in the transmission line
V_1 = the maximum voltage amplitude associated with the incident wave
I_1 = the maximum current amplitude associated with the incident wave

Figure A53a shows the tangent curve corresponding to the case of the short circuited line. The theoretical curve for the *modulus* of input impedance against frequency will therefore be as shown in *Figure A53b*.

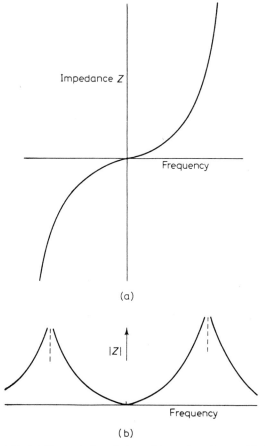

Figure A53. (a) Theoretical variation of input impedance with frequency for a short-circuited transmission line. (b) Theoretical variation of the modulus of input impedance $|Z|$ with frequency for a short-circuited transmission line. This curve should be compared with the experimental curves in *Figure A54*

The form of curve obtained in practice is shown in *Figure A54a* and differs from the theoretical curve because R, though small, is not exactly zero.

Figure A54b shows the case of the open-circuited line and it should be noted that the maxima of this curve coincide with the minima of the curve of *Figure A54a*.

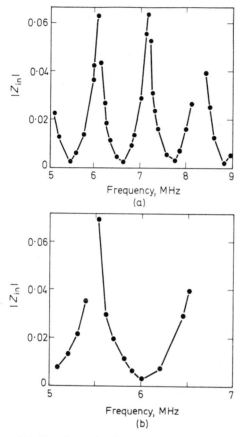

Figure A54. Experimental variation of the modulus of input impedance with frequency for (*a*) a short-circuited transmission line, and (*b*) for an open-circuited transmission line

When the characteristic impedance terminates the line no variation of input impedance with frequency is found.

Method

First find the values of the inductance per unit length and the capacity per unit length using a short piece of the coaxial cable (~150 cm long) which may be used as transmission line here.

To do this solder the central wire and sheath at one end of the line and connect the other ends to the inductor terminals of the Q meter. Set up the instrument carefully in accordance with specific operating instructions of the model being used and tune for the resonant frequency f_0 for a value of C of the Q meter condenser.

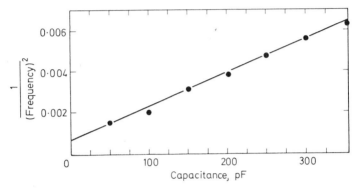

Figure A55. Graph to determine the inductance of a length of transmission line

Repeat the procedure for say six different values of C_1 and plot a graph (*Figure A55*) of $1/f_0^2$ against C_1 when the slope of the resulting straight line gives the inductance of the known length of line. Since

$$\omega^2 L_t(C_0 + C_1) = 1$$

where ω = pulsatance
 C_0 = capacitance of the line
 L_t = total inductance
 C_1 = selected capacitance on the Q meter,

i.e.

$$4\pi^2 L_t C_0 + 4\pi^2 L_t C_1 = \frac{1}{f_0^2}$$

and

$$\left(\frac{1}{f_0^2}\right) = 4\pi^2 L_t C_1 + 4\pi^2 L_t C_0$$

the slope of the graph = $4\pi^2 L_t C_1$

hence L_t can be found and knowing the length of the line L can be determined.

Now 'open circuit' the line with the wires of one end to the Q meter 'capacitor' terminals and place a known inductance ($\sim 5 \mu\text{H}$) across the 'inductor' connections. Set the Q-meter capacitor to

CHARACTERISTICS OF A TRANSMISSION LINE A1.10

(say) 50 pF and adjust for resonance. Disconnect the line and readjust the capacitor C_1 to re-establish resonance. The capacitance of the line may then be found $[(C_1 - 50) \text{ pF}]$.

By the theory given, Z_0 the characteristic impedance is then $(L/C)^{\frac{1}{2}}$ and the velocity constant of the wave in the line = $(C_s/C)^{\frac{1}{2}}$.

To obtain graphs of the variation of input impedance with frequency the longer piece of coaxial cable with which you will be supplied is used in a circuit such as shown in *Figure A56*.

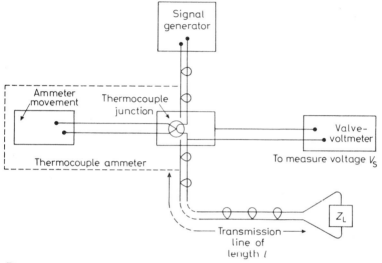

Figure A56. Schematic diagram of circuit used in the measurement of transmission line parameters

The inner conductor is 'tapped' as indicated so that the current may be measured by suitable meter.

The input voltage is noted from a suitable voltmeter and hence $Z_{in}(=V/I)$ can be determined for a series of frequencies from 5 to 10 MHz.

Graphs should then be plotted as shown in *Figure A54*.

Discussion of Results

The scatter of the points about the 'best fit' line of the graphs will enable valid inference concerning the accuracy of the experiment to be made. Remember that among other things it was assumed in the theory that the resistance can fall to zero when in fact this is not possible in practice. Also, contact resistances and meter errors are present and it has been assumed that no kinks exist in the conductors and no mechanical strain is introduced into the dielectric between them. (Was the coax cable in a smooth coil?)

Further Work

It is possible to estimate the length of the line from these results, for at minimum values of $|Z|$, using the short-circuited line, the line length $l = n\lambda_l/2$ for the electromagnetic radiation of wavelength λ_l in the line ($n = 1, 2, 3$, etc.).

Again

$$\lambda_l = \frac{v_l}{f} = \frac{pc}{f}$$

where v_l = phase velocity
        c = velocity of electromagnetic radiations
        p = velocity constant

therefore

$$l = \frac{npc}{2f}$$

and since p, c and l are constant a graph of n against frequency f is a straight line of slope $(2l/pc)$. A graph of n against f actually obtained by experiment is shown in *Figures A57* and *A58*.

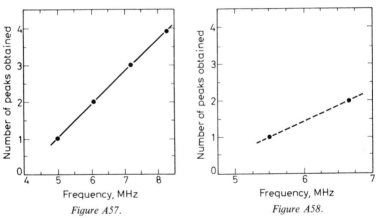

Figure A57.　　　　　　　　　　Figure A58.
Graphs enabling the length of a transmission line to be determined

A second estimate can be made plotting the maxima of the open-circuit case and this will again enable firm conclusions to be drawn about the accuracy of the experiment.

Reading and References

The theory of the transmission line may be found in a number of standard texts but in particular either Terman's *Radio Engineering* or W. J. Duffin's *Electricity and Magnetism* (McGraw-Hill, 1965) may be found helpful.

Experiment No. A1.11

TO PLOT EQUIPOTENTIALS USING THE ELECTROLYTIC TANK

Introduction

The electrolytic tank can be used as an effective analogue 'computer' for the solution of static electric (and other) field problems and the aim here is to obtain a series of experimental equipotentials and to compare them with the theoretical graphs derived from Laplace's equation.

The experiment may be extended to investigate the equipotentials associated with electrodes of different geometries which have important applications in electrical engineering. As an example a graph is given (*Figure A59*) of the equipotentials obtained with electrodes designed to simulate a high current shunt.

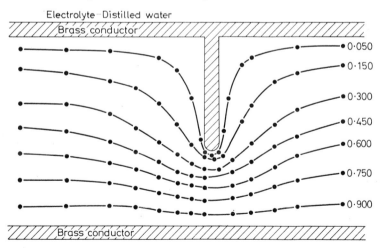

Figure A59. Equipotentials obtained using shunt simulator

Theory

If we consider the complex function

$$F(z) = F(x + iy) \tag{1}$$

where z represents a complex entity, then

$$\frac{\partial^2 F}{\partial x^2} + \frac{\partial^2 F}{\partial y^2} = 0 \tag{2}$$

i.e. F is a solution of Laplace's equation, and if

$$F = U + iV$$

then U and V are solutions of

$$\frac{\partial^2 F}{\partial x^2} + \frac{\partial^2 F}{\partial y^2} = 0$$

and from (1)

$$\frac{\partial F}{\partial y} = i \frac{\partial F}{\partial x}$$

$$\therefore \frac{\partial}{\partial y}\{U + iV\} = \left\{\frac{\partial U}{\partial y} + i\frac{\partial V}{\partial y}\right\} = i\left\{\frac{\partial U}{\partial x} + i\frac{\partial V}{\partial x}\right\}$$

from which

$$\frac{\partial U}{\partial x} = \frac{\partial V}{\partial y} \quad \text{and} \quad \frac{\partial U}{\partial y} = -\frac{\partial V}{\partial x}.$$

We may conclude from this that the surfaces represented by the equations

$$U(xy) = \text{constant} \quad \text{and} \quad V(xy) = \text{constant}$$

are orthogonal. Hence if V represents an equipotential surface, U represents the associated lines of force.

If we now consider the case of two cylindrical electrodes in the electrolytic tank, one carrying a charge of $+q$ and the other $-q$ then the potential V at a point P distance a from the first electrode and b from the second one is given by

$$V = +\frac{q}{2\pi\varepsilon_0}\{\log b - \log a\} = \frac{q}{2\pi\varepsilon_0} \log_e\left(\frac{b}{a}\right) \tag{3}$$

where ε_0 = permittivity of the dielectric considered. (Compare this with the standard bookwork of the case of two parallel current-carrying wires.)

V represents, in fact, a family of circles whose equations are

$$\left.\begin{array}{l} b^2 = (d + x)^2 + y^2 \\ a^2 = (d - x)^2 + y^2 \end{array}\right\} \quad (4)$$

where $2d$ is the distance between the centres of the two electrodes as shown in *Figure A60*.

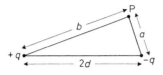

Figure A60

Now let

$$v = \frac{2\pi\varepsilon_0 V}{q}$$

then $v = \log_e (b/a)$ or $(b/a) = \exp(v)$ therefore

$$\frac{b^2}{a^2} = \frac{(d + x)^2 + y^2}{(d - x)^2 + y^2} \quad \text{[from (4)]}$$

and hence

$$\frac{(d + x)^2 + y^2}{(d - x)^2 + y^2} = \exp(2v) \text{ or } d^2 + 2dx + x^2 + y^2$$
$$= \exp(2v)\left[d^2 - 2dx + x^2 + y^2\right],$$

i.e.
$$(d^2 + x^2 + y^2)\{1 - \exp(2v)\} + 2\,dx\,\{1 + \exp(2v)\} = 0$$
or
$$(d^2 + x^2 + y^2)\{\exp(-v) - \exp(v)\} + 2\,dx\{\exp(-v) + \exp(v)\} = 0$$

this becomes
$$-2(d^2 + x^2 + y^2) \sinh v + 4\,dx \cosh v = 0$$
or
$$2dx \cosh v = (d^2 + x^2 + y^2) \sinh v,$$
i.e.
$$2dx \coth v = d^2 + x^2 + y^2$$

whence

$$(x - d \coth v)^2 + d^2 \left(1 - \frac{\cosh^2 v}{\sinh^2 v}\right) + y^2 = 0$$

$$(x - d \coth v)^2 + y^2 = -d^2 \frac{\sinh^2 v - \cosh^2 v}{\sinh^2 v} = +\frac{d^2}{\sinh^2 v}$$

$$\therefore y^2 + (x - d \coth v)^2 = (d \operatorname{cosech} v)^2.$$

For constant values of v, $d \operatorname{cosech} v = $ constant R (say) and $\coth v = $ constant B hence

$$y^2 + (x - dB)^2 = R^2.$$

This is the equation of a circle of radius R and with centre at (C, O) on the cartesian system where $C = dB$ (*Figure A62*).

Method

The large shallow watertight tank (*Figure A61*) should be filled to a depth of 6–7 mm with distilled water and the electrodes (in the first instance cylindrical brass ones) placed upright in it as shown in

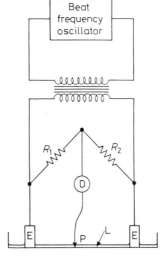

P = probe,   E = electrodes
D = detector,   L = liquid electrolyte

Figure A61

PLOTTING EQUIPOTENTIALS   A1.11

Figure A61. The sides of the tank may be of aluminium and graph paper beneath the glass base of the tank will enable the position of the electrodes to be set fairly accurately about 10 cm either side of the central y axis with both electrodes lying on the x axis.

The plane of water forms a resistance bridge with the two resistance boxes of the circuit acting as ratio arms R_1 and R_2 as shown

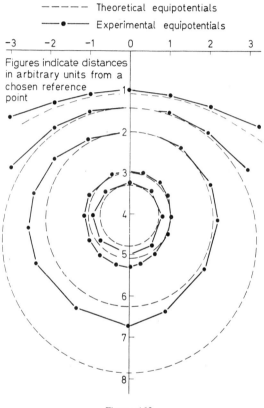

Figure A62

in *Figure A61*. The probe connects by way of a detector to a point between the resistance boxes. The detector itself may be an amplifier and c.r.o.

A simple wire probe is used and it should be fused into a long stem of glass tubing, taking care to see that in use it always remains upright.

The resistances should first be set at about 500 Ω each and the tank carefully levelled to produce a minimum spot on the c.r.o. when the probe passes along the y axis through the origin. The R_1/R_2 ratio should then be varied and points of zero deflection tabulated for each given value of the ratio. The equipotentials may then be plotted as indicated in the full line of *Figure A62* and the calculated equipotentials shown for comparison on the same graph.

Figure A59 shows the equipotential plot obtained with electrodes simulating the cut which is often made in the manganin bar of a high current shunt should its resistance, for practical use, require variation.

In this case the R_1/R_2 ratios are preset at values 0·1, 0·2, etc. up to 0·9 and the 'simulator' consists here of a T-piece of brass and a straight brass electrode forming the two sides of the rectangular 'tank', the other two sides and the base being of suitable insulator material. (The whole thing is made watertight using 'Bostik'.) The plot can then be made directly without need to tabulate the results if two holes are made in the base material outside the tank to enable the arrangement to be placed in a standard reference position so that no deflection on the galvanometer occurs along the chosen y axis and a pantograph style arm is used to carry the probe (*Figure A63*).

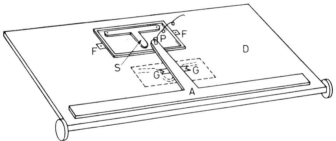

Figure A63. S represents model and tank; F, Locating pins or screws; A, Pantograph arm; P, Balance probe; D, Drawing board as mounting; G, Electrolytic tank plot made by using pencil in grooves

A number of standard sets simulating, for example, coaxial or three-core cable can be made up in a similar way.

Discussion of Results

Distortion of the equipotentials may occur due to one or more of the following factors.

PLOTTING EQUIPOTENTIALS

(1) In the first case, of the two cylindrical electrodes, the sides of the tank are not effectively at infinity although the theory assumes an infinite surface.

(2) Dirt on the base or sides of tank.

(3) Salts or contamination in the water (which should be *distilled*).

It is interesting here to take the result of *Figure A59* and to repeat the experiment using (*a*) tap water, (*b*) water containing small quantities of Teepol, (*c*) Teepol itself as the liquid in the tank. Note carefully the effect on the equipotentials especially near the electrode boundaries.

Further Work

Additional examples of electrode geometry to simulate physical conditions in many branches of physics and engineering are possible. *Figure A64* shows a configuration representing the cross-section of a three-core cable such as might be used in a three-phase electrical system. The plot shows the instantaneous equipotentials which exist when electrode A is at maximum potential and electrode B (and its mirror image) are both negative with respect to the earthed sheath C. (Since symmetry exists only half the 'cable' is shown.) *Figure A65* shows an appropriate circuit. *Figure A66* shows the plot that results in an attempt to simulate an electrode geometry similar to that which exists in a Nier ion source for a mass spectrometer. It displays the effect of ion focusing which takes place between the electrodes of the accelerator system and also the resultant divergence which can occur as the beam passes through the final earthed electrode of the gun.

If only the electrodes corresponding to the physical geometry of the source are used, then it is not possible to plot any points beyond the final accelerating electrode C. According to the result shown in *Figure A66* the ion beam is slightly diverging. The effect is much exaggerated in the experimental illustration since for clarity the slit widths have been drawn to a larger scale than that used for the electrode separations. The particles would of course, in practice, continue in a straight line, in the field free space beyond C.

However if a very small potential is placed upon the 'artificial' electrode D (shown in the diagram) the picture can be extended into the space beyond C. It is no longer a strictly true representation of what happens in practice but it illustrates the care that must be exercised in setting up and interpreting the electrolytic tank analogue.

A convenient and alternative form of this experiment is to use

Figure A64. Plot of the equipotential lines associated with a 'three-core cable' geometry obtained by the electrolytic tank analogue method. Note (1) that since the system is symmetrical only half the configuration is shown; (2) sheath of analogue is positive to simulate the condition where one conductor is negative with respect to earth

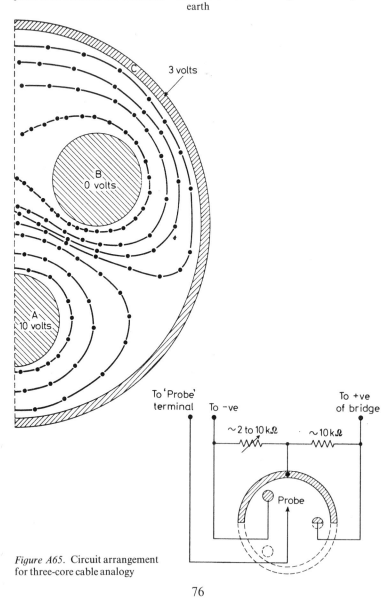

Figure A65. Circuit arrangement for three-core cable analogy

'Teledeltos' conducting paper in place of the watertight tank. Electrode geometries can then be painted on the paper using colloidal silver paint. (See also Experiment No. E1.7.) Generally, however, considerable accuracy is sacrificed in this way.

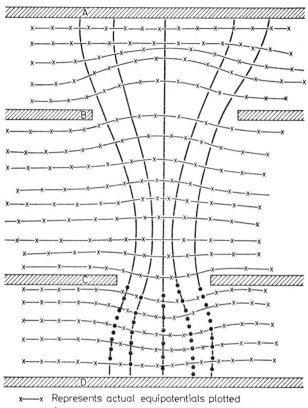

x———x Represents actual equipotentials plotted
———   Orthogonal system of lines showing focusing effect

Figure A66. Simulation of the geometry of a Neir ion source. A, Repelling electrode ($V \simeq 1\,205$ V in actual mass spectrometer); B, First focusing electrode ($V = 1\,200$ V); C, Second electrode ($V = 0$ V); D, 'Artificial' analogue electrode used to obtain equipotentials below C (see note in text)

Reading and References

Details of yet another type of set up (using ordinary paper) together with the necessary circuitry will be found in an article by

C. T. Murray and D. L. Hollway in *Journal of Scientific Instruments*, vol. 32, p. 481, December, 1955.

Details of the electrode geometry and the applied potentials appropriate to Nier ion sources may be found in *Modern Mass Spectrometry* by G. P. Barnard (Institute of Physics, 1953).

A note upon the electrolytic tank method, and the plotting of sets of orthogonal lines using yet another geometrical form may be found in *Electrical Machines* by A. Draper (Longmans, 2nd edition, 1967).

Experiment No. A2.1

TO DETERMINE THE ELECTRICAL CONDUCTIVITY OF A *p*-TYPE SEMI-CONDUCTING SILICON SPECIMEN

Introduction

Germanium and silicon are examples of intrinsic semiconductors. They are like insulators in having a full valence band, but the next allowed energy band lies only a small energy span above the valence band so that although the resistivity of the element is high at low temperatures it has dropped dramatically at room temperature and conduction is appreciable. The aim of this experiment is to measure this conductivity.

Theory

Considering the current density J, i.e. the current flowing across unit area of the specimen, one can write

$$J = \frac{I}{A} = nev \tag{1}$$

where I = current
   A = cross sectional area
   n = total number of carriers per unit volume
   v = the drift velocity of the carriers, and
   e = charge on the carrier.

CONDUCTIVITY OF A SEMICONDUCTOR

Note. 'Dimensionally' the left-hand side of the equation gives

$$\frac{\text{Current}}{(\text{Length})^2}$$

and the right-hand side

$$\frac{\text{Quantity}}{(\text{Length})^3} \times \frac{(\text{Length})}{(\text{Time})} \quad \text{i.e.} \quad \frac{\text{Current}}{(\text{Length})^2}$$

Again the mobility μ of the carrier is defined as the drift velocity acquired when unit electric field is applied, i.e.

$$\mu = \frac{v}{E} \tag{2}$$

Again the total number of carriers per unit volume

$$n = n_p + n_n$$

where subscripts n and p refer to negative and positive carriers respectively and thus the mobilities μ_p and μ_n must also be considered, i.e.

$$J = neE(\mu_p + \mu_n) \tag{3}$$

Looking now at the conductivity

$$\sigma = \frac{1}{S} \tag{4}$$

one can write (since $S = AR/l$) where R = resistance and l the length of the specimen:

$$\frac{V}{AR} = \frac{(El)}{A}\left(\frac{A}{Sl}\right) = \frac{E}{S} = E\sigma \tag{5}$$

Again

$$\left(\frac{V}{AR}\right) = \frac{I}{A} = J \text{ the current density}$$

where V = applied voltage
       I = current measured
       l = length between contacts

$$\sigma = \frac{J}{E} \tag{6}$$

therefore from equation (3)

$$\sigma = \frac{J}{E} = \frac{neE(\mu_p + \mu_n)}{E} = ne(\mu_p + \mu_n)$$

Method

The specimen consists of a thin bar of highly purified silicon, the ends of which are plated with rhodium in order to produce good contacts for the passage of current from a 6 V battery with a decade resistance box and ammeter in series as shown in *Figure A67*.

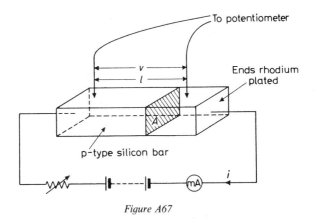

Figure A67

The voltage contacts are two gramophone needle like probes which are mounted a fixed distance l apart and are spring loaded onto the surface of the silicon specimen.

The voltage V between the probes is measured by a potentiometer and the distance between the probes by use of a travelling microscope. Plot a graph of the voltage against l (*Figure A68*). This should be linear and the slope of the graph should enable the conductivity or resistivity to be found.

Discussion of Results

The experiment is subject to a number of serious errors, perhaps the most important being that a true 'ohmic' contact between the silicon and the probes is unlikely to be achieved unless a specially prepared encapsulated specimen is used. (The contact region in fact acts as a poor quality junction rectifier, so that the resistance to passage of current in one direction is much larger than in the other.)

Secondly, with the values of the field encountered using this set up (of the order of a few volts per centimetre), Ohm's law is obeyed and it is not possible to demonstrate the dependence of the four parameters n_n, n_p, μ_n and μ_p upon electric field, which becomes important at very high field values.

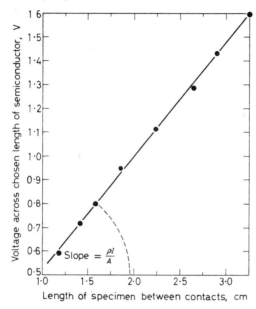

Figure A68. Graph used to determine the resistivity of a semiconductor.

$$R = \frac{\rho l}{A} = \frac{V}{I} \quad \therefore \rho = \frac{VA}{lI}$$

but V/l = slope of graph when A is constant and when I is constant

In assessing the accuracy of the experiment remember too that the measurement of the current is far less accurate than the measurement of the cross-sectional area of the specimen and the distance between the probes.

By far the least satisfactory part of the experiment is the cleaning and fixing of the wire leads.

It is very difficult to obtain true 'ohmic' contact particularly if the contact areas are very small.

Further Work

If a readily prepared specimen is not available it is possible to make one using germanium for which a standard preparation technique is as follows.

Clean the germanium specimen CAUTIOUSLY in concentrated nitric acid to which potassium fluoride has been added. To do this place the germanium on the bottom of a small beaker, add a small quantity of potassium fluoride and then slowly add the nitric acid. After a quick 'swill' round pour off the solution into a LARGE quantity of water and then wash the specimen throughly in distilled water before dipping it in shellac which has been dissolved in chloroform. Repeat the dipping to produce a good protective coat, and then scrape four small areas free of shellac at appropriate points on the edge of the germanium. Dip the specimen quickly into the cleaning mixture (being careful not to dissolve the main coat of shellac), then dip it into distilled water before transferring quickly to a nickel plating solution consisting of 2·5 g of sodium citrate, 1·25 g of ammonium chloride, and 0·75 g of nickel chloride in 25 ml of distilled water.

The solution is then gently heated to 65°C before adding 0·25 g of sodium hypophosphite and heating further to 78°C. At this temperature sufficient ammonium hydroxide is added to turn the whole contents blue and the temperature raised to about 90°C to facilitate plating. After removal the shellac is dissolved away using industrial methylated spirits and a 38 s.w.g. (~ 0.152 mm) copper wire is soldered to each of the plated areas.

Reading and References

A simple outline of the band theory of intrinsic semiconductors may be found in *Physics for Electrical Engineers* by W. P. Jolly (English Universities Press, 1961).

Detail of a method of measuring resistivity of semiconductors of arbitrary shape is given by L. J. Van der Pauw in *Phillips Technical Report*, vol. 13, No. 1, pp. 1–9 (February, 1958).

Typical Results

$V = 0.033$ V $\qquad A = 5 \times 5$ mm
$I = 0.5$ mA $\qquad l = 1.25$ cm

Experiment No. A2.2

TO MEASURE THE HALL COEFFICIENT, THE NUMBER OF CHARGE CARRIERS PER UNIT VOLUME AND THE CARRIER MOBILITY IN A DOPED SEMICONDUCTOR

Introduction

When a current flows in a semiconductor located in a magnetic field such that a component of the field is perpendicular to the current, then a voltage is developed across the specimen mutually at right angles to both the current and the field. This is known as the Hall effect. An important application of this is in magnetic field measuring probes which replace the search coil for this purpose.

Thus if we consider a rectangular slice of semiconducting material (*Figure A69*) of length l, breadth b and thickness t placed with its

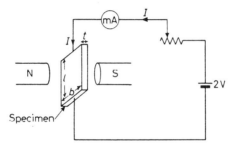

Figure A69

large face perpendicular to a magnetic field H (or more strictly to the flux density B) and imagine a current I is passed along the length of the specimen then a potential difference V appears in a direction at right angles to that of current flow and field. This is called the Hall voltage.

Theory

The Hall voltage arises because the negative (or positive) current carriers in a semiconductor are deflected by the magnetic field so that opposite charges accumulate on the two faces across which the voltage is measured (*Figure A70*).

When charge ceases to accumulate further and equilibrium is

reached, the force on an electron (to consider the simplest carrier) just balances the force due to the magnetic field,

$$Bev = Ee$$

where E = electric field intensity V/b in volts cm^{-1}
  B = flux density in the specimen
and e = charge on the electron
  v = velocity of the electron.

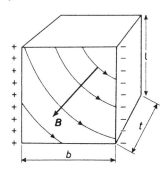

Figure A70

If there are n carriers cm^{-3} then

$$I = nevt.b$$

and

$$\frac{eV}{b} = \frac{IB}{ntb}$$

or

$$\frac{Vt}{IB} = \left(\frac{1}{ne}\right)$$

The quantity (Vt/IB) is known as the Hall coefficient K_H of the semiconductor material, i.e.

$$K_H = \left(\frac{1}{ne}\right).$$

It is negative for those semiconductors where conduction is predominantly by electrons but positive for those which have an excess of electron deficiencies, i.e. positive 'holes' moving in the material lattice. Hence determination of K_H tells us something of the nature of the process of conduction in the substance as well as the actual concentration of the carriers, i.e. the number per unit volume.

A more rigorous theoretical treatment (taking account of the tendency for positive holes and negative electrons to 'combine' and cancel the effects of one another) actually gives

$$\frac{3\pi}{8}\left(\frac{1}{ne}\right) = K_H$$

We shall adopt this form but the simple formula is sufficiently accurate for many purposes ($3\pi/8 = 1.18$).

The mobility μ of the charge carriers may be defined as the drift velocity acquired upon the application of unit electric field and we can write

$$i = E_{App}\, e\mu n$$

[N.B. The applied field E_{App} must NOT be confused with the Hall field E.]

where

$$i = \text{current density} = \frac{I}{\text{Area}} = \frac{I}{bt}$$

(remember n = no. of carriers cm^{-3})
hence

$$\mu = \left(\frac{I}{bt}\right)\frac{1}{E_{App}}\left(\frac{1}{ne}\right)$$

but

$$\frac{1}{ne} = K_H \quad \text{and} \quad \frac{1}{E_{App}} = \frac{l}{V}$$

therefore

$$\mu = \frac{I}{V}\left(\frac{l}{bt}\right)K_H$$

$$\frac{I}{V} = \frac{1}{\text{Resistance}} = \frac{\text{Area}}{\rho l} \quad \text{where } \rho = \text{resistivity}.$$

Hence

$$\mu = \left(\frac{bt}{\rho l}\right)\left(\frac{l}{bt}\right)K_H$$

or taking account of the constant previously mentioned

$$\mu = \frac{8}{3\pi} \frac{K_H}{\rho}$$

i.e. the mobility of the charge carriers can be calculated. Finally if we assume the charge on the electron to be $1·6 \times 10^{-19}$ coulombs we have

$$n = \frac{3\pi}{8} \frac{1}{K_H e}$$

hence the number of charge carriers cm^{-3} can be calculated.

Method

Set up the semiconductor sample (a germanium one may be obtained quite readily) in a simple series circuit consisting of a battery (~ 2 V) a decade resistance box and a milliameter (0–5 mA) (*Figure A69*).

The leads across which the Hall voltage is to be measured are connected to an accurate potentiometer and first the readings are taken WITHOUT the magnetic field on. (This voltage is due to the fact that the probe leads may not be directly opposite one another.) The specimen is then placed in a magnetic field of 1 500 oersted ($1·2 \times 10^5$ ampere-turns m^{-1}) or more and the voltage tabulated once again for various current values from say 1 to 5 mA.

The true Hall voltage is then obtained by subtraction of the first reading from the second at corresponding current values and a graph of Hall voltage V against current I should give a straight line (*Figure A71*). Its slope (V/I) enables the Hall coefficient K_H, the number of carriers cm^{-3} and the carrier mobility μ to be calculated if the magnetic field H (or flux density B) the resistivity ρ and the dimensions of the specimen are known.

The resistivity of the germanium can be determined by tabulating the voltages *across the current leads* for those values of I which were previously used (*Table A1*), Resistance $= \rho(l/A)$, where

l = length of specimen
A = appropriate face area $= b.t.$

Results

Table A1

Current I mA	Voltage across no field (a)	Hall probes with field (b)	True Hall voltage (a − b)	Voltage across current leads
0·5	0·00174	0·00331	0·00157	0·06
1·0	0·00351	0·00679	0·00328	0·120
5·0	0·00751	0·03372	0·02621	0·620

Hall plate dimensions: length l = 1 cm
breadth b = 0·5 cm
thickness t = 0·04 cm.

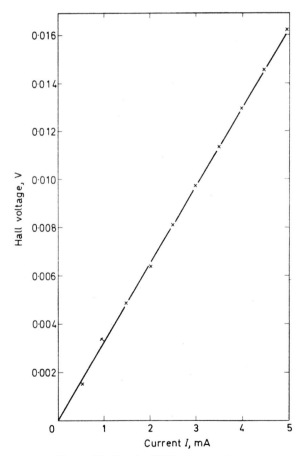

Figure A71. Graph of Hall voltage against current

ELECTRONICS AND SOLID STATE PHYSICS

Discussion of Results

The magnetic field can be measured using a search coil and fluxmeter but this is perhaps the least accurate measurement in the experiment. Use of a substance of standard susceptibility is preferable. To do this, finely powdered ferrous ammonium sulphate may be used which has a magnetic susceptibility of 32.3×10^{-6} e.m.u. cm^{-3} or 4.05×10^{-4} in terms of S.I. units.

The powder is filled to a chosen reference mark in a Pyrex tube some 10 to 15 cm long, the tube is gently tapped to ensure that the powder is well shaken down. Tube and contents are suspended from one arm of a balance with the lower end in the region of maximum magnetic field strength and the force on the specimen measured. (The specimen is strongly paramagnetic and experiences a pull down towards the strongest part of the field.) The force (upward) on the empty Pyrex tube is then measured separately and using the well known Gouy formula

$$mg = \tfrac{1}{2}(k - k_0) A\mu_0 (H^2 - H\text{o}^2)$$

the magnetic field H can be found,

$mg =$ the force on the specimen alone
$k =$ the magnetic susceptibility of the chosen standard, i.e. ferrous ammonium sulphate
$k_0 =$ the known magnetic susceptibility of air (3.76×10^{-7} in S.I. units or 0.03×10^{-6} e.m.u. cm^{-3})
$A =$ the cross-sectional area of the tube
$\mu_0 =$ the permeability of free space, and
$H_\text{o} =$ the magnetic field strength at the other end of the specimen, well outside the region of the pole pieces. This should be negligible compared with H.

If an electromagnet is used to provide the field, care should be taken that no remanent field exists should any 'no-field' readings be retaken.

Note. The answer obtained does depend upon the degree of doping which is found in the germanium and particularly so at room temperature. A graph showing the variation of conductivity with temperature in doped and undoped germanium can be found in the third reference below.

Reading and References

Two useful books for background reading here are:
Fundamental Principles of Transistors by J. Evans (Heywood, 1962).

Semiconductor Devices by J. N. Shive (Van Nostrand, 1959).
A very simple treatment of semiconductors and the Hall effect is given in *Physics for Electrical Engineers* by W. P. Jolly (English Universities Press, 1961), p. 119 and pp. 148–167. (Note particularly the example given on pp. 55–56.)
The Gouy method of measurement of magnetic susceptibilities may be found in *Modern Magnetism* by L. F. Bates (Cambridge University Press, 4th edition, 1961).

Experiment No. A2.3

TO DETERMINE THE ENERGY GAP OF A SPECIMEN OF INTRINSIC SEMICONDUCTING N-TYPE GERMANIUM

Introduction

The band theory of solids shows that when atoms or molecules combine to form solids, 'allowed' and 'forbidden' energy levels exist analagous to those appropriate to the electron in the theory of the atom but that the allowed levels form a band of preferred energies separated by zones of forbidden levels.

The forbidden region between a full band where all the allowed levels are occupied and the empty (conduction) band, is often referred to in semiconductor technology as the 'energy gap' and the aim of this experiment is to determine this for a specimen of n-type germanium.

Theory

The resistivity of a semiconductor depends significantly upon the temperature in accordance with the equation

$$\rho_T = \rho_0 \exp(B/T)$$

where ρ_T = the resistivity at temperature T
ρ_0 = the resistivity at temperature $T = 0$
B = the constant.
At temperature T_0

$$\rho_{T_0} = \rho_0 \exp(B/T_0)$$

hence
$$\frac{\rho_T}{\rho_0} = \exp\left\{B\left(\frac{1}{T} - \frac{1}{T_0}\right)\right\}$$
or
$$\rho_T = \rho_{T_0} \exp\left\{B\left(\frac{1}{T} - \frac{1}{T_0}\right)\right\}$$
where T_0 is a suitable reference temperature, say 273° K. Thus we may write
$$R_T = R_0 \exp\left\{B\left(\frac{1}{T} - \frac{1}{T_0}\right)\right\}$$
and taking logs
$$\log_e R_T = \log_e R_0 + B\left(\frac{1}{T} - \frac{1}{T_0}\right)$$
when, converting to logarithms to the base 10
$$\log_{10} R_T = \log_{10} R_0 + \frac{B}{2\cdot 303}\left(\frac{1}{T} - \frac{1}{T_0}\right)$$
or
$$\log_{10} R_T = \left(\log_{10} R_0 - \frac{B}{2\cdot 303 T_0}\right) + \frac{B}{2\cdot 303}\left(\frac{1}{T}\right)$$
whence plotting log R on the y axis and $1/T$ on the x axis and comparing $y = mx + c$ the slope of the graph gives
$$\frac{B}{2\cdot 303}$$
and $B = E/2k$

where E represents the energy gap appropriate to the semiconductor under test and $k =$ Boltzmann's constant.

Method

The specimen is enclosed in a thin walled tube (*Figure A72*) and connecting leads positioned from the semiconductor through the bung; the whole arrangement forming one arm of a Wheatstone network.

The resistance R may be found from
$$R = \frac{R_2 R_3}{R_1} \quad \text{(see Figure A72)}$$

and as the temperature of the bath is slowly raised the bridge is balanced at convenient temperature intervals and the results tabulated (*Figure A73*). A graph is then plotted of $\log_{10} R$ against $1/T$

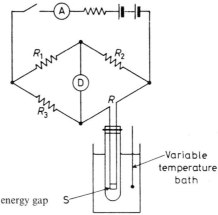

Figure A72. Determination of the energy gap of n-type germanium

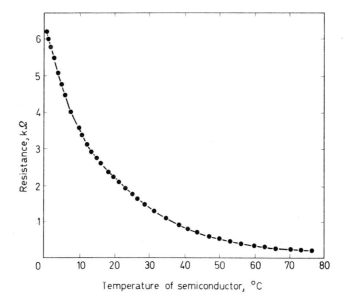

Figure A73. Graph of resistance against temperature for an intrinsic semiconductor

(*Figure A74*), from the slope of which $E = 4.606k \times$ (Slope of graph).

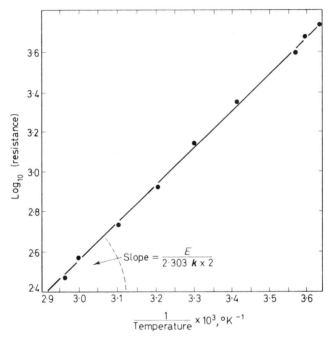

Figure A74. Graph of $\log_{10}$ (resistance) against 1/temperature

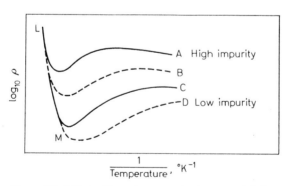

Figure A75. Effect of impurity concentration on resistivity as (1/temperature) varies

Discussion of Results

In assessing the accuracy of the experiment it should be recalled that the resistances rather than resistivity have been plotted and therefore it has been assumed that the dimensions of the specimen have not varied with temperature.

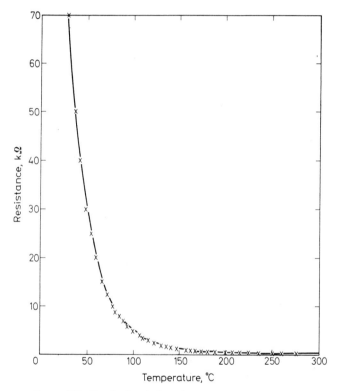

Figure A76. Graph of variation of resistance with temperature

Readings should be taken not only as the temperature rises but also as it cools when it is probable that better results are obtained since more stable conditions prevail.

Good ohmic contacts should be made between the lead and the specimen and it is necessary to check that the resistances due to these connections are low and of the same order of magnitude irrespective of the way current flows in the leads.

Further Work

If possible a number of different samples should be investigated since at low temperatures the resistivity depends to a very large degree upon the impurity content, the effect becoming less prominent at high temperatures where all samples should approach a common line representing the conductivity region LM of *Figure A75*.

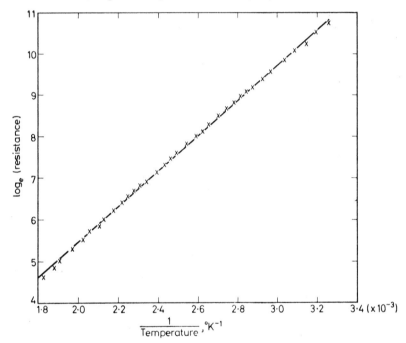

Figure A77. Graph of $\log_e$ (resistance) against 1/temperature

Reading and References

Elementary background reading upon the electrical classification of solids can be found in *Physics of the Atom* by M. R. Wehr and J. A. Richards (Addison-Wesley, 1960).

A detailed treatment of intrinsic and extrinsic semiconductors is to be found in *Semiconductor Devices* by J. N. Shive (Van Nostrand, 1959).

Note. *Figure A76* shows the results obtained with a second specimen and *Figure A77* the logarithm of resistance to the base e plotted directly against (1/temperature).

Experiment No. A2.4

TO INVESTIGATE THE CHARACTERISTICS OF A THERMISTOR

Introduction

Thermistors are semiconducting devices which have large negative temperature coefficients of resistance. This property enables them to be used as 'thermometers' over restricted ranges and makes them valuable as resistance control devices in conditions of changing temperature. Other applications are found in 'surge' suppression, trigger circuitry, power measurement and in the stabilization of low-frequency oscillators.

Theory

The equation governing the variation of resistance R of a thermistor with temperature T may be written

$$R = a \exp(b/T) \quad (1)$$

where a and b are constants depending upon size and mode of mounting and upon thermistor material respectively. A typical thermistor resistance temperature graph is shown in *Figure A78*. By definition the temperature coefficient of resistance α is given as

$$\alpha = \frac{1}{R}\left(\frac{dR}{dT}\right) \quad (2)$$

and since

$$R = a \exp(x) \quad \text{where} \quad x = bT^{-1}$$

$$\frac{dR}{dx} = a \exp(x) \quad \text{and} \quad \frac{dx}{dt} = -\frac{b}{T^2}$$

therefore

$$\frac{dR}{dT} = \frac{dR}{dx}\frac{dx}{dT} = a \exp(x)\left(-\frac{b}{T^2}\right) = -\left(\frac{ab}{T^2}\right)\exp\left(\frac{b}{T}\right)$$

therefore

$$\alpha = \left\{\frac{1}{a \exp(b/T)}\right\}\left(-\frac{ab}{T^2}\right)\exp\left(\frac{b}{T}\right) = -\frac{b}{T^2} \quad (3)$$

ELECTRONICS AND SOLID STATE PHYSICS

If logs are taken

$$\log R = \log a + \frac{b}{T} \quad \text{from equation (1)}$$

hence a graph of log R against $1/T$ gives a straight line whose slope yields the value of the constant b and whose intercept on the y axis

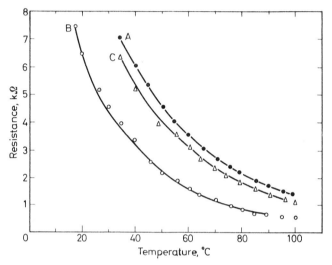

Figure A78. A typical thermistor resistance-temperature graph. Three specimens are shown. Ordinate values should be multiplied by 1 for specimen A, by 10 for specimen B and by 0·4 for specimen C

enables a to be found (*Figure A79*), whence the temperature coefficient of resistance α is given by equation (3).

Method

The thermistor is placed in one arm of a Wheatstone bridge such as shown in *Figure A80*. The 1 kΩ ratio arms can be the usual plug type boxes and the variable arm R_1 a dial decade box.

The thermistor should be mounted in such a way that it can be immersed in an oil bath and the bridge should be balanced whilst the bath is at room temperature. The value of R_1 is then noted.

The process is repeated over suitable temperature intervals up to say 100° C and the graphs plotted as indicated above.

Do not raise the temperature of the oil bath too high. The experiment should preferably be done in a fume cupboard.

Discussion of Results

The graphs given indicate that the experiment is capable of yielding good results, but it must be realized that variations in temperature within the bath do occur unless adequate means of stirring are provided.

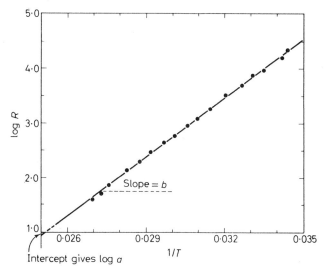

Figure A79

Further Work

The graphs shown are in fact those for 'home made' thermistors made in accordance with the instructions given in Experiment D10.9. It will be worthwhile to compare the results obtained in this way with those obtained with a commercial thermistor, e.g. S.T.C. Sentercel F.15.

Reading and References

Background reading on semiconductors may be found in *Physics for Electrical Engineers* by W. P. Jolly (English Universities Press, 1961). *A Simple Explanation of Semiconductor Devices* published by

Mullard Ltd. (1965), *An Introduction to Semiconductors* by W. C. Dunlap (Wiley, 1957), or *Fundamental Principles of Transistors* by J. Evans (Heywood, 2nd edition, 1962).

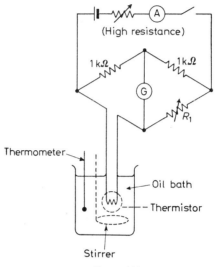

Figure A80

A brief account of thermistors may also be found in *Electronic Processes in Materials* by L. V. Azaroff and J. J. Brophy (McGraw-Hill, 1963). An elementary account of thermistors, which gives a number of applications, is given by G. J. King in *Practical Electronics* for March 1968. Detail is also given of a simple bridge circuit for building an 'electronic thermometer' suitable for the temperature range from $-70°$ C to $200°$ C in which the response is rendered approximately linear.

Experiment No. A2.5

TO PLOT THE CURRENT/VOLTAGE CHARACTERISTIC OF A JUNCTION DIODE

Introduction

The quantum concept of discrete energy levels within the atom may be extended to solids, provided that the levels are considered as bands of finely spaced levels rather than as single levels themselves. On this basis one may subdivide the solids into: (*a*) insulators, where electrons fill the lower band of energy levels (the valence band) leaving the levels in the upper band (the conduction band) (*Figure A81*) empty but with a forbidden gap between the two bands so

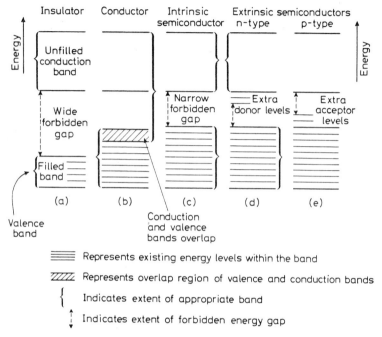

Figure A81. Energy-level diagrams showing the band structure of various types of materials

large that electrons cannot cross it; (*b*) conductors, where electrons already exist in the conduction band and on being given energy simply move to higher levels within the band; (*c*) semiconductors, which are materials possessing a small forbidden energy gap, such that application of thermal energy results in electrons crossing the gap. The small gap may exist naturally, e.g. as in germanium (0·7 eV, $1·1214 \times 10^{-19}$ J) when we have an intrinsic semiconductor, or it may be created by 'doping' the chosen material to create extra energy levels in what was normally the forbidden gap (*Figure A81*) giving an impurity or extrinsic semiconductor.

If the extra level is occupied by an electron we have an n-type semiconductor.

If the new impurity level lies near the valence band electrons are enabled to pass from the full valence band leaving positively charged 'holes'—a p-type semiconductor.

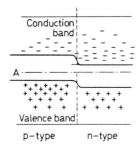

Figure A82. Schematic diagram of the energy levels within a junction of n- and p-type semiconducting materials. The shift of energy levels takes place within the junction system so that a single Fermi level characteristic of the system (and common to both types of junction material) results

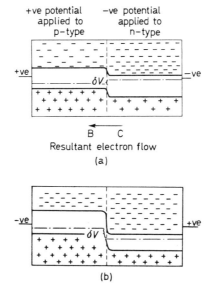

Figure A83. Application of potential difference to a junction diode. (a) The condition of forward bias. Biasing in this way increases the energy of electrons in the n-type material more than that of the electrons in the p-type with resultant electron flow in direction CB. (b) The condition of reverse bias. Resultant electron flow small and tending to zero across junction region

If p- and n-type materials are placed in conjunction as in *Figure A82*, the device functions as a diode rectifier, for on applying a potential difference with the positive on the left (*Figure A83*) positive

CHARACTERISTIC OF A JUNCTION DIODE   A2.5

'holes' drift from left to right and electrons from right to left, resulting in a large 'forward current'.

If the positive is applied on the right the charge carriers drift away from the junction leaving what is effectively an 'insulating' area so that little or no current will flow.

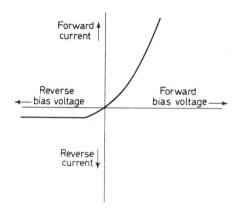

Figure A84. Idealized characteristic of semiconductor diode

Figure A84 shows the ideal form of the solid state diode characteristic but the actual one deviates markedly from this, as *Figures A85* and *A86* demonstrate.

Theory

To pass a current through the diode in the direction of easy flow (*Figure A87a*) only a small voltage ($\sim$ a few volts) is required. In the reverse direction a large voltage is needed.

It is necessary to check the polarity most carefully otherwise the diode may short-circuit with the application of the high voltage and damage the microammeter. For this reason the check is made using the low voltage cell as supply (*Figure A87a*) when, if the polarity is incorrect, it will only result in zero readings without damaging the measuring instruments.

Method

Wire up the circuit as shown in *Figure A87a* and using a small value of voltage V, note the milliameter reading (i). Increase the

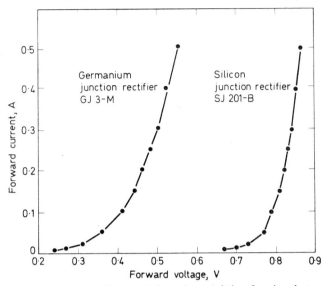

Figure A85. Forward current/voltage characteristics of semiconductor diodes

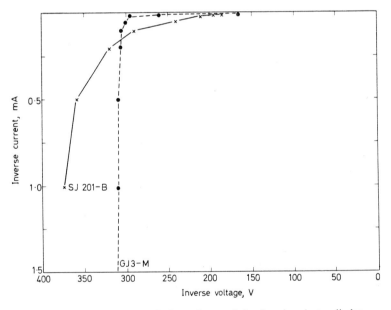

Figure A86. Inverse current/voltage characteristic of semiconductor diodes

CHARACTERISTIC OF A JUNCTION DIODE   A2.5

voltage in small steps up to the maximum value recommended for the device being used.

Now wire the diode in the reverse sense using a different rheostat and voltmeter in the circuit.

Adjust the voltage tapping to give a low voltage reading (~ 10 V) and if the milliameter reading is too small to be read conveniently, a change to the more sensitive microammeter may be made.

Increase the voltage slowly in steps of 5 V and note the corresponding values of current.

Plot the graph of current on the y axis against voltage on the x axis as shown in *Figures A85* and *A86*.

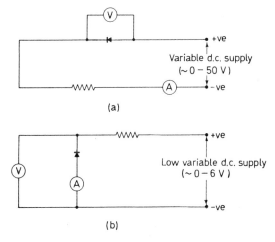

Figure A87. Circuit arrangements for measurement of the forward (*a*) and inverse (*b*) characteristics of a junction diode

Discussion of Results

It is important to note the peak inverse voltage that the device will withstand without breaking down and also the maximum permissible temperature at which it can operate. The forward resistance should be low if power wastage is to be avoided and the inverse resistance high so that the current in this direction is low. For the germanium junction rectifier for which the characteristic is plotted (*Figures A85* and *A86*), the inverse resistance is seen to be very high, giving an inverse current of less than 10 µA up to the breakdown point.

It is of interest to note that the experimental breakdown point is just above 300 V, whilst the manufacturer's specified peak inverse voltage in this case is 200 V.

Further Work

The characteristics of a number of different semiconducting rectifier devices should be compared. *Figures A85* and *A86* show not only those of germanium but also that of the more robust silicon junction device less easily damaged by overheating, used for power rectifiers.

Reading and References

An elementary account of band theory and diodes may be found in *Electricity* by J. Goodier and J. W. Meynell (Mills and Boon, 1964).

A deeper treatment is given in *The Electrical and Magnetic Properties of Solids* by N. Cusack (Longmans, 1958).

Experiment No. A2.6

TO PLOT THE COLLECTOR AND BASE CHARACTERISTICS OF A JUNCTION TRANSISTOR

Introduction

The junction transistor consists of a thin slice of n-type material (usually germanium) sandwiched between two pellets of p-type material. There are therefore three distinct regions, the emitter, the collector and the base. There are thus three different ways in which the transistor may be wired in circuit, but the mode chosen here is the common emitter connection, i.e. the emitter lead is common to input and output signals, where the base current (not the emitter current) is the controlling factor and change in this causes corresponding change in the output circuit.

CHARACTERISTICS OF A JUNCTION TRANSISTOR A2.6

Theory

Let I_B = base current of the transistor
I_C = collector current of the transistor
and β = the current gain appropriate to common emitter connection, then

$$\beta = \left(\frac{I_C}{I_B}\right) = \left(\frac{\delta I_C}{\delta I_B}\right)$$

over the linear portion of the characteristic having collector voltage V_C constant. A graph showing how the gain β varies with the current I_C may be plotted (*Figure A88*).

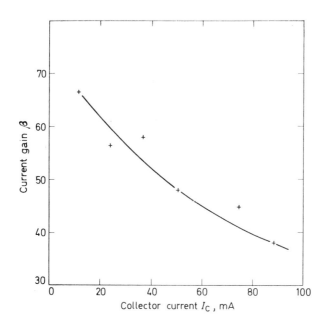

Figure A88. Characteristic of a junction transistor showing the variation of current gain with collector current

Theoretically eight different characteristic curves, each involving three parameters, may be plotted from the four variables, base current, control current, output voltage and input voltage, and two are required to give a full specification for the junction.

Method

Connect up the circuit as in *Figure A89* taking the utmost care to check the polarity of each connection before switching on the electrical supplies, otherwise damage to the transistor will result.

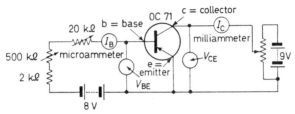

Figure A89. Circuit used to investigate the characteristics of a pnp transistor connected in the common emitter mode

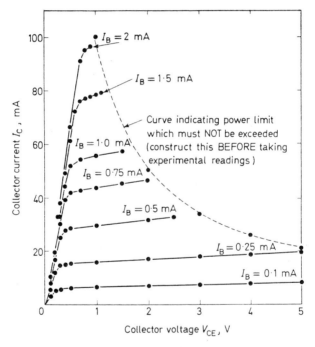

Figure A90. The collector characteristics of a junction transistor connected in the common emitter mode

Having done this set the base to emitter voltage V_B and the collector to emitter voltage V_C to zero.

CHARACTERISTICS OF A JUNCTION TRANSISTOR   A2.6

Increase the voltage V_C in convenient steps up to the maximum recommended voltage of the device being used (Mullard or Ediswan provide suitable transistors) and tabulate corresponding values of collector current I_C.

Reset the base current to a small convenient value and keeping this constant repeat the procedure. Plot the graphs of collector current against collector voltage to display the family of collector characteristics (*Figure A90*).

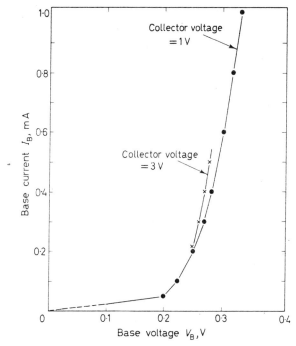

Figure A91. The base characteristics of a junction transistor connected in the common emitter mode

Now set the collector voltage to a small convenient value (~ 1 V). Increase the base current I_B in very small steps (see *Figure A91*) for typical values) and note the base voltage V_B. Again do not exceed the maximum power recommended for the junction.

Repeat this for a different constant value of collector voltage and plot the base characteristic of the transistor.

Finally calculate the current-gain appropriate to the common emitter mode of connection $\beta = \delta I_C / \delta I_B$ at constant V_C (*Figure A88*).

ELECTRONICS AND SOLID STATE PHYSICS

Discussion of Results

It is wise before plotting the actual characteristics to tabulate a convenient series of collector voltages V_C (from say 1 to 5 V, in 1 V steps) and to calculate the collector current determined by the recommended power limit which will be stated for the junction used. Then a graph representing the power limit, which must not be exceeded, may be plotted (shown dotted in *Figure A90*).

Notice that whilst analogies between the junction transistor and the thermionic valve may be made, the transistor is a current amplifier, not a voltage amplifier as is the thermionic triode.

Also the collector characteristics of the solid state transistor junction resemble those of the pentode valve rather than the triode.

Further Work

As pointed out above, it is possible to plot other characteristics for the transistor using different modes of connection and to link the current gain associated with the other possible ones.

Reading and References

A very useful elementary introduction to transistors is found in *Simple Transistor Measurements* (Mullard, 1963). This publication gives brief details of transistor manufacture as well as a useful bibliography.

Transistor characteristics and circuits are discussed in *Semiconductors* by H. Teichmann (Butterworths, 1964).

Experiment No. A2.7

TO MEASURE THE CURRENT GAIN OF A TRANSISTOR AMPLIFIER CONNECTED IN THE COMMON EMITTER MODE

Introduction

Investigation of the transistor characteristics (Experiment A2.6) showed that analogies between the pnp transistor and the thermionic triode could be drawn, the emitter, base and collector of the solid state device corresponding to the cathode, grid and anode respectively of the valve.

CURRENT GAIN OF A TRANSISTOR AMPLIFIER A2.7

The transistor may thus be used as an amplifier and the aim of this experiment is to plot the graphs of current gain, voltage gain and power gain against load resistance for the transistor connected in the common emitter mode.

Theory

Figure A92a shows the basic circuit to be used. It makes use of a potential divider $R_1 R_2$ and an emitter resistance R_4 in order to give the negative feed back which is necessary (otherwise emitter current

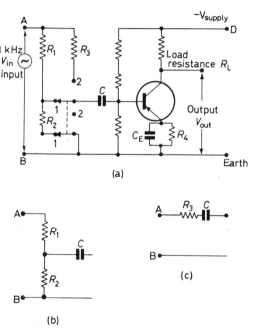

Figure A92. Circuit for determination of the gain of a transistor amplifier. (*a*) The switch is ganged and in position 1 gives circuit (*b*) thus making use of the potential divider $R_1 R_2$. With the switch in position 2, the circuit becomes effectively as shown in (*c*). Typical component values for an OC 71 transistor are as follows: R_1, 3 kΩ; R_2, 0·1 kΩ; R_3, 100 kΩ; C, 5 µF; C_E, 5 µF; R_L, 0·5–5 kΩ

would tend to produce positive bias on the base and to reduce the emitter current). The capacitor C_E ensures that for the a.c. signal the emitter is effectively at earth potential.

The net (alternating) base to emitter voltage V_{be} can be calculated

$$V_{be} = V_{in} \frac{R_2}{R_2 + R_1}$$ (assuming $R_2 <$ the input impedance to the transistor)

since the input voltage V_{in} and the resistances of the potential divider are known.

The voltage gain (V_{out}/V_{be}) can thus be calculated, so may the base current

$$I_b = \frac{V_{in} - V_{be}}{R_3}$$

and the collector current $I_c = V_{out}/R_L$ where V_{out} represents the output voltage.

The current gain I_c/I_b can thus be found and the power gain becomes

$$\frac{V_{out}I_c}{V_{be}I_b}$$

Taking a series of values of the resistance load R_L in this way enables graphs of current gain, voltage gain and power gain to be plotted against the load (*Figure A93*).

Method

The circuit is assembled as shown in *Figure A92a*, where a typical circuit is given as example. The ganged switch is turned to position 1 giving effectively the circuit of *Figure A92b* and a cathode ray oscilloscope is connected across the output in order to examine the waveform of the output voltage.

A suitable oscillator giving a signal of 1 kHz is connected across the input together with suitable direct voltage supply (about 6 V).

The amplitude of the input signal is slowly increased to that maximum value which will still give an undistorted output waveform on the oscilloscope. The d.c. input is now adjusted to give maximum amplitude of the undistorted output waveform and first the alternating input voltage and then the alternating output voltage are measured using a suitable valve voltmeter.

The a.c. signal is now reduced to zero and the d.c. supply current is measured.

Now restore the a.c. input signal and after checking that the output voltage is of the same value as previously read, put the switch to position 2 (*Figure A92*).

CURRENT GAIN OF A TRANSISTOR AMPLIFIER A2.7

The circuit of *Figure A92c* is now effectively in operation and the input voltage is adjusted to a new value V'_{in} in order to keep the output voltage as it was previously. Note the new input voltage.

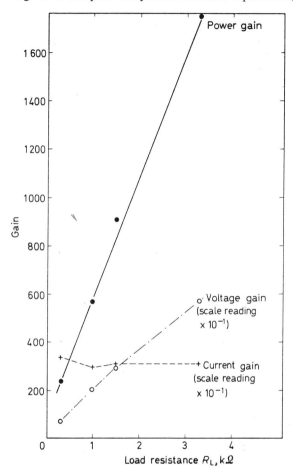

Figure A93. Graph of the gain of a transistor amplifier against load-resistance

Tabulate the base voltage, the base current, the collector current, the voltage gain, the current gain, and the power gain calculated as indicated above.

Repeat the procedure for different values of the load resistance and plot the graphs of *Figure A94*.

Discussion of Results

The results bring out a number of criteria of importance to the design engineer, e.g. the graph of *Figure A94* shows that a maximum undistorted output voltage occurs for a particular value of load resistance (about 1 kΩ in the example quoted).

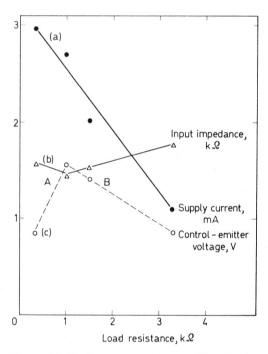

Figure A94. The junction transistor amplifier. Graphs showing variation of (*a*) supply current, (*b*) input impedance and (*c*) control emitter voltage with load

If however a large voltage gain for a small input signal is required then a higher load resistance is indicated (*Figure A93*).

To obtain a more accurate estimate of optimum load, a large number of different load resistances must be taken, corresponding to the region AB of the curve, when it would be found that the graph does not rise sharply to a point but passes slowly through a maximum.

Perhaps the most difficult part of the experiment is assessing the point at which distortion first occurs on the oscilloscope waveform.

Further Work

It is of interest to calculate the input impedance $Z_{in} = V_{be}/I_b$, and to determine the variation of this quantity with load resistances as shown in *Figure A94b*.

Reading and References

The theory behind the transistor amplifier may be found in *Principles of Electronics* by M. R. Gavin and J. E. Houldin (English Universities Press, 1959). The different modes of connection are treated here, together with equivalent circuits and load-line theory.

A more elementary treatment may be found in *Handbook for Science Masters and Lecturers* by I. L. Muler and K. E. J. Bowden (Advance Components Ltd., 1963).

Experiment No. A2.8

TO DETERMINE THE ENERGY GAP OF GERMANIUM FROM INVESTIGATION OF THE VARIATION OF THE SATURATION CURRENT WITH CHANGE IN TEMPERATURE UNDER THE CONDITIONS OF REVERSE BIAS FOR A JUNCTION DIODE

Introduction

As was seen in Experiment A2.5, a pn junction is in a condition of reverse bias when in circuit such as shown in *Figure A97*, so that the p side of the junction is held more negative than the n side.

For any fixed temperature the current across the junction rapidly reaches saturation (*Figure A96*). At room temperature this value is quite low (~microamps) but if the junction is heated in a constant temperature bath the reverse saturation current increases dramatically (*Figure A96*).

Theory

The semiconductor diode characteristic may be represented fairly closely by the equation

$$I = I_s\{\exp(eV/kT) - 1\}$$

where I = general value of current
 I_s = reverse saturation current
 e = fundamental charge on the electron
 V = applied voltage
 T = absolute temperature, and
 k = Boltzmann's constant.

The saturation current I_s is itself a function of (a) electron charge, (b) diffusion lengths and diffusion coefficients of both holes and electrons as well as of (c) the number of holes and electrons present under conditions of equilibrium.

We shall assume that it is of the form

$$I_s = A \exp(-E_g/2kT)$$

where A is a constant, and E_g is the energy gap corresponding to germanium, i.e.

$$\log_e I_s = \log_e A - \frac{E_g}{2k}\left(\frac{1}{T}\right)$$

or

$$\log_{10} I_s = \frac{\log_e A}{2\cdot 303} - \frac{E_g}{4\cdot 606 k}\left(\frac{1}{T}\right).$$

Thus a plot of $\log I_s$ against $1/T$ enables an estimate to be made of the energy gap E_g knowing the Boltzmann constant k.

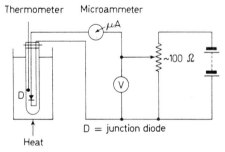

Figure A95. Variation of reverse saturation current with temperature for a germanium junction diode

Method

First place the diode in a thin-walled test tube together with a thermometer and put the whole arrangement into an ice bath.

ENERGY GAP OF GERMANIUM   A2.8

Connect up the circuit as shown in *Figure A95*, and increase the reverse bias in suitable steps such as indicated by *Figure A96*, noting the reverse current at each step *and also* the temperature.

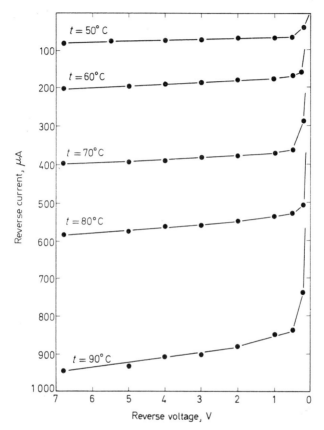

Figure A96. Graph showing the change in reverse bias characteristic with change in temperature

Repeat the procedure as the water bath is gently heated and plot graphs such as *Figure A96*, in order to determine the saturation current at each chosen value of temperature.

Then plot the logarithm of this value against the reciprocal of absolute temperature, when a straight line graph should result (*Figure A97*) the slope of which enables the energy E_g of germanium to be calculated.

Discussion of Results

At the outset it may be found that the current is too small to measure accurately at 0° C in which case one must commence reading at the lowest convenient value. The answer obtained by this method is generally very high. This is due to a number of causes.

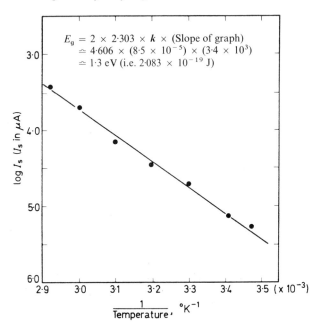

Figure A97. Graph showing the variation of saturation reverse bias current with temperature

First of all the form of the equation used in this experiment is only very approximately true. It would be nearer the truth to write

$$I_s = AT^{\frac{3}{2}} \exp(-E_g/2kT)$$

but analysis of a plot of $\log(I_s/T^{\frac{3}{2}})$ against $1/T$ would show that in practice the effect of the $T^{\frac{3}{2}}$ term is 'swamped' by that of the exponential term and may for practical purposes be assumed to be constant in this case.

Secondly the term $(E_g/2)$ used in the index of the law is an erroneous one, which assumed that the Fermi level lies in the middle of the forbidden gap (*Figure A98*). It would be more accurate to write

$$I_s = AT^{\frac{3}{2}} \exp(E_F - E_B/kT)$$

but $(E_F - E_B)$ is in fact not strictly constant, since as the temperature rises the Fermi level tends to rise until, at the upper range of temperatures quoted here, it lies close to the level of the conduction band. It is not surprising therefore that the answer obtained here is much higher than the true value.

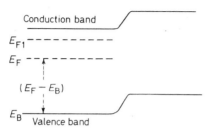

Figure A98. Schematic diagram showing the shift of the Fermi level upwards with rise in temperature.

- E_B represents the lowest level of the valence band
- E_F represents the Fermi level at the lower temperature
- E_{F1} represents the position of the Fermi level at higher temperature where it has moved upwards towards the conduction band

For the purposes of illustration in the results quoted, the temperature has been taken up to 100° C but the upper operational temperature of a germanium device is in the region of 70° C and it is not advisable to take the junction above this.

Further Work

It would be gratifying to carry out the same experiment with a silicon junction, which can operate at a higher temperature than the germanium, but the currents are of very low order and beyond the range of the conventional microammeter. One might however compare the results obtained here with those of the Zener diode (Experiment No. 2.11) and also look at those of the point contact diode and the tunnel diode.

Reading and References

Background reading on pn junctions may be found in *Modern Physics* by R. L. Sproull (Wiley, 1964), as well as in *Solid State Physics* by A. J. Decker (Macmillan, 1960), and in *Solid State Physics* by C. Kittel (Wiley, 1966).

ELECTRONICS AND SOLID STATE PHYSICS

Experiment No. A2.9

THE USE OF A SEMICONDUCTING THERMOELECTRIC MATERIAL (BISMUTH TELLURIDE) TO DETERMINE THE DEW POINT BY OBSERVATION OF THE PELTIER EFFECT

Introduction

Bismuth telluride, when prepared from its constituent elements in the correct stoichiometric proportions, is always a p-type semiconductor. The n-type material however may be produced by the addition of one of the halogens. If the p-type material is made one leg of a thermojunction with the n-type the other leg, thermoelectric effects are observed to a marked degree, and when current passes through the junction Peltier heating (or cooling) occurs.

In this experiment the cooling of such a bismuth telluride junction is used to determine the dew point and hence the relative humidity of the atmosphere.

Control of humidity is for various reasons an important factor in many industrial processes. For example, such materials as cotton need careful control of the atmospheric conditions in which they are produced; tobacco and many other commodities have to be stored under optimum humidity conditions, air conditioning is important for personnel comfort in factories, laboratories and other working environments, electrical components may suffer in operation unless the atmospheric conditions about them are strictly controlled. Thus hygrometry (the study and measurement of humidity) assumes an industrial importance which may not at first sight be obvious.

Theory

When current passes through the junction, change in temperature will take place due to two effects: (1) 'Joule' heating ($=I^2R$), where I represents current, and R resistance. This effect is independent of the way in which current flows. (2) The Peltier effect, where the heat H involved is given by $\Pi I t$ (Π being the Peltier coefficient and t the time for which the current flows). This effect is reversible and produces rise or fall in temperature depending upon the direction of current flow.

In this experiment one junction is in contact with a metal mirror

DETERMINATION OF DEW POINT A2.9

and the second with a series of fins which act as a heat sink and effectively maintain this junction at a constant (reference) temperature.

The dissimilar materials forming the junction are the p- and n-type forms of bismuth telluride.

The total potential difference across the thermal junction will, for the reasons noted above, consist of two components, one due to Peltier effect (V_p) and one (V) obeying Ohm's law ($=IR$), i.e.

$$V_{total} = V_p + V.$$

Method

Room temperature at the beginning of the experiment is first carefully noted. The apparatus is set up as shown in the diagram (*Figure A99*). (A suitable bismuth telluride junction with readily

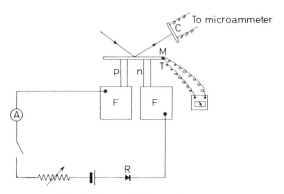

Figure A99. Circuit used in dew point measurement
p and n represent p- and n-type bismuth telluride components forming the thermojunction
F, Cooling fins maintaining one thermojunction effectively at constant temperature
M, Metal mirror upon which mist forms at dew point
R, Rectifying semiconductor diode (for safety)
C, Photocell by which dew point is detected
T, Conventional thermocouple used to measure the temperature of the mirror surface

mounted heat sink and mirror may be obtained from Salford Instruments, Ltd.)

The beam from a suitable white light source is directed onto the mirror and reflected to impinge on a photocell as indicated in the diagram, when a maximum deflection of the photocell microammeter should be obtained.

The rheostat is then adjusted to give a current of say 0·2 A when the potential difference (which may be measured on a suitable standardized potentiometer or recorder) is seen to rise instantaneously in obedience to Ohm's law, i.e. $V = IR$. However, the thermal e.m.f. in the bismuth telluride junction which is superimposed upon this (IR) voltage, takes a little time to stabilize and not until this is seen to be so from the constancy of the conventional

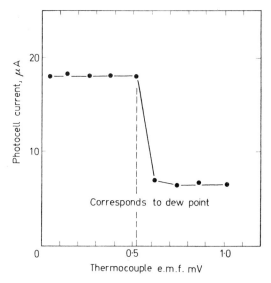

Figure A100. Use of a semiconducting thermoelement to determine the dew point by means of the Peltier effect

thermocouple reading (indicating that the mirror temperature has reached equilibrium with its surroundings), should the thermocouple galvanometer and photocell microammeter readings be tabulated. The current is then increased in steps of say 0·1 A and the procedure repeated.

It is noted that at one point of increase of current the photocell microammeter reading suddenly falls (*Figure A100*). This corresponds to the condition where the formation of 'dew' on the polished mirror surface has drastically reduced the amount of light reflected.

Discussion of Results

It is essential to allow sufficient time for stabilization of the thermoelectric e.m.f. before readings are taken. It is wise to screen the area of the mirror since breath on the surface, or the presence of draughts about it, can falsify the results.

Having determined dew point from the graph (*Figure A100*) the percentage relative humidity of the atmosphere may be obtained from standard tables.

Further Work

If the thermojunction is in firm contact with a small thin copper rod or wire then the junction provides a means of heating (and cooling) the wire and by keeping the conditions at the junction steady, the thermal conductivity of the wire may be determined or, alternatively, if the junction temperature is allowed to fluctuate alternately hot and cold then the thermal diffusivity of the wire may be determined (see Experiment No. E1.6 on Ångström's method).

Reading and References

The performance of bismuth telluride thermojunctions is dealt with in detail in a paper of that title by H. J. Goldsmid, A. R. Sheerd and D. A. Wright in the *British Journal of Applied Physics*, vol. 9, p. 365, Sept. 1958.

The background reading of the Peltier and other thermoelectric effects is contained in the many standard degree texts, in particular, *Electricity and Magnetism* by J. H. Fewkes and T. M. Yarwood (University Tutorial Press, 1965).

Experiment No. A2.10

TO DETERMINE THE VOLTAGE-CURRENT CHARACTERISTIC OF A VARISTOR

Introduction

A varistor, as its name implies, is a voltage dependent resistor, the value of whose resistance varies with applied voltage in such a way that above a certain critical operational value of voltage, the current rises very rapidly so that the voltage-current characteristic is non-linear. The varistor is basically silicon carbide with a ceramic binder and the non-linearity of the voltage-current dependence is due to contact resistance between the crystallites of silicon carbide.

The aim of this experiment is to investigate the nature of the law governing the voltage-current characteristic.

The devices are used as protective 'surge' arrangements.

Theory

Assuming the law has a form

$$I = aV^n$$
$$\log I = n \log V + \log a$$

and a straight line graph results whose slope gives the value of n. Both a and n are constants whose values depend upon the physical dimensions, the composition and the processes of manufacture of the varistor.

Method

The varistor is connected in the circuit as shown in *Figure A101*, and at the outset the potential divider is set to give zero volts across

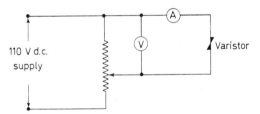

Figure A101. Circuit used in the determination of the voltage-current characteristic of a varistor

the varistor. The voltage is slowly increased in steps of say 10 V and at each point the current through the device (in milliamps) is noted.

A graph may be plotted such as *Figure A102* showing the rapid rise of the current above the point A.

To investigate the form of the law appropriate to this curve log I is plotted against log V when a straight line results (*Figure A103*).

Discussion of Results

As *Figure A103* shows it is possible to obtain a fairly accurate value of the index n of the law, the 'scatter' of the points in the case

CHARACTERISTIC OF A VARISTOR A2.10

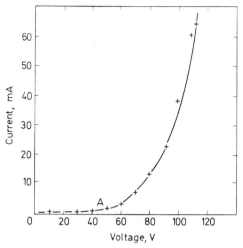

Figure A102. The voltage-current characteristic of a varistor

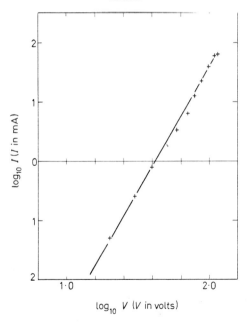

Figure A103. Graph showing the relationship between $\log_{10} I$ and $\log_{10} V$ for a varistor. I, current through varistor; V, voltage across varistor

shown indicating an accuracy to the order of ±5% in the value obtained. It is not possible however to achieve such accuracy in the estimation of log a since a very small variation in the value of the slope can give rise to a large change in the value of log a. Note too that displacement (up or down) of the best fit line can affect the value of log a.

Further Work

The value of the index may vary widely (between about 2 and 5) for varistors of different specifications and a number should be investigated other than the one used here ('Metrosil').

It is also worthwhile to vary the temperature at which the (encapsulated) varistor operates to discover the effect of change in temperature upon the log I – log V characteristic.

Reading and References

Details of several voltage dependent resistors are given in the *Mullard Technical Handbook*, vol. 5, under 'Non-linear resistors'.

The Characteristics and some Applications of Varistors by F. R. Stansel in *Proceedings of the Institute of Radio Engineers*, vol. 39, No. 4, April, 1951, forms useful background reading.

Experiment No. A2.11

TO INVESTIGATE THE EFFECT OF CHANGE IN TEMPERATURE DOWN TO LIQUID NITROGEN TEMPERATURE UPON THE BREAKDOWN POTENTIAL OF A VOLTAGE REFERENCE DIODE

Introduction

Electrical breakdown is known to occur in an insulator if a large electric field exceeding a certain critical value is applied to the insulator.

With a semiconductor, breakdown can take place in two ways:

(1) By Zener breakdown, where a sufficiently large electric field

BREAKDOWN POTENTIAL OF A REFERENCE DIODE   A2.11

causes some electrons to move from the valence band to the conduction band.

(2) By avalanche breakdown where a free electron gains kinetic energy from the applied field, greater than that which it loses in collision with atoms of the material in which it moves.

This excess energy can then be lost by collision with a valence electron and results in the production of an electron-hole pair. The *two* electrons can in their turn produce more electrons and an avalanche effect is rapidly created.

Theory

The inverse voltage current characteristic of a voltage-reference diode is as shown in *Figure A104*. (Compare also the graphs of semi-

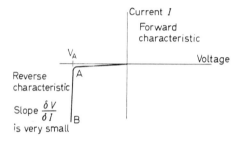

Figure A104. Idealized characteristic of a voltage reference diode. Note that in region AB the voltage across the diode is effectively constant for small changes in current

conductor diodes shown in Experiments No. A2.5 and A2.8.) Over the region A to B the slope dV/dI is very small and for practical purposes V_A is effectively constant and may therefore be used as a voltage reference provided the temperature at which the device operates remains constant.

However, if the temperature varies the reference voltage varies and breakdown of the pn junction changes as shown in *Figure A105*.

The breakdown as stated above may either be Zener breakdown or avalanche breakdown, but it is found that when V_A is greater than 5 V or so, the avalanche effect predominates, whilst if V_A is less than 5 V the Zener effect is the one chiefly responsible for breakdown.

In the case of avalanche breakdown, the breakdown voltage is found to increase with temperature, i.e. it has a positive temperature coefficient, whereas for Zener breakdown the slope is negative.

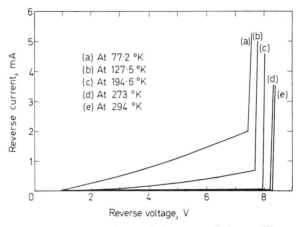

Figure A105. Characteristics of the Zener diode at different temperatures under reverse bias

In this experiment a graph of breakdown voltage against temperature as the latter is reduced to liquid nitrogen temperature is obtained, and from the nature of the slope of the graph the mechanism responsible for breakdown is deduced.

Method

The voltage reference diode is placed in a circuit such as shown in *Figure A106*, so that it is in the condition of reverse bias (i.e. conventional current flow is against the direction of easy flow through the junction). The voltage across the junction may be measured using a suitable valve-voltmeter or as is indicated in the diagram by a suitable direct recorder.

The current may be conveniently measured in terms of the voltage developed across a standard 1 Ω resistance. The rheostat resistance is increased slowly from zero to its maximum value, the reverse current and reverse voltage are noted and a graph (*Figure A105*) plotted.

The diode is then placed in turn in melting ice, solid carbon dioxide (which sublimes at a temperature of 194·6° K) and liquid nitrogen (contained in a suitable Dewar and boiling off at 77·2° K), similar graphs showing breakdown being plotted.

Finally the pn junction is suspended just above the liquid nitrogen and the breakdown voltage carefully noted.

The breakdown voltages obtained from this first graph are then

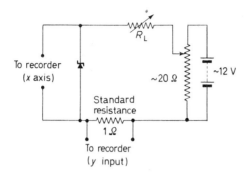

Figure A106. Investigation of avalanche or Zener breakdown in a junction diode

plotted against the known temperatures used in the experiment when a graph such as *Figure A107* results.

Using the graph the temperature corresponding to that of the vapour just above the liquid nitrogen may be estimated.

Discussion of Results

Notice that the graph of *Figure A107* is nearly linear. The slope at the mid-point of the chosen range should be found, when it is seen that the device might be used as a thermometer. It should also be noted that the breakdown voltage is greater than 5 V in the example given and that the slope of the graph of *Figure A107* is positive, indicating that the avalanche breakdown effect is the predominant one here.

Further Work

A voltage reference diode of a rating below 5 V should be chosen and the experiment repeated, when a negative slope should result corresponding to Zener breakdown. It is also of interest to look at the region about 5 V where the temperature coefficient passes from positive to negative.

Reading and References

Background reading on voltage reference diodes and the temperature dependence of breakdown voltage may be found in *Introduction to the Theory and Practice of Transistors* by J. R. Tillman and F. F. Roberts (Pitman, 1961), also in R. H. Mattson's *Electronics* (Wiley, 1966).

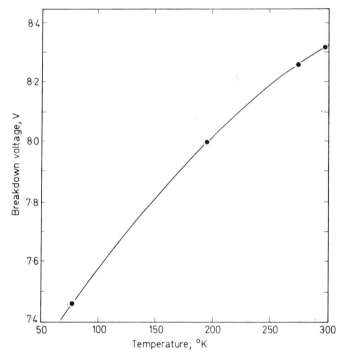

Figure A107. Variation of breakdown voltage of a Zener diode with temperature

A note on voltage reference diodes of various ratings together with detailed graphs showing their performance are to be found in a pamphlet of that name No. 5865–6. (B.T.-H. Ltd., 1959). The graphs apply to much higher temperatures than those involved in this experiment, but it is of interest to compare values of temperature coefficient in the two ranges.

Section B. Nuclear and Radiation Physics

Experiment No. B1.1

TO PLOT THE CHARACTERISTIC CURVE OF A GEIGER COUNTER AND TO DETERMINE THE RESOLVING TIME OF THE COUNTER

Introduction

The aim of the experiment is (*a*) to plot the variation of count rate with E.H.T. voltage applied to the Geiger-Müller tube, noting the length and slope of the plateau region (*Figure B1*) and (*b*) to determine the period during which the system fails to register pulses

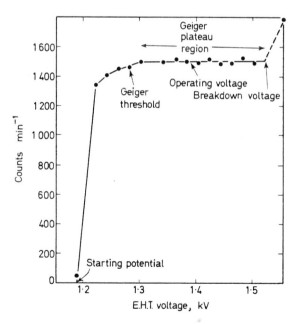

Figure B1. Graph showing the Geiger plateau

NUCLEAR AND RADIATION PHYSICS

due to particles entering the tube—the so called 'resolving time' or paralysis time of the counter.

Theory

Basically the Geiger-Müller tube consists of a cylindrical cathode surrounding a central wire anode, the space between containing a suitable gas at low pressure.

An ionizing particle entering the gas gives rise to avalanches of electrons which cause the potential on the anode to fall below the level needed to operate the tube, thus registering the entry of the ionizing particle as a pulse in the external circuit (*Figure B2*).

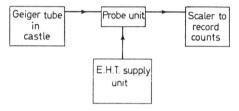

Figure B2. Block diagram of Geiger counting set up

The count rate is found to depend upon the voltage applied between the anode and cathode of the tube and it is first necessary to plot the characteristic graph of this variation. A graph such as *Figure B1* is obtained and the starting potential (below which pulses are too small to be detected) and the plateau region (in the centre of which lies the correct operating voltage for the tube) should be noted. Beyond this the counter begins to 'race' since continuous discharge takes place and this should be avoided for it is harmful to the tube.

After the pulse is registered a 'sheath' of positive ions that gradually increases in radius remains about the central wire. This effectively decreases the potential gradient near the wire and not until this space charge has drifted sufficiently far from the anode will the counter become sensitive again. The time taken to do this is called the dead time of the tube (*Figure B3*). A further time must elapse (the recovery time) before the detection level reaches its fully sensitive state once more. The total time involved is sometimes referred to as the resolution time or the paralysis time and since with the Geiger counter the dead time is not truly constant, it is general practice to swamp the counter paralysis time with a longer

CHARACTERISTICS OF A GEIGER COUNTER B1.1

electronic-circuit resolving time which may be selected by suitable design of the scaling unit.

This paralysis time t means that for each single count recorded the system is inoperative for t seconds. Thus if we have n recorded counts registered per second the lost time in one second is nt seconds and the effective operating time is thus $(1 - nt)$ seconds; so that n

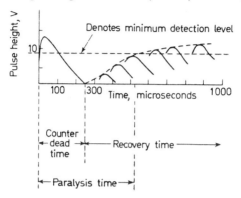

Figure B3. The paralysis time of a Geiger counter

counts actually take place in $(1 - nt)$ seconds and if we assume the corrected count rate is N per second then

$$\frac{N}{n} = \frac{1}{1 - nt}$$

from which $N = n/1 - nt$.

If chosen values of the paralysis time are taken in turn on the instrument selector switch, then whilst the selected time is less than that of the counter it will not affect the count rate and a plot of 1/count rate against paralysis time will be approximately linear. Once above the tube paralysis time the count rate will fall and hence the graph of 1/count rate rises steeply (*Figure B4*) and this marks the paralysis time of the counter.

Method

Essentially the apparatus will consist of a unit to supply E.H.T. voltage to the Geiger tube (which is housed in a suitable lead castle) and a scaler unit for counting the particles (and upon which the chosen paralysis time may be set) (*Figure B2*).

(a) To plot the characteristic, switch on the equipment and allow ten minutes or so for the equipment to 'warm up' making sure that the voltage output from the E.H.T. supply is at its minimum value.

Meanwhile using tweezers, place the planchette source on the castle shelf just below the tube window. Increase the volts applied to the tube until counting starts (the threshold voltage) and at this point count for 3 minutes to determine the count rate per minute.

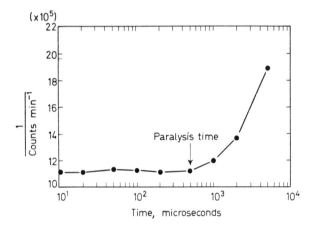

Figure B4. Graph used to determine the paralysis time of a Geiger counter assembly

Increase the applied voltage by the smallest convenient amount and repeat the counting. Continue to do this until the complete curve is obtained (*Figure B1*).

DO NOT increase the voltage beyond the point at which the count rate once again begins to increase rapidly, for this marks the end of the plateau region (the breakdown voltage).

From the resulting curve note the slope of the plateau, the threshold potential, the breakdown voltage, the operating voltage of the tube, the Geiger threshold (i.e. the point at which the plateau commences) and the length of the plateau region.

(b) To determine the paralysis time of the counter, set the E.H.T. supply to the correct working voltage of the tube and the paralysis time on the scale to its minimum setting. Count for 5 minutes and repeat three times. Increase the paralysis time to its next convenient value and repeat the counting procedure. Do this for each paralysis time setting and plot a graph (*Figure B4*) of (1/count rate) against

CHARACTERISTICS OF A GEIGER COUNTER   B1.1

paralysis time. Note the point at which the graph begins to rise, which marks the paralysis time of the counting tube.

Discussion of Results

A good Geiger counter will have a plateau at least 150 V in extent and a slope of not more than $0.1\% \, V^{-1}$ but as the tube 'ages' the plateau tends to grow shorter and steeper.

The dead time of the tube is not strictly constant but shows some variation so that in the graphical method of determining the paralysis time of the counting tube the (theoretical) straight line portion may show quite a wide scatter of points about it. For this reason several readings of count rate at any one setting should be made using as long a counting interval as conveniently possible.

The result obtained should be checked by using the 'two source' method. In this a second source is counted in the same way as the one above and then both sources together.

It was seen earlier that

$$N - n = Nnt \quad \left\{ \text{or } n = \frac{N}{1 + Nt} \right\}$$

and considering two sources, for the first source

$$N_1 - n_1 = N_1 n_1 t \tag{1}$$

for the second source

$$N_2 - n_2 = N_2 n_2 t \tag{2}$$

for both sources together

$$N_3 - n_3 = N_3 n_3 t$$

but

$$N_3 = N_2 + N_1 = \frac{n_1}{1 - n_1 t} + \frac{n_2}{1 - n_2 t}$$

$$= \left\{ \frac{n_1(1 - n_2 t) + n_2(1 - n_1 t)}{1 - (n_1 + n_2)t + n_1 n_2 t^2} \right\} \tag{3}$$

133

Substituting into

$$n_3 = \frac{N_3}{1 + N_3 t}$$

$$\therefore n_3 = \left\{\frac{n_1 + n_2 - 2n_1 n_2 t}{1 - (n_1 + n_2)t + n_1 n_2 t^2}\right\}$$

$$\times \left\{\frac{1}{1 + t\left(\dfrac{n_1 + n_2 - 2n_1 n_2 t}{1 - (n_1 + n_2)t + n_1 n_2 t^2}\right)}\right\}$$

i.e. $n_3 = \dfrac{n_1 + n_2 - 2n_1 n_2 t}{\{1 - t(n_1 + n_2) + n_1 n_2 t^2\} + t\{(n_1 + n_2) - 2n_1 n_2 t\}}$

$= \dfrac{n_1 + n_2 - 2n_1 n_2 t}{1 - 2n_1 n_2 t^2}$

whence for

$$n_1 n_2 t^2 \ll 1 \quad \therefore t \simeq \frac{n_1 + n_2 - n_3}{2 n_1 n_2}$$

where n_1 = observed count rate from first source
n_2 = observed count rate from second source
n_3 = observed count rate from both sources.

Note that $(n_1 + n_2)$ is very nearly equal to n_3 and hence long counting intervals should be taken if the result is to be an accurate one.

Further Work

It is worthwhile to determine and compare the efficiency of the Geiger tube for different types of radiation (Radium-226 provides suitable gamma emission and thallium gives beta particles). The intrinsic efficiency ε is defined by the equation

$$\varepsilon = \frac{4\pi N}{\omega S}$$

where ω = solid angle which the tube 'window' subtends at the source (assumed to approximate to a point) and S the source strength.

To calculate the solid angle the Geiger tube dimensions and the distance of the source from the window must be known. Then

CHARACTERISTICS OF A GEIGER COUNTER B1.1

$$\frac{\text{Number of ionizing particles passing through the window}}{\text{Total number emitted}}$$

$$= \frac{\omega}{4\pi} = \frac{\text{Area } A}{4\pi R^2} = \frac{2\pi Rh}{4\pi R^2} = \frac{h}{2R}$$

but

$$R = (d^2 + r^2)^{\frac{1}{2}} \quad \text{and} \quad h = (d^2 + r^2)^{\frac{1}{2}} - d$$

$$\therefore \frac{\omega}{4\pi} = \frac{1}{2}\left\{1 - \frac{d}{(d^2 + r^2)^{\frac{1}{2}}}\right\}$$

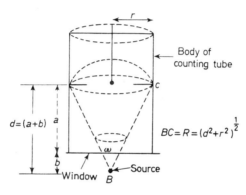

Figure B5. Geiger-Müller tube geometry considered in calculating intrinsic efficiency

where d represents the distance from the tube probe to the source $= (a + b)$ on the diagram and $r =$ radius of the Geiger tube (*Figure B5*).

Reading and References

Background reading on the Geiger counter and upon resolving time may be found in *Nuclear Radiation Physics* by R. E. Lapp and H. L. Andrews (Pitman, 3rd edition, 1963). The text also includes illustrative worked examples including one on the double source method.

NUCLEAR AND RADIATION PHYSICS

Experiment No. B1.2

THE STATISTICS OF COUNTING OF THE PARTICLES EMITTED FROM A RADIOACTIVE SOURCE

Introduction

It will already have been noted that using a single constant radioactive source, the recorded counts of emission over a fixed interval of time will show some fluctuation and will seldom repeat themselves exactly. One would expect these statistical fluctuations, since nuclear disintegration is a random process and the aim of this experiment is to investigate the statistical laws which govern the emission.

Theory

The probability P of N particles being emitted in time t from the radioactive source is given by

$$P = \frac{\bar{N}^N}{N!} \exp(-\bar{N}) \tag{1}$$

where $\bar{N}$ = mean number of particles emitted in time t, i.e. it obeys the law of Poisson distribution.

For a large number of particles (i.e. large N) this distribution approximates to the normal (or Gaussian) distribution which may be expressed in the form

$$P = \frac{1}{(2\pi\bar{N})^{\frac{1}{2}}} \exp\left\{\frac{-(N-\bar{N})^2}{2\bar{N}}\right\} \tag{2}$$

It differs from the Poisson distribution in that it is symmetrical about the maximum corresponding to the mean number of particles $\bar{N}$.

The probability may alternatively be expressed in terms of the standard deviation σ which is itself given by

$$\sigma = \left\{\frac{\Sigma(N-\bar{N})^2}{n-1}\right\}^{\frac{1}{2}} \simeq \bar{N}^{\frac{1}{2}} \tag{3}$$

STATISTICS OF PARTICLE COUNTING   B1.2

for the Poisson distribution, where n represents the number of experimental counts taken, thus

$$P = \frac{1}{2\pi\sigma^2} \exp\left\{\frac{-(N-\bar{N})^2}{2\sigma^2}\right\} \tag{4}$$

The first thing to do therefore is to construct a histogram and to calculate σ from the results.

The Gaussian law may then be written

$$y = \frac{\text{Area of histogram}}{(2\pi\sigma^2)^{\frac{1}{2}}} \exp\left\{\frac{-(N-\bar{N})^2}{2\sigma^2}\right\} \tag{5}$$

where y = frequency distribution of the Gaussian curve.

For our purpose it is convenient to use the probability rather than frequency distribution and since the sum of all probabilities is unity, the area under the normal probability distribution curve is normalized, i.e. considered as unity (see equation (2)) so that the probability P is given by

$$P = \frac{1}{\sigma(2\pi)^{\frac{1}{2}}} \exp\left\{\frac{-(N-\bar{N})^2}{2\sigma^2}\right\} \tag{6}$$

hence

$$y = k \exp\left\{\frac{-(N-\bar{N})^2}{2\sigma^2}\right\} \tag{7}$$

where

$$k = \frac{\text{The area of the histogram}}{\sigma(2\pi)^{\frac{1}{2}}}$$

Method

A suitable radioactive source (e.g. a 0·5 microcurie thallium beta-emitter) is placed within the lead castle and the Geiger counter operated at its working voltage in the usual way (see Experiment B1.1). Choose a suitable time interval (say half a minute or one minute) and take as large a number of readings as possible (100 would be advisable) of the counts registered in the chosen interval. Correct the readings for counter paralysis time using the equation

$$N_0 = N_c(1 - \beta t)$$

where N_0 = observed number of counts
       N_c = corrected number of counts
       t = counter paralysis time
       β = average number of particles emitted in unit time

(see also Experiment No. B1.1).

Choose convenient 'class intervals' for the experimental counts obtained, subdividing them so that the frequency f with which the count falls within the chosen subdivision, is obtained (see for example *Table B1*).

Table B1

Selected range of count rate (counts min^{-1})	Frequency with which counts fall within the particular range
710 to 729	1
730 to 749	2
750 to 769	2
770 to 789	5
790 to 809	9
810 to 829	9
830 to 849	10
850 to 869	5
870 to 889	3
890 to 909	1
910 to 929	0
930 to 949	2
950 to 969	1

Hence construct the histogram and frequency polygon as shown in *Figure B6*.

Finally using equation (7) plot the theoretical Gaussian curve and superimpose it over the histogram.

Discussion of Results

The Gaussian distribution assumes that n tends to infinity and deviation from this condition will yield discrepancies between the general trend of the histogram and that of the Gaussian curve.

The Poisson distribution, which is strictly appropriate to the random process of radioactive emission, does not yield a symmetrical curve and this asymmetric tendency is clearly seen in the histogram of *Figure B6*.

The secondary pip is typical of the sampling error often obtained in the student laboratory experiment. Possible reasons for this may

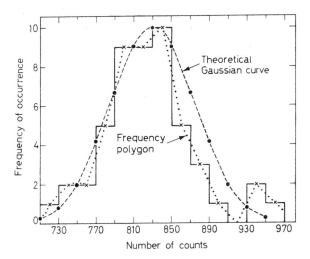

Figure B6. The random nature of the emission from a radioactive substance. An experimental histogram with theoretical Gaussian curve and the frequency polygon superimposed

lie in the fact that more than one type of radioactive emanation is present or in a wrong choice of class interval (or of course in the electronic counting equipment).

Further Work

Using a very weak source (or even the background count of the laboratory) one can in fact investigate the Poisson distribution itself rather than the Gaussian curve. For these small values of counts the distribution is such that the mean value $\overline{N}$ lies somewhat to the right of the maximum (*Figure B7*).

Reading and References

An elementary *Statistics* by M. R. Spiegel (Schaum, 1961) deals with histograms, the normal and Poisson distributions and standard deviations. Worked examples illustrate all these topics.

An alternative treatment of the normalized probability distribution and its uses is to be found in *Basic Statistical Methods for*

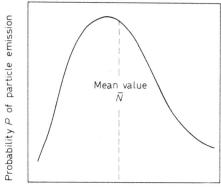

Figure B7. Graph of Poisson distribution showing (for low N values) the mean value $\bar{N}$ of the number of radioactive particles emitted lying to the right of the maximum

Engineers and Scientists by A. M. Neville and J. B. Kennedy (International Text Book Co., 1964).

Experiment No. B1.3

TO INVESTIGATE THE ABSORPTION OF BETA RAYS FROM A RADIOACTIVE SUBSTANCE BY ALUMINIUM

Introduction

The aim of this experiment is to measure the mass absorption coefficient for β rays and to estimate the maximum range of the rays, from which the approximate energy of the beta particles may be determined.

ABSORPTION OF BETA RAYS BY ALUMINIUM

Theory

If one assumes that the absorption of β rays in any medium follows an exponential law

$$I = I_0 \exp(-\mu x)$$

where I = the intensity of activity after absorption
I_0 = the initial intensity
μ = linear absorption coefficient
x = thickness of the absorber

then a plot of $\log I$ against x theoretically results in a straight line whose negative slope gives μ.

In practice the linear absorption coefficient μ will vary widely from one material to another but the quantity (μ/ρ) where ρ = density is approximately constant for all types of absorbers and hence the use of this quantity, the mass absorption coefficient, is to be preferred. This necessitates expressing the thickness of absorber not in units of length, but in gram cm^{-2} thus

$$I = I_0 \exp\left\{-\left(\frac{\mu}{\rho}\right)(\rho x)\right\}$$

and

$$\log I = \log I_0 - \left(\frac{\mu}{\rho}\right)(\rho x)$$

Hence plotting the logarithm of the count rate against absorber thickness, the slope of the graph gives the mass absorption coefficient and the extrapolation of the linear region to the x axis enables the maximum range of the β's in terms of gram cm^{-2} to be found (*Figure B8*). Then using one of the empirical formulae given below, relating energy and range of β particles, their energy may be calculated.

If the energy E is above 0·8 MeV {1 MeV = 1·60 × 10^{-13} joules}

$$\text{Range } R = 0.542E - 0.133$$

If the energy lies between 0·15 and 0·8 MeV, $R = 0.407E^{1.38}$ (see references).

Note the 'tail' on the graph (*Figure B8*) due to background count rate.

Method

A suitable particle emitter is placed in the lead castle and the count rate determined in the usual way using a Geiger counter arrangement (Experiment No. B1.1).

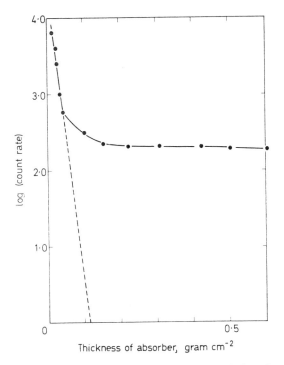

Figure B8. The absorption of beta particles: variation of the log of the count rate with absorber thickness. Extrapolation shows how the maximum range of the β's may be estimated. The slope of the linear portion of the graph gives the absorption coefficient

Absorbers of known thickness (in terms of gram cm^{-2}) are introduced in turn and the rate corresponding to each is tabulated (*Figure B9*).

Correct the count rate for lost counts, and for background counts, then plot the graph in the manner suggested above.

Calculate the range by extrapolating the linear region of the graph and hence find the beta particle energy.

Discussion of Results

In assessing the accuracy of the result remember that no allowance has been made in this simple treatment for the effect of the Geiger tube 'window' thickness, nor for the air-gap between the source

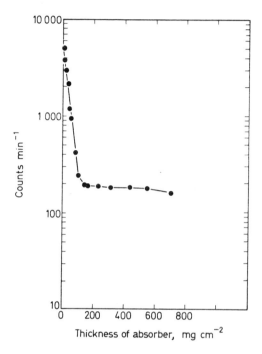

Figure B9. The absorption of β rays emitted from a radioactive source

and the window. Scattering of the beta particles from the atoms of air, the absorbers themselves from the material supporting and covering the source, has been ignored while generally too the collimation of the emitted beam is not good.

The 'tail' of the curve (*Figure B8*) indicates the presence of gamma radiation or of *Bremsstrahlung* (electromagnetic radiation of the same nature as gamma) produced by interaction of beta particles with the target atoms in the absorber. (See also Experiment No. B1.11.)

Further Work

As in a later experiment (number B2.5) it is possible to calculate the half value thickness t of the absorbing material and one should do so using the formula $t = 0.69/(\mu/\rho)$.

Reading and References

Absorption of beta rays and the reasons why the mass absorption coefficient may be assumed approximately constant are dealt with in *Introduction to Reactor Physics* by D. J. Littler and J. F. Raffle (Pergamon, 1957).

The empirical formulae due to Glendenin relating energy and range are quoted in *Textbook of Nuclear Physics* by C. M. A. Smith (Pergamon, 1965).

Tables showing the energies of beta particles emitted from various radioactive nuclides are to be found in *Nuclear Physics* by I. Kaplin (Addison-Wesley, 2nd edition, 1965), p. 350.

Experiment No. B1.4

TO DETERMINE THE CHARACTERISTICS OF A SCINTILLATION COUNTER

Introduction

It has already been seen in the case of a Geiger counter that the count rate recorded by the counting system is a function of the voltage applied to the tube. It was necessary therefore, before bringing it into use, to determine the optimum operational voltage and to do this the characteristic curve of the counter, which displayed a plateau region, was plotted.

The same procedure must be adopted with the scintillation counter and the aim here is to plot curves showing the variation of counts min^{-1} on a logarithmic scale with applied voltage (*Figure B10*).

This is done in the case of (*a*) the total counts due to source and background, (*b*) source counts alone and (*c*) background counts alone.

CHARACTERISTICS OF A SCINTILLATION COUNTER B1.4

Theory

It will be recalled that in Rutherford's earlier experiments upon alpha particles the scintillations produced when the charged particles struck a zinc sulphide screen were used to 'count' the particles. In the scintillation counter the same principle is employed, the ionizing radiations being allowed to fall on a light sensitive surface so that electrons are ejected from the surface. A photomultiplier arrangement is then employed to amplify the effect and to convert the electron 'pulses' to voltage pulses which may then be counted using a scaler in the usual way.

If R represents the count rate and subscripts s, b and t represent source, background and total (i.e. due to source and background) respectively then

$$R_s = R_t - R_b$$

Since we are dealing with a large number of emitted particles variation in count rate response will occur and the measure of statistical uncertainty is given by the standard deviation σ and this deviation in the counting rate

$$R = \left(\frac{\text{number of ionizations } N \text{ occurring}}{\text{time of observation } t}\right)$$

is shown by statistical theory to be given by $\sigma = N^{\frac{1}{2}}/t$. Also in the case of a sum or difference such as above, where

$$R_s = R_t - R_b$$
$$\sigma_s = (\sigma_t^2 + \sigma_b^2)^{\frac{1}{2}}$$

but from above

$$\sigma_t^2 = \frac{N_t}{t_t^2} \quad \text{and} \quad \sigma_B^2 = \frac{N_B}{t_B^2}$$

and since

$$R = \frac{N}{t}, \quad \sigma_t^2 = \frac{R_t}{t_t} \quad \text{and} \quad \sigma_B^2 = \frac{R_b}{t_b}$$

$$\therefore \sigma_s = \left(\frac{R_t}{t_t} + \frac{R_b}{t_b}\right)^{\frac{1}{2}}$$

For convenience in this experiment the scintillation counter is usually set to count for equal time periods, so that $t_t = t_b = t$ and

$$\sigma_s = \left(\frac{R_t + R_b}{t}\right)^{\frac{1}{2}}$$

or

$$t = \frac{R_t + R_b}{\sigma_s^2}$$

For practical measurements the standard deviation is better expressed as a fraction of the quantity being measured (N) and we define a coefficient of variation

$$V = \frac{\sigma}{R} = \left(\frac{N^{\frac{1}{2}}/t}{N/t}\right) \quad \text{i.e. } \sigma = RV$$

therefore

$$t = \frac{R_t + R_b}{R_s^2}$$

or in terms of source and background since $R_t = R_s + R_b$

$$t = \frac{R_s + 2R_b}{R_s^2}$$

If a weak source is employed such that $R_b > R_s$ then

$$t \simeq \frac{2R_b}{R_s^2}$$

The more responsive the counter the smaller the total time required to make representative measurements and hence (R_s^2/R_b) becomes a 'figure of merit' of the counter.

For this reason graphs of the ratio {(source count)^2/background count} against the applied voltage are plotted (*Figure B10*). The maximum of the resulting curve then gives the optimum E.H.T. value for that particular bias and attenuation setting of the counter.

Method

The apparatus is set up as shown in *Figure B11*. The voltage to the tube is set at the minimum for the particular apparatus being used and sufficient time is allowed for 'warming up' before actually making measurements (20 minutes or so).

The bias setting is put at its minimum value (~ 5 V) and the chosen source (a 0·1 μcurie gamma emitter is suitable) placed in position in the lead castle.

CHARACTERISTICS OF A SCINTILLATION COUNTER B1.4

Set the 'attenuation' to minimum and slowly increase the applied voltage until counting commences. Observe the count rate at slowly increasing values until 'racing' occurs on the counter. At this point stop immediately and reduce the E.H.T. to minimum once more.

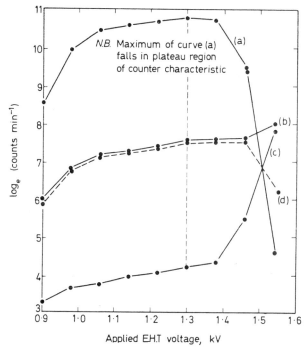

Figure B10. To determine the characteristics of a scintillation counter. (*a*) Graph of $\log_e$ {(source count rate)^2/background count rate} to estimate optimum operating conditions; (*b*) Source and background counts against E.H.T.; (*c*) Background counts against E.H.T.; (*d*) Source counts against E.H.T.

The source is now removed and the measurements taken again for the background count alone.

Plot the graphs of counts (on a log scale) against applied E.H.T. voltage as shown in *Figure B10*.

Compute the ratio {(source count)^2/background count} as suggested in the theory and note the maximum value when the quantity is plotted against applied voltage.

Discussion of Results

It is important to note that the theory given is only strictly true if the count rate due to the background is greater than that due to the source. The chosen source must therefore be a weak one and the results given were obtained using a caesium-137 0·1 microcurie gamma emitter.

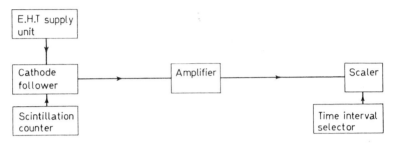

Figure B11. Block diagram of apparatus used in the determination of the characteristics of a scintillation counter

Considerable improvement could be effected if smaller E.H.T. intervals could be taken in the region about the maximum of the curve of the {(source count)^2/background count} against E.H.T. This unfortunately is often limited by the instrument settings.

Further Work

Notice that the method of equal counting time intervals chosen here is not the only one possible. The conditions appropriate to minimal total counting time and to setting for equal total counts might be investigated and compared with the above.

Reading and References

A theoretical account of possible modes of counting may be found in *Nuclear Radiation Physics* by R. I. Lapp and H. L. Andrews (Pitman, 3rd edition, 1963).

A briefer description of this experiment may be found in *Experimental Nucleonics* by B. Brown (Iliffe, 1963) whilst details of different types of phosphors for both beta particles and gamma radiation are given in *Practical Nucleonics* by F. J. Pearson and R. R. Osborne (Spon, 1960).

Experiment No. B1.5

TO DETERMINE THE ENERGY RESOLUTION OF A SCINTILLATION COUNTER USING GAMMA RAYS FROM ^{137}Cs

Introduction

Energy conversion by photoelectric absorption of gamma photons is so rapid that the resulting light pulses produced by the scintillation counter are proportional to the energy of the original gamma photons. Thus the instrument acts as a gamma ray spectrometer, and the spectrum can be investigated using a pulse height

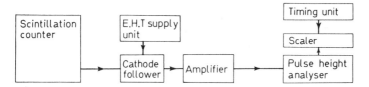

Figure B12. Block diagram of a typical scintillation counter assembly

analyser (*Figure B12*) which responds to pulses whose magnitude lie between chosen upper and lower limits, i.e. they define a chosen 'channel width'.

Theory

There are three main processes by which gamma rays give up their energy to matter. These are:

(1) by the photoelectric effect
(2) by the Compton effect
(3) by pair production.

In using the scintillation counter here as a detector of gamma rays from ^{137}Cs we are concerned with radiation having energy between 0·5 and 1 MeV $\{1 \text{ MeV} = 1·6 \times 10^{-13} \text{ joule}\}$ so that the third process of absorption by pair production is not important since it has a threshold energy of 1·02 MeV.

The resulting graph when ^{137}Cs, which emits monoenergetic gammas of 0·66 MeV, is used thus displays:

(1) a photoelectric peak, whose position is proportional to the energy of the incident gammas, and

(2) a Compton distribution of recoil electrons with a well-defined 'Compton Edge' (see *Figure B13*).

Notice that in fact the photoelectric peak has finite width which limits the resolution of peaks of different (but closely grouped) gamma ray energies.

By convention the energy resolution of the scintillation spectrometer is taken as

$$\frac{\text{The width of photopeak at half its maximum height}}{\text{Pulse height at the photopeak maximum}}$$

and usually multiplied by 100 to express as a percentage.

Method

As in the previous experiment, switch on the apparatus, set the E.H.T. control to minimum and allow the apparatus to warm up.

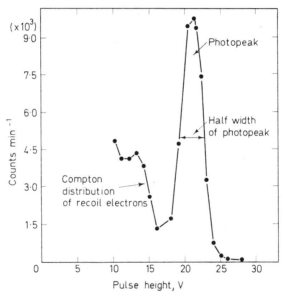

Figure B13. Photoelectric peak of caesium-137 obtained using a scintillation counter

Place the ^{137}Cs source in position in the lead castle, set the attenuation control to its maximum (i.e. employ minimum amplifier gain), and set the E.H.T. at the working voltage as determined in the experiment on scintillation counter characteristics.

Set the voltage on the pulse height analyser to its minimum value (~ 5 V) and the channel width to say 0·5 V.

Observe the number of counts over (say) 3 minutes at the pulse height of 5 V, then repeat the counting as the pulse height is increased in steps of 5 V until the whole spectrum has been investigated.

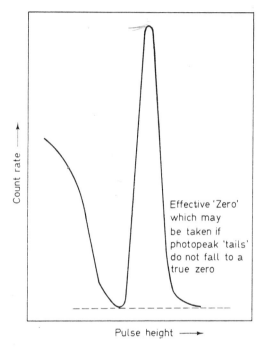

Figure B14. Graph showing possible mode of treatment when photopeak 'tails' do not fall to zero

Plot the graph of (counts minute^{-1}) against pulse height in volts as shown in *Figure B13*.

Note the pulse height corresponding to the maximum of the photoelectric peak and also the 'half width' as indicated on *Figure B13*. Hence the resolution of the counter may be determined using

$$\text{Energy resolution} = \frac{\text{Half width} \times 100}{\text{Pulse height at the photo peak maximum}}$$

Discussion of Results

The resolution will probably have a value of between 10 and 20% although with a large clear crystal hermetically sealed and surrounded by magnesium oxide (which effectively reflects back stray scintillations) it is possible to obtain resolutions as good as 7 or 8% if the optical contact between the crystal and the photomultiplier window is good (silicone grease is used).

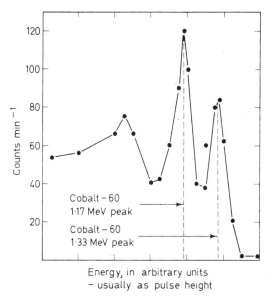

Figure B15. Energy resolution of a scintillation counter using cobalt-60

When in the region of the photopeak it is advisable in the interests of accuracy to choose smaller pulse-height intervals than the 5 V steps, for remember unless the peak position is accurately known a significant error is introduced into the estimation of the 'half width'.

If the tails of the photopeak do not fall to zero, accuracy is enhanced by taking the horizontal through the lower tail as the effective zero, as indicated by the dotted line in *Figure B14*.

Correction should also be made for background counts and for the paralysis time of the counter.

Further Work

The procedure should be repeated using cobalt-60 or sodium-22 as the gamma sources (graphs for which are shown in *Figures B15* and *B16*).

The use of cobalt-60 affords an alternative method of measuring the resolution of the counter which is more frequently quoted by manufacturers of scintillation counting equipment, for cobalt-60

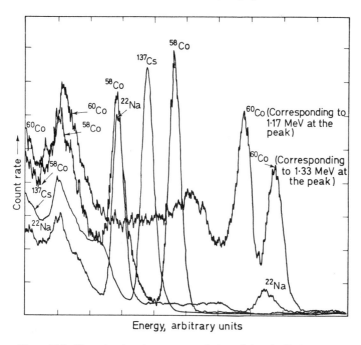

Figure B16. Chart showing the energy resolution of the scintillation counter. N.B. Count rates of the various curves are not all on the same scale

emits gamma rays of two energies (1·17 and 1·33 MeV respectively) and the resolution can be defined by the ratio

Amplitude (i.e. Count rate) of the maximum of the 1·33 peak
―――――――――――――――――――――――――――――
Amplitude (i.e. Count rate) at the lowest point of
              the preceding valley

For a good large crystal this should be about 4:1 but it may be considerably lower (e.g. *Figure B16*). (Notice that using the first convention the better the resolution the lower the percentage figure obtained but by this second convention the better resolution the bigger the ratio.)

Using the second method the 1·17 and 1·33 MeV peaks enable the

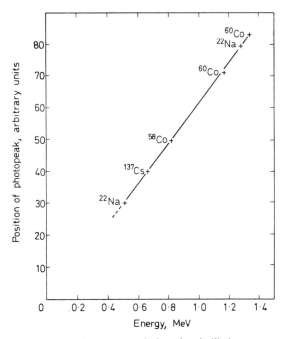

Figure B17. The energy resolution of a scintillation counter. Graph showing that a linear relationship holds between the position of the photopeak and its energy

pulse height axis to be calibrated in terms of energies in MeV. Photopeak position against photopeak energy for several gamma emitters is plotted in *Figure B17*.

Reading and References

The theory of the three processes of γ absorption may be found in many standard texts on nuclear physics as well as in *Nuclear Radiation Physics* by R. E. Lapp and M. L. Andrews (Pitman, 1963), p. 96. *Practical Nucleonics* by F. J. Pearson and R. R. Osborne (Spon, 1960) discusses counters, ancillary equipment and counting techniques in some detail.

Comparative data for a number of scintillation crystals and on the two methods of measuring resolution may be found in the UKAEA Research Group User Guide, Issue 2, October 1965, entitled 'Thallium Activated Sodium Iodide Scintillation Crystals'.
'Modern Developments in Radiation Detection Techniques' are dealt with in an AERE Harwell publication PGEC/L.35 of that title by W. Abson, R. B. Owen, W. R. Loosemore and P. Iredale, whilst a paper by J. D. Ridley in *Nuclear Engineering*, vol. 13, No. 141, Feb. 1968, p. 109, gives an account of germanium gamma ray spectrometers which use the ion drift technique.

Experiment No. B1.6

TO INVESTIGATE THE ABSORPTION OF GAMMA RADIATION BY LEAD AND TO DETERMINE THE MASS ABSORPTION COEFFICIENT AND THE HALF VALUE THICKNESS OF THE ABSORBER

Introduction

The theory behind this experiment may be compared with that given in a previous Experiment (No. B1.3) and of Experiment No. B1.11.

Theory

As in the experiments quoted above one may write

$$I = I_0 \exp\{(-\mu/\rho)(\rho x)\}$$

where I = intensity of gamma activity after absorption
I_0 = initial gamma intensity
μ/ρ = the mass absorption coefficient of the absorber
ρx = thickness of absorber expressed in gram cm^{-2}.

Hence the half value thickness $x_{\frac{1}{2}}$ corresponds to a resulting intensity $I_0/2$
i.e.

$$\frac{I_0}{2} = I_0 \exp(-\mu x_{\frac{1}{2}})$$

where μ is the linear absorption coefficient.
Therefore

$$\log_e \tfrac{1}{2} = -\mu x_{\frac{1}{2}} \quad \text{or} \quad x_{\frac{1}{2}} = 0\cdot 69/\mu$$

Thus if the mass absorption coefficient (μ/ρ) is first found from the slope of the log I against (ρx) graph, the half value thickness of the absorber is readily obtained knowing the density ρ of the material. Note that if logarithms to the base 10 are used then $(\mu/\rho) = 2\cdot 303$ times the slope of the graph.

Method

A fairly strong gamma source (e.g. 10 microcurie Radium-226) is placed in a suitable lead castle.

Set up the Geiger counter as described in Experiment B1.1 and determine the number of counts per second corresponding to the

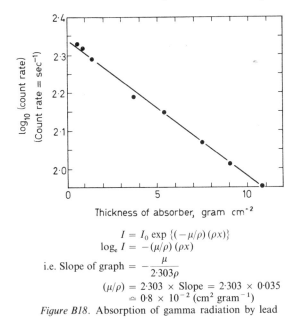

$$I = I_0 \exp\{(-\mu/\rho)(\rho x)\}$$
$$\log_e I = -(\mu/\rho)(\rho x)$$
i.e. Slope of graph $= -\dfrac{\mu}{2\cdot 303\rho}$

$(\mu/\rho) = 2\cdot 303 \times$ Slope $= 2\cdot 303 \times 0\cdot 035$
$\simeq 0\cdot 8 \times 10^{-2}$ (cm^2 gram^{-1})

Figure B18. Absorption of gamma radiation by lead

gamma emission from the chosen source. (See *Figures B18* and *B19* for order of magnitude of count rate.)

Insert a thin foil of lead between the source and the counting tube window and once more obtain the count rate.

Repeat this for a number of different thicknesses of lead and plot the logarithm of the count rate against the absorber thickness expressed in gram cm^{-2}.

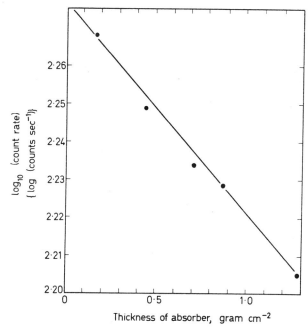

$(\mu/\rho) = 2\cdot303 \times \text{Slope} \simeq 2\cdot303 \times 0\cdot055$
$(\mu/\rho) = 1\cdot26 \times 10^{-2}$ (cm^2 gram^{-1})

Figure B19. Absorption of gamma radiation by aluminium

Assuming logarithms to the base ten are used the slope of the resulting graph gives $\{(\mu/\rho)/2\cdot303\}$.

Discussion of Results

The 'scatter' of the points about the best fit line of the graph gives a good indication of the order of experimental accuracy involved but one should refresh one's thoughts concerning the accuracy of the Geiger counting system by reading again Experiment No. B1.1.

Compare the results obtained for the half value thickness with the accepted values and try to account for the discrepancies which may arise.

Further Work

If possible, the experiment should be repeated using a scintillation counter rather than a Geiger counter.

Using the same source and housing of similar geometry the scintillation counter should be found to give a shorter resolving time and a much higher and more stable counting efficiency.

The experiment should also be carried out using other absorbers (e.g. aluminium, *Figure B19*, and steel).

Reading and References

The experimental scripts upon the Geiger counter and the scintillation counter should be read in conjunction with this experiment. Both instruments and their use are dealt with in *Practical Nucleonics* by F. J. Pearson and R. R. Osborne (Spon, 1960).

Experiment No. B1.7

TO DETERMINE THE SLOWING DOWN LENGTH OF NEUTRONS IN PARAFFIN WAX

Introduction

In this experiment a number of indium foils of identical known dimensions are placed in the flux of neutrons at known positions and irradiated for a period of several hours. This has the effect of rendering the foils radioactive and on removal they are found to emit beta rays which may be counted in the usual way.

The foils are placed in a moderator which surrounds the neutron source (e.g. paraffin wax, water or graphite) and the fast neutrons of energy of the order of several MeV are first 'slowed down' by collision with the moderator atoms with a consequent reduction in energy to 'thermal' energies (0.025 eV or 4×10^{-21} joule). Appropriate to any particular moderator and monoenergetic group of neutrons, there is a slowing down length L_s which characterizes this process.

The aim of this experiment is to determine the slowing down length of neutrons in the chosen moderator and it is assumed that the activity induced in the foils is a direct measure of the number of neutrons per second slowing down past a given energy level (in this case 1·44 eV, 2·307 × 10^{-19} J, and not true 'thermal' energy level 0·025 eV, 0·04 × 10^{-19} J).

Theory

Let N = number of neutrons per unit volume slowing down past the given level in 1 second. Then considering a spherical shell of radius r through which the neutrons are emitted,

the surface area of the shell = $4\pi r^2$

and

the volume of the shell = $4\pi r^2 \delta r$

hence the total number of neutrons slowing down in one second within the shell

$$= \int_0^\infty 4\pi N_1 r^2 \, dr$$

integrating over the whole sphere.

The mean square distance $\bar{r}^2$ travelled by a neutron as it is slowed to this level is therefore given by

$$\frac{\int_0^\infty 4\pi N_1 r^4 \, dr}{\int_0^\infty 4\pi N_1 r^2 \, dr}$$

Hence if it is assumed that the count rate (or intensity of activity I) is a direct measure of N_1 then

$$\int_0^\infty 4\pi N_1 r^4 \, dr$$

may be represented by the area under the graph of Ir^4 against r and

$$\int_0^\infty 4\pi N_1 r^2 \, dr$$

by the area under the graph of Ir^2 against r. The ratio of the two areas thus gives an estimate of $\bar{r}^2$.

In activating the foils only those neutrons emitted over a small solid angle ω are utilized whereas the source emits neutrons over the solid angle 4π. By comparison with elementary kinetic theory it is not difficult to appreciate therefore that the slowing down length L_s is given by

$$L_s^2 = \bar{r}^2/6$$

and hence

$$L_s^2 = \left(\frac{\text{Area under the } Ir^4 \text{ graph}}{\text{Area under the } Ir^2 \text{ graph}}\right)$$

from which L_s may be determined.

Assuming that neutrons slow down in accordance with the natural exponential decay law (the 'Fermi age' theory).

$$I = \frac{S \exp(-r^2/4T)}{(4\pi T)^{\frac{3}{2}}}$$

where T is the Fermi age of the neutrons $= L_s^2$.

Hence

$$I = \frac{S \exp(-r^2/4L_s^2)}{(4\pi L_s^2)^{\frac{3}{2}}} \quad \text{or} \quad \log_e I = \log_e \left\{\frac{S}{(4\pi L_s^2)^{\frac{3}{2}}}\right\} - \frac{1}{4L_s^2} r^2.$$

Thus the graph of $\log_e I$ against r^2 should have a negative slope $= -(1/4L_s^2)$.

Method

The initial activity of the indium foils sheathed in cadmium is first determined by Geiger counter arrangement in the usual way (Experiment No. B1.1.) This is necessary since if used continuously the formation of indium-114, whose half life is 50 days, results in a significant initial activity which one does not want. If possible use fresh foils or foils that have not been used for some considerable time.

Load the sheathed foils into the slots in the paraffin wax plug (*Figure B20*), noting carefully the distance of each slot from the base of plug, so that the distance from the source to each particular foil may be computed.

The blank plug is removed from the neutron source tank and the loaded plug lowered into position over the source, which should emit of the order of 10^6 neutrons second^{-1}. (500 millicurie Americium-beryllium may be recommended as suitable.)

The time of loading should be noted and the foils should be irradiated for several hours, after which the plug is removed, the time of removal from the tank again being carefully noted. The blank plug is replaced and the foils removed from their slots. (Use tweezers and only do this under supervision.) Store the foils in a suitable box whilst the count rate of each one, starting with the weakest, which lay furthest from the source, is measured. (If the foils have been mounted in aluminium allow 15 minutes to elapse after removal in order for short-lived β activity in the aluminium to decay.)

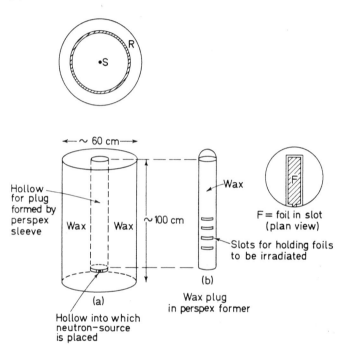

Figure B20. Shaded region R represents shell in which $4\pi r^2 I \, dr$ neutrons are slowing down past the chosen energy level

Tabulate in each case the time of starting to count, the time of ending the count, and the mean time which has elapsed from the time of removal of the foils. If possible repeat the process of measuring the count rate of the foils in the same sequence.

The following corrections must now be applied:

(a) Correction for the paralysis time of the counter equipment (Tables will be provided). This is necessary since the number of particles per second entering the counter is higher than the rate recorded. (The instrument takes ~500 μs to operate and two particles entering within this time will be recorded as one.)

$$\text{Paralysis time} = \frac{1}{\text{Observed count}} - \frac{1}{\text{True count}}$$

therefore

$$\text{True count} = \frac{\text{Observed count}}{1 - \text{Paralysis time} \times \text{observed count}}.$$

(b) Correction for background activity. Some of the counts come from other events than activity in the foil. A separate background

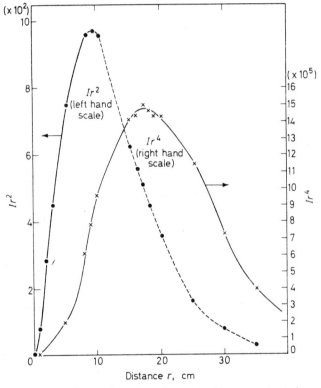

Figure B21. Ir^2 and Ir^4 graphs against r (paraffin wax moderator)

count is made at the beginning and end of the experiment and the result deducted from the observed foil activity.

Count due to specimen = True count (above) − Background.

(c) Correction for decay in indium activity between the time at which counts were made and the time of removal from source. The specimen count above is divided by the appropriate decay factor which is given in tables to obtain the activity at the time of removal from the region of the source.

Having corrected each foil activity in this way, plot Ir^4 against r and Ir^2 against r giving graphs of the type shown in *Figure B21*. A planimeter may be used to find the area under each curve and hence the ratio

$$\left(\frac{\text{Area } Ir^4 \text{ curve}}{\text{Area } Ir^2 \text{ curve}}\right) = \frac{\int_0^\infty Ir^4 \, dr}{\int_0^\infty Ir^2 \, dr} = \bar{r}^2 \quad \text{from the original theory.}$$

Then

$$L_s^2 = \bar{r}^2/6.$$

Discussion of Results

It will probably be found that the observed results do not continue far enough to get an accurate estimate of area in which case make use of the equation given, viz.

$$I = \frac{S \exp(-r^2/4L_s^2)}{(4\pi L_s^2)^{\frac{3}{2}}}$$

whence taking logs

$$\log_e I + \tfrac{3}{2} \log_e (4\pi L_s^2) = -\left(\frac{1}{4L_s^2}\right) r^2 + \log S$$

which is of the form $y = mx + c$.

So that the slope (m) of the graph gives $-1/4L_s^2$ and hence L_s. Typical graphs are shown in *Figure B22a* and *b* (plotted on 3 cycle log-linear paper).

The estimate of area under the Ir^4 and Ir^2 curves proves difficult unless the foils are many and their positions in the plug well chosen (this will have been predetermined of course by the apparatus) since the peak of the curve and the extremes are seldom well defined (*Figure B22*). Careful interpolation of Ir^4 and Ir^2 values about the peak and the extremities approaching 0 and infinity (the theoretically specified limits) is required from the I against r graph (*Figure B23*) in order to obtain reasonable graphs. Improved accuracy may be obtained by plotting I against r on semi-log graph paper and extrapolating to obtain the value of I at large values of r (*Figures B23* and *B24*).

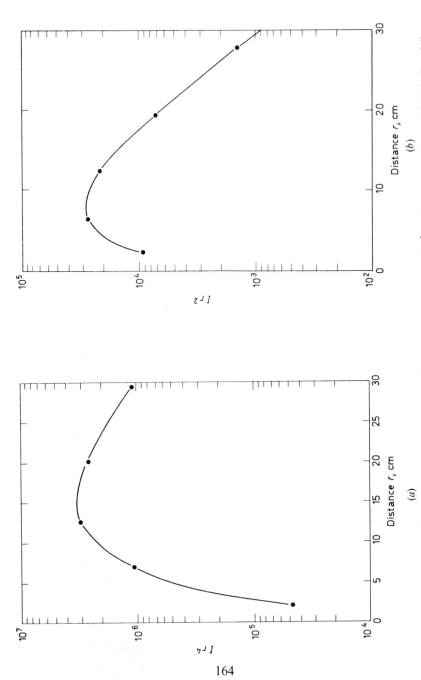

Figure B22 (*a*) Graph of Ir^4 against r for cadmium-sheathed indium foils; (*b*) Graph of Ir^2 against r for cadmium-shielded indium foils

If a graphite moderator is used then a check upon the accuracy of the first method may be made by use of the 'Fermi age' equation and the resulting straight line plot suggested in the theory (*Figure B25* shows an actual result) but with paraffin wax the loss of energy

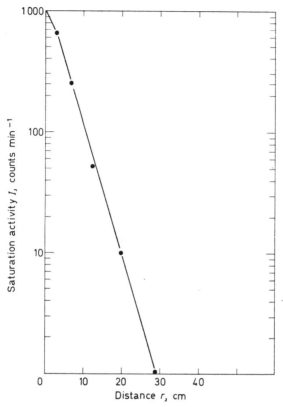

Figure B23. Graph of saturation activity I (on log scale) against r to obtain extrapolation values for Ir^2 and Ir^4 curves using a paraffin wax moderator

takes place in a succession of sharp changes which occur at each elastic collision between a neutron and a target nucleus. These do not approximate to the continuous exponential energy loss suggested by the Fermi formula and the resulting graph is not a straight line. However, it may approximate to this over the central region of readings, enabling an estimate of the slowing down length to be

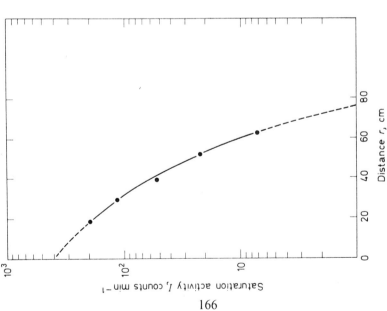

Figure B24. Saturation activity I (on log scale) against r using a graphite moderator

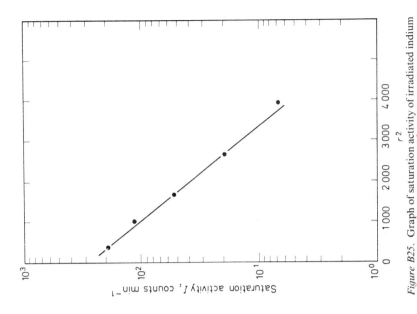

Figure B25. Graph of saturation activity of irradiated indium foils against (distance from source)^2 used in the determination of the slowing-down length of neutrons in graphite

made. (The energy losses in graphite and such heavier moderators are relatively much smaller than in hydrogenous media and do approximate to the theory of exponential loss.)

Other factors to be considered are:

(1) The location of the source for the true measurement of r is not exactly known.

(2) The log I against r^2 graph is not strictly linear at large r values.

(3) No allowance has been made for the fact that we have slowed down to indium resonance level (1·44 eV) not to true thermal energy

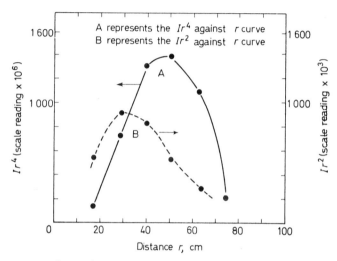

Figure B26. Ir^2 and Ir^4 against r curves to determine the slowing-down length of neutrons in graphite

level ($=0.025$ eV) when about 20% more collisions must take place. An approximate correction for this may be made by multiplying the result for L_s by $(1·2)^{\frac{1}{2}}$.

Note. The graphs (*Figures B23, B24,* and *B25*) show 'Saturation activity' and this is defined as the resultant activity if the foil were irradiated for infinite time. It is needed only if comparison of results with those of other observers is required and is found by dividing I by $(1 - F)$ where F is the appropriate factor given for the particular time of irradiation, in standard tables.

Although the planimeter has been suggested as the method of measuring area, it is instructive to determine the area by alternative means, e.g. by direct counting of squares on graph paper or by cutting out the areas and weighing, making sure that allowance is made for any different scales used. It would not be possible to cut out the areas on the examples given in *Figures B21* and *B26* since one scale has to be multiplied by 10^6 and the other by 10^3. The discrepancies brought to light in this way will emphasize clearly the large inaccuracies involved in the experiment.

Further Work

If a number of identical indium foils are placed in the sleeve at one given distance from the source and irradiated for 2, 4, 6, 8 and 10 minutes respectively a graph of the build up of activity within the foils may be plotted. It is also possible to plot the decay curve of the most active foil and from it to determine the half-life of the activity (see also Experiment No. B1.10).

Reading and References

The mathematical background behind the equations given in the theory may be found in *Nuclear Reactor Engineering* by S. Glasstone and A. Sesonske (Van Nostrand, 1963).

Detail of the cross-sections of indium and cadmium showing the 1·44 electron volt resonance in indium and the high capture cross-section of cadmium for thermal neutrons may be found in *Nuclei and Particles* by E. Segre (Benjamin, 1964).

Experiment No. B1.8

TO MEASURE THE DIFFUSION LENGTH OF NEUTRONS IN A MODERATING MEDIUM, e.g. GRAPHITE OR PARAFFIN WAX

Introduction

Experiment No. B1.7 was concerned with the slowing down of fast neutrons to thermal energies. This experiment deals with those neutrons sufficiently distant from a point source to have been thermalized. They then diffuse through the moderating medium, constantly colliding with the target nuclei and their motion may be represented by a general diffusion equation, the analysis of which may be compared with that treated in Experiment No. E1.6 on thermal diffusivity.

Theory

Once a steady state of diffusion has been established one may write

$$\nabla^2 \phi - \left(\frac{1}{L^2}\right)\phi = 0$$

where ϕ represents the neutron flux, and L the diffusion length. L may be defined by the equation

$$L^2 = (\bar{r}^2/6),$$

i.e. the diffusion length is equal to the square root of one sixth of the mean square (or shortest) distance from the point at which the neutron was thermalized to the point at which it was absorbed by the moderator (leakage and consequent escape are neglected in this simplified treatment).

For a moderator of cylindrical geometry

$$\nabla^2 \phi = \left(\frac{\partial^2 \phi}{\partial r^2} + \frac{1}{r}\frac{\partial \phi}{\partial r} + \frac{\partial^2 \phi}{\partial z^2}\right)$$

and hence the diffusion equation may be written

$$\frac{\partial^2 \phi}{\partial r^2} + \frac{1}{r}\frac{\partial \phi}{\partial r} + \frac{\partial^2 \phi}{\partial z^2} - \frac{1}{L^2}\phi = 0$$

where r = radius of the cylinder and z = distance along the axis.

Now it may be assumed that the neutron flux ϕ is a function of both the axial distance z from the source and the radial distance r from the central axis of the cylinder, i.e.

$$\phi = ZR$$

where Z represents the function of z and R the function of r. Then the diffusion equation becomes

$$\frac{\partial^2}{\partial r^2}(RZ) + \frac{1}{r}\frac{\partial}{\partial r}(RZ) + \frac{\partial^2(RZ)}{\partial z^2} - \frac{1}{L^2}(RZ) = 0$$

or

$$Z\frac{d^2 R}{dr^2} + \frac{Z}{r}\frac{dR}{dr} + R\frac{d^2 Z}{dz^2} - \frac{1}{L^2}(RZ) = 0$$

and dividing by (RZ) and rearranging

$$\frac{1}{R}\left(\frac{d^2R}{dr^2}\right) + \frac{1}{(Rr)}\left(\frac{dR}{dr}\right) - \frac{1}{L^2} = -\frac{1}{Z}\frac{d^2Z}{dz^2}$$

It should now be noted that the left-hand side of the equation is a function of r only and the right-hand side a function of z only. For this condition to be realized each side of the equation in turn must be equal to a constant, i.e.

$$\frac{1}{R}\frac{d^2R}{dr^2} + \frac{1}{rR}\left(\frac{dR}{dr}\right) - \frac{1}{L^2} = k = -\beta^2 \tag{1}$$

and

$$\frac{1}{Z}\frac{d^2Z}{dz^2} = -\beta^2 \tag{2}$$

where k and β are constants. (For a non-periodic solution the constant in equation (1) must be negative hence the negative sign is included since β^2 is essentially positive, i.e.

$$\frac{d^2Z}{dz^2} = \beta^2 Z \quad \text{and} \quad Z = A\exp(+\beta z) + B\exp(-\beta z) \tag{3}$$

Differentiate twice to check this solution.)

Since leakage is neglected and it is assumed that no neutrons escape from the system $Z = 0$ at the extrapolated boundary of the moderator where $z = b_1$, hence using equation (3)

$$A\exp(\beta b_1) + B\exp(-\beta b_1) = 0 \quad \text{or} \quad A = -B\exp(-2\beta b_1)$$

and

$$\therefore Z = -B\exp(-2\beta b_1)\exp(+\beta z) + B\exp(-\beta z)$$

or

$$Z = B\exp(-\beta b_1)\{\exp(-\beta z)\exp(+\beta b_1)$$
$$- \exp(+\beta z)\exp(-\beta b_1)\}$$
$$= C[\exp\{-\beta(z - b_1)\} - \exp\{\beta(z - b_1)\}]$$

This represents the solution governing the z variation only, and it will be assumed that the solution governing the r variation can be considered independently in the same way. Thus we had

$$\left(\frac{1}{R}\right)\frac{d^2R}{dr^2} + \left(\frac{1}{rR}\right)\frac{dR}{dr} - \frac{1}{L^2} = -\beta^2$$

DIFFUSION LENGTH OF NEUTRONS B1.8

and rearranging

$$r^2 \frac{d^2R}{dr^2} + r \frac{dR}{dr} + \left(\beta^2 - \frac{1}{L^2}\right) r^2 R = 0$$

This is Bessel's equation of zero order (see references), for if

$$r \left(\beta^2 - \frac{1}{L^2}\right)^{\frac{1}{2}} = r_1$$

$$dr \left(\beta^2 - \frac{1}{L^2}\right)^{\frac{1}{2}} = dr_1$$

and substituting in the above equation

$$r_1^2 \frac{d^2R}{dr_1^2} + r_1 \frac{dR}{dr_1} + r_1 R = 0$$

Hence

$$R = J_0(r_1) = J_0 r \left\{\left(\beta^2 - \frac{1}{L^2}\right)^{\frac{1}{2}}\right\}$$

but when $r = a_1$, $\phi = 0$, and when $\phi = 0$, $J_0(r_1) = 0$.

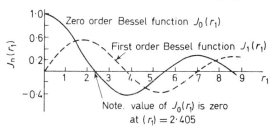

Figure B27. Graph of the zero order and first order Bessel functions $J_0(r_1)$ and $J_1(r_1)$

From the graph (*Figure B27*) $J_0(r_1) = 0$ when $r_1 = 2\cdot405$, i.e.

$$J_0 \left\{a_1 \left(\beta^2 - \frac{1}{L^2}\right)^{\frac{1}{2}}\right\} = 0$$

therefore

$$a_1 \left(\beta^2 - \frac{1}{L^2}\right)^{\frac{1}{2}} = 2\cdot405$$

or

$$\beta^2 - \frac{1}{L^2} = \left[\frac{2\cdot405}{a_1}\right]^2 \quad \text{and} \quad R = J_0 \left[\frac{2\cdot405}{a_1} r\right]$$

so that taking account of the dependence of flux variation upon z as well as upon r

$$\phi = AJ_0\left(\frac{2\cdot 405 r}{a_1}\right)[\exp\{-\beta(z-b_1)\} - \exp\{+\beta(z-b_1)\}].$$

It may be assumed that the geometry of the moderator tank is such that $z < b$, hence $(z - b_1)$ is negative and the term $\exp\{+\beta(z-b_1)\}$ may be neglected in comparison with $\exp\{-\beta(z-b_1)\}$. Therefore plotting $\log \phi$ against z results in a straight line, the slope of which is $-\beta$ (*Figures B28* and *B30*).

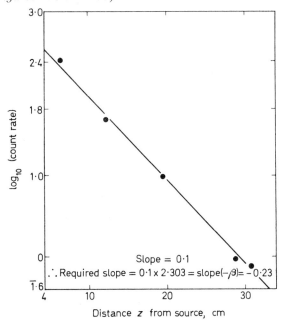

$$\frac{1}{L^2} = (\text{Slope }\beta)^2 - \left(\frac{2\cdot 405}{a_1}\right)^2$$

a_1 = extrapolated radius of cylindrical moderator
    = 28·3 cm

Hence $\dfrac{1}{L^2} = (0\cdot 1 \times 2\cdot 303)^2 - \left(\dfrac{2\cdot 405}{28\cdot 3}\right)^2$

Note: 2·303 is introduced since the slope of a $\log_e$ graph is required and $\log_{10}$ is plotted

$$L \simeq 4 \text{ cm}$$

Figure B28. Graph of log (count rate) against distance from source, the slope of which enables the diffusion length in paraffin wax to be calculated

Method

Six or so identical indium foils are irradiated as in experiment B1.7 in the fixed positions within the neutron flux from an americium-beryllium source of about 500 millicuries for several hours but this time they are not sheathed in cadmium. They are (as before) removed and allowed to decay for 10–15 minutes before the weakest one is placed in a suitable castle and the count rate from it measured. The others are then taken in sequence and the count rate noted.

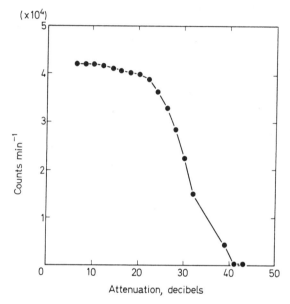

Figure B29. Attenuation of count rate in the determination of the diffusion length of neutrons in graphite

Corrections for the paralysis time of the counter, the background counts, and for the delay before counting takes place are made in just the same way as in experiment B1.7.

The logarithm of the corrected count rate is then plotted against the distance z along the axis of the moderator and the slope of the graph enables the diffusion length of the neutrons in the paraffin wax to be determined from the equation

$$\frac{1}{L^2} = (\text{Slope})^2 - \left(\frac{2 \cdot 405}{a_1}\right)^2$$

NUCLEAR AND RADIATION PHYSICS

since a_1, the extrapolated radius of the cylindrical moderator, is known.

Discussion of Results

Note that only distances beyond the point of thermalization of neutrons may be considered and also points for which $z < b_1$.

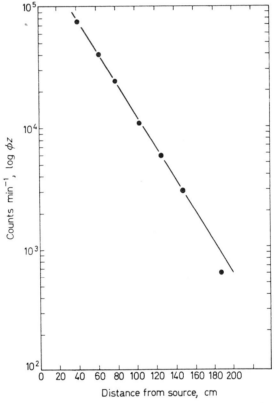

Figure B30. Graph of count rate (on log scale) against distance z from the source for neutrons in graphite

Hence the linearity occurs only over a restricted central portion of the graph. Strictly speaking the dimension

$$b_1 = b + 0.71\lambda_t \quad \text{and} \quad a_1 = a + 0.71\lambda_t$$

but λ_t the transport mean free path is of the order of 0.5 cm for paraffin wax and may be neglected compared with the dimensions of the usual moderator tank.

The values obtained may show serious error since the following assumptions have been made which are not strictly true:

(1) that the neutrons are monoenergetic
(2) that the scattering takes place from a free particle and this is of course not so.

Further Work

Figures B29 and *B30* show the graphs obtained using a graphite stack of cubic geometry. The diffusion length in graphite is much higher than in paraffin wax, so is the transport mean free path (~ 3 cm) and this latter parameter cannot in this case be neglected. The mathematics appropriate to a moderating stack of rectangular geometry should be carefully noted (see references), even if it is not possible to carry out the experiment in a moderator of this type.

Reading and References

The mathematical background of the zero order Bessel equation and a detailed treatment of the graph of *Figure B27*, will be found in *Mathematical Methods of Physics* by J. Mathews and R. L. Walker (Benjamin, 1965).

Greater detail concerning neutron diffusion and the theory of cylindrical and cubic stacks is given in *Introduction to Reactor Physics* by D. J. Littler and J. F. Raffle (Pergamon, 1957).

Tables of slowing down lengths, diffusion length and transport mean free path of neutrons in various moderating materials may be found in *Nuclear Reactor Engineering* by S. Glasstone and A. Sesonske (Van Nostrand, 1963).

Experiment No. B1.9

TO DETERMINE THE RANGE OF ALPHA PARTICLES EMITTED FROM A RADIOACTIVE SOURCE

Introduction

The range of alpha particles in air at atmospheric pressure is small and of the order of a few centimetres. As the pressure falls the alpha range over which ionization occurs increases.

NUCLEAR AND RADIATION PHYSICS

In this experiment a flask of air is evacuated until the point is reached when the alpha particles from a radioactive source located at the centre of the flask reach the conducting walls of the vessel. The voltage established between the central wire holding the radioactive substance and the walls then falls dramatically and thus one can estimate the range of the particle at this particular pressure.

Theory

Let R = range of the emitted alpha particles
r = radius of the conducting sphere
P_A = pressure of the atmosphere
P_c = the critical pressure at which the range R of the alpha particle equals r the radius of the conducting sphere.

As the pressure decreases the alpha range increases, and it may be assumed that

$$R_\alpha = \frac{1}{\text{Pressure } P}$$

i.e.

$$\frac{R_\alpha}{r} = \frac{P_c}{P_A} \quad \text{or} \quad R_\alpha = \frac{rP_c}{P_A}$$

Method

A weak alpha source (10 μcurie ^{210}Po source is suitable) is mounted on the tip of a brass rod which passes through an insulating stopper to lie at the centre of a brass or copper (conducting) spherical flask as shown in *Figure B31*.

To ensure that all the alphas are attracted towards the wall of the flask a 'saturation' voltage is applied between the walls of the flask and the electrode. This will be of the order of a few hundred volts (see note of warning below). *Figure B32* shows a typical initial 'control' graph to determine a suitable 'saturation' voltage.

The sphere is evacuated by means of a rotary pump and the pressure noted from readings on a mercury manometer attached to the reservoir of the system, as indicated (*Figure B33*).

The effective ionization current should be noted as the pressure is reduced from atmospheric to about 5 cm mercury pressure.

As long as all the alphas are stopped within the gas the ionization remains constant and a graph of the variation against pressure in this region (AB *Figure B34*) shows that it is independent of pressure.

Ionization current may be plotted directly. To do this a suitable electrometer is required for the measurement of the ionization current which will be of the order of 10^{-11} amperes.

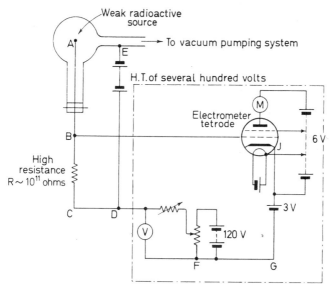

Figure B31. Measurement of the range of alpha particles. The broken line encloses the circuit used for measuring ionization current (a suitable electrometer might be substituted)

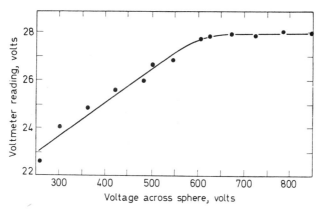

Figure B32. Determination of the range of alpha particles—control graph

The actual graph shown is plotted in terms of voltage, and the readings taken when a constant anode current flowed in an electrometer tetrode, using a circuit of the type illustrated in *Figure B31*. This constant value of current used, is that which exists when no voltage difference is applied between the electrode and the sphere,

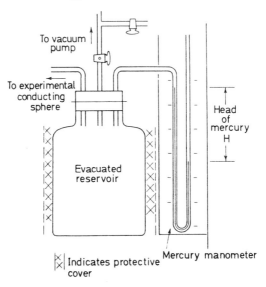

Figure B33. Arrangement for reducing the pressure in the experimental sphere and measuring the resulting pressure

i.e. when no current flows in the circuit ACDE which contains the high resistance R ($\sim 10^{11}$ Ω). However, grid current does flow in the valve circuit BCDFGJ and the value of this determines the standing anode current through the microammeter M.

On establishing the high voltage across AE current flows in the circuit ABCDE thus increasing the voltage across R (negatively). This in turn affects the grid of the valve and current through M varies. Restoring this current to its original value is done by varying the voltage V (*Figure B31*) from its original value when $(V - V_0)$, which is equal to ir, is a direct measure of the ionization current.

As indicated by the theory, below the critical pressure P_c, the ionization current or voltage falls fairly rapidly in accordance with a linear law (region BC *Figure B34*), thus knowing P_c, the alpha range may be determined as shown in the theory.

RANGE OF ALPHA PARTICLES   B1.9

Warning Note: The voltage applied across the sphere is of a high order and a guard ring or other protective device should be used in the experiment. Read carefully any instructions given in this regard with respect to the particular apparatus being used.

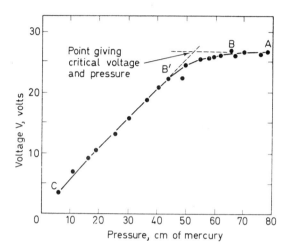

Figure B34. Determination of the range of alpha particles

Discussion of Results

As the graph clearly shows, the 'turn over' point is not a sharp one. This is due to the fact that random variations in the range of alphas of one particular energy occur. Depending upon the radioactive source used, it may also indicate the presence of alphas of different energies (i.e. ranges).

The data of the graph shown in *Figure B34* are investigated in greater detail in *Figure B35*. This seems to indicate that three distinct alpha energies are present in the emission from the particular source chosen in this case.

Further Work

Knowing the ionization current and the range one can determine the number of alpha particles emitted per second by the radioactive source and compare this with the accepted value. If the resulting experimental value differs markedly from the textbook value it should be recalled that for long paths in air, at the higher pressures involved in the experiment, some recombination of electron pairs may take place yielding a low value for the ionization current.

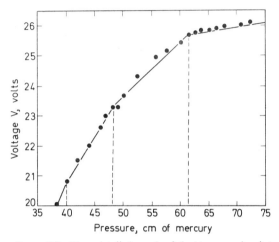

Figure B35. More detailed graph of the 'turn-over' point, indicating the presence of alphas of more than one single range

Reading and References

Background reading upon the properties of alpha particles may be found in *Electricity, Magnetism and Atomic Physics*, vol. 2, by J. Yarwood (University Tutorial Press, 1958).

A brief description of this experiment is given, together with comparative ionization current curves for various different radioactive sources.

Experiment No. B1.10

TO DETERMINE THE HALF-LIFE OF NEUTRON-IRRADIATED INDIUM FOIL

Introduction

The half-life of a radioactive substance is defined as the time taken for the initially observed intensity of activity in the substance to decay to half that original value.

HALF-LIFE OF INDIUM FOIL   B1.10

Some nuclides have extremely short half-lives, of the order of fractions of a second, others have very long ones ($\sim 10^9$ years) and to measure these, sophisticated experimental methods are required.

Irradiated indium however has a half-life of about 54 minutes, which makes it an ideal choice of substance for the general laboratory determination of half-life.

Theory

From the definition of half-life given above one may write

$$I = \frac{I_0}{2}$$

where I = intensity of activity at time t
       I_0 = initial intensity of activity

and since the decay of a radioactive substance is governed by a law of the form

$$I = I_0 \exp(-\lambda t)$$

where λ is the decay constant, then at the half-life τ

$$\frac{I_0}{2} = I_0 \exp(-\lambda \tau) \quad \text{and} \quad \tfrac{1}{2} = \exp(-\lambda \tau)$$

so that

$$\log_e 2 = +\lambda \tau \quad \text{and} \quad \tau = \frac{0 \cdot 69}{\lambda}.$$

When the indium is irradiated an (n, γ) reaction takes place thus

$$^{115}_{49}\text{In} + ^{1}_{0}\text{n} = ^{116}_{49}\text{In} + \gamma.$$

The product nucleus ($^{116}_{49}\text{In}$) is itself radioactive and decays with the emission of a negative beta particle (together with further gamma emission) so that the 116 isotope of tin is produced.

$$^{116}_{49}\text{In} \rightarrow ^{116}_{50}\text{Sn} + ^{0}_{-1}\beta + \gamma$$

This product $^{116}_{50}\text{Sn}$ is stable and remains (in traces) in the body of the indium foil.

If the beta decay of the foil is measured, then a graph of log I against t should be a straight line since $\log_e (I/I_0) = -\lambda t$ and I_0 is a constant.

Using logarithms to the base ten the slope of the line is $-\lambda/2\cdot303$ so that the half-life

$$\tau = \frac{0\cdot69}{-2\cdot303 \times (\text{Slope})}$$

Method

Load the indium foil (it will probably be mounted in aluminium for mechanical strength) into the neutron-tank sleeve under the direction of the laboratory demonstrator, placing it in the slot which will lie nearest the source during irradiation. Remove the blank plug and using the tongs provided, lower the loaded sleeve into position. Irradiate the foil for a period of about 1 hour and then remove the foil and replace the blank plug. Note carefully the time of removal of the foil and put the foil aside for 5 minutes to allow for the very short-lived 'aluminium activity' to decay to negligible proportions.

Set up the Geiger counting equipment in the manner described in Experiment No. B1.1 and determine the beta count rate of the foil. Note the times at the beginning and end of counting and take the average value to compute the time of foil decay. Repeat at convenient intervals (see *Figure B36*) and from the slope of the resulting graph calculate the half-life of the irradiated indium in the manner indicated in the theory.

Throughout the experiment background count rates should be observed and the necessary corrections for background activity and counter paralysis time should be made (see Experiment No. B1.1).

Discussion of Results

Provided fresh foils (or alternatively foils that have NOT been in use for some time) are used the experiment generally gives good results for the half-life of the irradiated indium, in close agreement with the accepted value of 54 minutes. The graph of *Figure B36* shows little scatter of individual points so that the position of the 'best fit' line is well defined.

Further Work

If after irradiating a single foil, stacks of 2, 3, 4 and 5 foils are irradiated and the activity of each separate foil counted (in rotation) the ratio of activity of each stack relative to the single foil may be found. A fall in relative activity is observed due to self-absorption in the foils of the stack. This may be calculated and compared with the appropriate theory.

Reading and References

The theory of the half-life of a radioactive substance may be found in any of the standard degree texts. The abbreviated derivation given

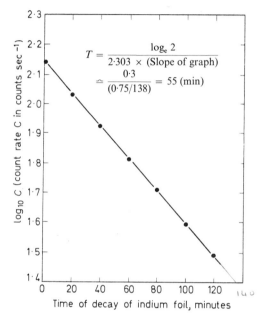

Figure B36. Decay of beta activity on irradiated indium: graph of the logarithm of count rate against time of decay, used to determine the half-life of the activity

above should be compared with that given in Experiment No. B2.5 on the half value thickness of x-ray absorbers and also with the notes on absorption of light (Experiment No. C2.3).

Self-absorption in thick samples is dealt with in *Nuclear Radiation Physics* by R. E. Lapp and H. L. Andrews (Pitman, 3rd edition, 1963) where a typical graph is given.

NUCLEAR AND RADIATION PHYSICS

Experiment No. B1.11

TO INVESTIGATE THE SIMULTANEOUS EMISSION OF BETA AND GAMMA ACTIVITY FROM SODIUM-22 AND TO VERIFY THE EMPIRICAL FORMULA RELATING THE MASS ABSORPTION COEFFICIENT TO THE MAXIMUM ENERGY IN THE BETA SPECTRUM

Introduction

Previous experiments (Nos. B1.3 and B1.6) have dealt with the absorption of the emission of beta particles alone and of gamma radiation alone, respectively, and it was noted that in the emission of beta particles a '*Bremsstrahlung* tail' could appear on the beta absorption curve. Many radioactive disintegrations take place, however, where beta particles and gamma rays are emitted simultaneously and this experiment is concerned with the analysis of the data from the absorption curve resulting from the use of such simultaneous emission.

Theory

When beta particles are emitted, a 'spectrum' of beta energies is found to exist and since the scattering process is the predominant one, the beta track is tortuous so that particles having passed through the absorbing medium may emerge in different directions after traversing paths of widely different magnitude.

These two effects result in the experimentally observed facts that (1) the emitted particles do not strictly have a specified range as do alpha particles (although as seen from Experiment No. B1.3 an empirical estimate of 'range' is often used), and (2) that the ionization—and hence the absorption—follows the exponential law assumed in Experiment No. B1.3, namely

$$I = I_0 \exp\{-(\mu/\rho)(\rho x)\}$$

where I = the intensity of activity after passing through an absorber of thickness (ρx) (gram cm^{-2})
I_0 = initial intensity of activity
(μ/ρ) = the mass absorption coefficient.

BETA AND GAMMA ACTIVITY FROM SODIUM-22 B1.11

A number of empirical formulae has been produced to link the mass absorption coefficient with the maximum energy of the beta particles in the emitted spectrum, of which one is

$$\left(\frac{\mu}{\rho}\right) = \frac{22}{E_{\max}^{1\cdot 33}}$$

where $E_{\max}$ represents the maximum energy in the beta spectrum.

From the one curve resulting from simultaneous emission of both beta particles and gamma rays, the linear and mass absorption coefficients for the beta and gamma rays may be determined from the slopes AB and CD of the curve (*Figure B37*) and one may define the limits of exponential decay of ionization for the beta particles as well as verify the empirical formula quoted above.

Method

The isotope of sodium (^{22}Na) forms a suitable source for this experiment since it emits gamma rays and positive beta particles simultaneously.

The source should be placed in a locating slot a fixed distance from the window of the Geiger tube, with just sufficient space to insert absorbers on a shelf above them.

As in previous Geiger counting experiments, the mean count rates, first without an absorber inserted and later corresponding to the insertion of aluminium absorbers of various thicknesses, are recorded. These readings are then corrected for the paralysis time of the counter as well as for background radiation and a graph is plotted of the logarithm of the corrected count rate against the absorber thickness expressed in gram cm^{-2}.

Extrapolate the regions AB and CD of the graph as shown in *Figure B37* and from the slopes of the two portions the linear absorption coefficient for the betas and the gammas can be found.

Assuming that $\mu/\rho = 22/E_{\max}^{n}$ and knowing the maximum energy in the beta spectrum (in this case 1·28 MeV) the value of the index n can be found and compared with the accepted value.

Discussion of Results

The theory relevant to the betas assumes good geometry, i.e. a thin highly collimated beam of particles and this of course is not strictly true in practice. Three separate measurements of count rate under each of the specified conditions and careful note of the scatter of the points obtained on the graph, will show that errors of the order of 5% can occur in the experiment.

Owing to the shallow slope of the CD region of the curve the value obtained for the absorption coefficient of the gamma radiation

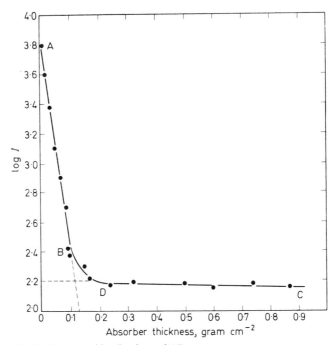

For the betas consider the slope of AB.

$$\text{From graph } \frac{\mu}{\rho} = \frac{1\cdot 6}{120} \times 2\cdot 303 \simeq 31$$

$$\text{Using the formula } \frac{\mu}{\rho} = \frac{22}{E_{max}^n}$$

where E_{max} = maximum energy of betas in the spectrum = 1·28 MeV

$$31 = \frac{22}{(1\cdot 28)^n}$$

or $n \log 1\cdot 28 = \log 0\cdot 715$

$$n \simeq \frac{-1\cdot 46}{1\cdot 1} = 1\cdot 3$$

Similarly for the gamma radiation the slope of the portion CD enables the absorption coefficient to be found

Figure B37. Curve resulting from simultaneous absorption of beta particles and gamma rays. I = intensity of activity in terms of count rate

is subject to much larger errors than are present in the beta calculation, and great care is required in plotting the CD region with as many points being chosen as possible.

Compare the values obtained with the accepted values found in the literature to help confirm the findings of the experiment.

Further Work

It is interesting to construct the graph due to beta particle absorption alone, by extrapolating the gamma 'tail' back to the condition of zero thickness so that the initial intensity of gamma radiation alone may be computed and subtracting the value obtained from the experimental corrected count rate value for total initial intensity of activity, thus giving the initial intensity due to betas alone. This may be done at several different points on the curve to construct the graph appropriate to the absorption of betas alone.

Reading and References

The absorption of both beta and gamma radiation may be found in *Introduction to Reactor Physics* by D. J. Littler and J. F. Raffle (Pergamon, 2nd edition, 1957). The formula $(\mu/\rho) = (22/E^{1.33})$ is also quoted.

Experiment No. B2.1

TO ANALYSE AN X-RAY 'POWDER' PHOTOGRAPH FOR A SUBSTANCE HAVING A CUBIC CRYSTAL LATTICE

Introduction

If a powdered crystalline material is bombarded by a monochromatic beam of x-radiation the randomly oriented crystallites diffract the beam in such a way that a series of lines is obtained (*Plate B1*) corresponding to those angles satisfying the Bragg relationship

$$n\lambda = 2d_{(h, k, l)} \sin \theta_n$$

where $n = 1, 2, 3$, etc.
$d_{(h, k, l)}$ = interplanar spacing
θ_n = angle of diffraction.

From measurements on the photograph the discrete values of the interplanar spacings (the d's) can be calculated.

Theory

Figure B38 represents the photograph and the way in which it is wrapped in the camera. The distance between corresponding lines

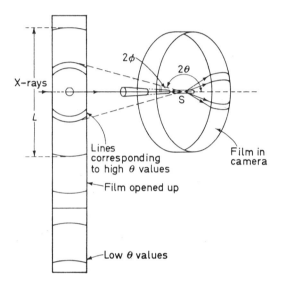

Figure B38. Diagram of the arrangement of the film in the camera for taking x-ray powder diffraction photographs

on either side of the collimator hole in the film enables the angle to be calculated since the radius R of the camera is known ($\phi = L/4R$).

From the diagram we also see that $\theta = (\pi/2 - \phi)$, hence placing values in the Bragg equation and remembering that for copper K_α radiation $n = 1$, d values can be tabulated (*Table B2*) and hence $(1/d^2)$.

For a cubic lattice

$$d = \frac{a}{(h^2 + k^2 + l^2)^{\frac{1}{2}}}$$

where a is the lattice parameter and (h, k, l) are the Miller indices.

ANALYSING AN X-RAY POWDER PHOTOGRAPH B2.1

Now

$$\frac{1}{d^2} = \frac{1}{a^2}(h^2 + k^2 + l^2)$$

i.e.

$\frac{1}{d^2}$ is proportional to $(h^2 + k^2 + l^2)$

Table B2. X-ray powder diffraction pattern

L	$h^2 + k^2 + l^2$	d	$\log d$	$\log(h^2 + k^2 + l^2)$	$1/d^2$
4·075	2	2·21	0·344	0·3010	
5·875	4	1·57	0·1956	0·6021	
7·375	6	1·282	0·1079	0·7782	
8·775	8	1·11	0·0414	0·9031	
10·125	10	0·996	$\bar{1}$·9983	1·00	
11·575	12	0·909	$\bar{1}$·9586	1·0792	
13·225	14	0·842	$\bar{1}$·9253	1·1461	
15·475	16	0·789	$\bar{1}$·8932	1·2041	

N.B. The above results are not for the particular plate shown (B2), a camera of different radius and film insertion was used.

For low order reflections, e.g. at the (100), (110) and (111) planes, $(h^2 + k^2 + l^2)$ has small integral values, viz. 1, 2 and 3 and we look for a common divisor in the $1/d^2$ values such that the *quotients* are integers.

One should then tabulate a greater number of (h, k, l) and $(h^2 + k^2 + l^2)$ values as in *Table B3*.

Table B3. Possible values

$h^2 + k^2 + l^2$	h, k, l	$h^2 + k^2 + l^2$	h, k, l
1	1, 0, 0	17	4, 1, 0 or 3, 2, 2
2	1, 1, 0	18	4, 1, 1 or 3, 3, 0
3	1, 1, 1	19	3, 3, 1
4	2, 0, 0	20	4, 2, 0
5	2, 1, 0	21	4, 2, 1
6	2, 1, 1	22	3, 3, 2
–	– – –	–	– – –
8	2, 2, 0	24	4, 2, 2
9	3, 0, 0 or 2, 2, 1	25	5, 0, 0 or 4, 3, 0
10	3, 1, 0	26	5, 1, 0 or 4, 3, 1
11	3, 1, 1	27	5, 1, 1 or 3, 3, 3
12	2, 2, 2	–	– – –
13	3, 2, 0	29	5, 2, 0 or 4, 3, 2
14	3, 2, 1	30	5, 2, 1
–	– – –		
16	4, 0, 0		

In the case of the simple cubic crystal lattice note that h, k and l may have *any* integral values. This is not true for a face centred cubic lattice where the integers will be all odd or all even whilst for a body centred cubic lattice the *sum* of (h + k + l) must be even. It is then possible to calculate the lattice constant a for each value of d. This should of course yield a single value.

Method

Powder specimens may be mounted by dipping a thin glass fibre first into Canada balsam and then into the crystalline powder to be

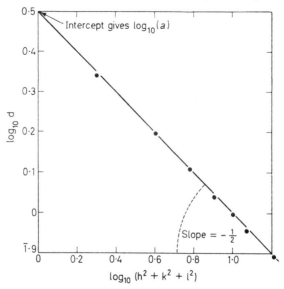

Figure B39. Graph of log d against log $(h^2 + k^2 + l^2)$

investigated or else by mixing the powder to a paste with water and gum tragacanth and rolling the mixture to a long thin thread. A suitable specimen here, however, is tungsten in the form of thin wire (~ 0.02 mm diameter) about six threads being twisted together. The specimen is then placed in the camera chuck, great care being exercised in aligning it so that it is upright and does not 'wobble' when in the x-ray beam. A small turntable should be provided for this purpose and alignment carried out through a low power eyepiece containing two cross wires (to define the close limits outside which the specimen must not fall) BEFORE being placed in the x-ray camera.

ANALYSING AN X-RAY POWDER PHOTOGRAPH   B2.1

Using a template, cut the collimator hole through the centre or other appropriate position of the film and mount it (complete with its black paper cover) in the camera.

After consultation with the operator mount the camera in the x-ray set (see safety notes—Experiment No. B2.3).

Expose the film for about 40 minutes. Switch off and process the negative in accordance with instructions given with the film.

Discussion of Results

By far the most important factor in obtaining sharp lines on the film is the alignment of the specimen in the beam and failure to do this correctly can result in 'fuzzy' lines.

The graph (*Figure B39*) shows an alternative way of treating the results. Note that

$$d^2 = a^2/(h^2 + k^2 + l^2)$$

so that

$$2 \log_e d = 2 \log_e a - \log_e (h^2 + k^2 + l^2)$$

or

$$\log_e d = -\tfrac{1}{2} \log_e (h^2 + k^2 + l^2) + \log_e a$$

and thus plotting $\log d$ against $\log (h^2 + k^2 + l^2)$ yields a graph of slope $(-\tfrac{1}{2})$ and the intercept on the y axis gives $\log a$. (The 'scatter' of the points about the 'best fit' line should enable conclusions to be drawn concerning the accuracy of the experiment.)

Figure B40 shows a typical computer print out for the evaluation of the function

$$\frac{1}{2}\left(\frac{\cos^2 \theta}{\sin \theta} + \frac{\cos^2 \theta}{\theta}\right)$$

The most accurate value of a has been shown to be given when $\theta = 90°$, i.e.

$$f(\theta) = \frac{1}{2}\left(\frac{\cos^2 \theta}{\sin \theta} + \frac{\cos^2 \theta}{\theta}\right) = 0$$

Thus by plotting a against $f(\theta)$ and extrapolating back to the y axis, optimum accuracy in the determination of a can be achieved.

Further Work

Having analysed the pattern corresponding to a known element of cubic structure it is worth while to try to identify the elements in a

NUCLEAR AND RADIATION PHYSICS

mixture of two compounds such as nickel and copper oxides, to grow familiar with the use of the cards of the ASTM Index.

```
LATTICE CONSTANTS FROM X RAY DIFFRACTION
              OCT 66
FREE STORE= 4630- 6433
BEGIN REAL A,S,R,LAMBDA,D
INTEGER N,H,K,L'
SWITCH SS:=NEXT'
READ LAMBDA,R'
PRINT££L5S2?LAMBDA=?,SAMELINE,FREEPOINT(5),LAMBDA,
£(A)     R=?,R,£(MM)?'
PRINT ££L2S2?S(MM)    N  H  K  L    D(A)     A(A)?'
NEXT:READ S,N,H,K,L'
D:=(N*LAMBDA)/(2*SIN(S/(2*R)))'
A:=D*SQRT((H*H)+(K*K)+(L*L))'
PRINT ££L2??,SAMELINE,ALIGNED(3,2),S,£ ?,DIGITS(2),N,H,K,

L,£ ?,ALIGNED(2,3),D,£ ?,A'
GOTO NEXT'
END'
LAMBDA= 1.5390(A)    R= 57.300(MM)
```

S(MM)	N	H	K	L	D(A)	A(A)
40.75	1	1	1	0	2.210	3.126
58.75	1	2	0	0	1.569	3.138
73.75	1	2	1	1	1.282	3.141
87.75	1	2	2	0	1.110	3.140
101.25	1	3	1	0	0.996	3.148
115.75	1	2	2	2	0.909	3.148
132.25	1	3	2	1	0.842	3.149
154.75	1	4	0	0	0.789	3.154

Figure B40. Computer print out of lattice constants from x-ray diffraction

Reading and References

More detailed theory concerning the Bragg equation and the Miller indices may be found in standard physics degree texts, e.g. *Atomic Physics* by J. Yarwood (University Tutorial Press, 2nd edition, 1964).

Suitable background reading concerning x-ray diffraction methods may be found in *Diffraction Methods in Materials Science* by J. B. Cohen (Macmillan, 1966).

Experiment No. B2.2

THE USE OF ASTM INDEX CARDS TO IDENTIFY SUBSTANCES IN A MIXTURE BY THE X-RAY POWDER METHOD

Introduction

Experiment B2.1 was concerned with the determination of the inter-atomic spacing of a known element (tungsten) from measurements taken on the powder photograph.

Each crystalline substance has its own characteristic pattern and an index of patterns has been built up by the American Society for Testing Materials (the ASTM Index).

The use of this index is of great importance in the identification of crystalline substances in industry and whilst a comprehensive index may not always be available it is of interest to familiarize oneself with the analysis of a simple compound or mixture of compounds and to identify the constituents from a small chosen range of cards.

Theory

The cards give data on the d spacings of the indexed substances and upon the relative intensities of the lines on the resulting powder patterns of each (*Figure B42*).

As was seen in Experiment No. B2.1, the Bragg equation was appropriate to x-ray diffraction, viz.

$$\sin \theta = \frac{n\lambda}{2d}$$

where n = an integer
  θ = angle of diffraction
  λ = wavelength of the x-rays used, and
  d = crystal 'spacing'.

For each value of θ a cone of diffracted rays exists (*Figure B41*) and corresponding to each cone two lines will be produced on the film.

The position of the lines from the centre of the system is characteristic of the crystal at which diffraction occurs, as is the intensity of the lines. (Lines characteristic of other crystals may of course overlap which will complicate the interpretation of the pattern.)

The spacing *l* between the lines may then be related to the angle of diffraction and the radius of the powder camera chamber as described in Experiment No. B2.1.

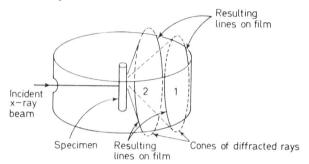

Figure B41. Diagram showing the relation between the lines on the powder-method film and the cones of diffracted x-rays. Cone 1 corresponds to angle of diffraction θ_1 and semi-angle $2\theta_1$. Cone 2 corresponds to angle of diffraction θ_2 and semi-angle $2\theta_2$

By reference to the ASTM index cards (*Figure B42*) it is possible to identify the substance or substances forming the pattern from the characteristic spacing between the various sets of lines.

Method

The specimen is mounted as a thin cylindrical thread at the centre of the camera chamber. This is usually done by first thoroughly cleaning a thin rod, or tube, of glass about 3 cm long (and less than 1 mm diameter) and fitting it into the chuck in a rotatable mount, which will lie at the centre of the photographic chamber during the actual experiment.

Using an optical method (as in Experiment No. B2.1) care is taken to see that on rotation the chuck remains upright, so that when placed in the x-ray beam the specimen will not move out of the beam at any point during the rotation.

The rod in the detachable part of the chuck is dipped in Canada balsam, then gently and carefully turned over in the specimen powder contained in a test tube. The chuck is replaced and a check made to see that no displacement of the alignment has occurred.

In the darkroom the film is placed round the inside of the camera, care being taken to see that alignment is made where necessary with any fiducial marks or stops in the apparatus and the light shield then placed over the chamber.

USE OF ASTM CARDS B2.2

The camera is positioned in the apparatus and the alignment checked by (or in the presence of) the supervisor.

9 – 17 MINOR CORRECTION

d	3·40	3·21	1·95	3·40	CuCl (HIGH TEMP)		★
I/I_1	100	80	80	100	COPPER (I) CHLORIDE		

Rad. CuK λ 1·542 Filter Dia. 19 CM	d Å	I/I_1	hkl	d Å	I/I_1	hkl
Cut off I/I_1 VISUAL	3·395	100	100			
Ref. LORENZ AND PRENER, ACTA CRYST. 9 538 (1956)	3·208	80	002			
	2·997	10	101			
Sys. HEXAGONAL S.G.	2·334	30	102			
a_0 3·91 b_0 c_0 6·42 A C 1·64	1·953	80	110			
α β γ Z Dx	1·805	70	103			
Ref. IBID.	1·668	50	112			
	1·639	20	201			
$\xi\alpha$ n$\omega\beta$ $\xi\gamma$ Sign						
2V D mp Color						
Ref.						
HIGH TEMPERATURE MODIFICATION, STABLE FROM 407°C TO MP. DIAGRAM TAKEN AT 410°C.						

4 – 0686 MAJOR CORRECTION

d 4-0678	2·86	2·48	1·49	2·855	PB		★
I/I_1 4-0686	100	50	32	100	LEAD	(LEAD)	

Rad. CuKα_1 λ 1·5405 Filter Ni	d Å	I/I_1	hkl	d Å	I/I_1	hkl
Dia. Cut off Coll.	2·855	100	111			
I/I_1 G. C. DIFFRACTOMETER d corr. abs.?	2·475	50	200			
Ref. SWANSON AND TATGE, JC FEL REPORTS, NBS (1951)	1·750	31	220			
	1·493	32	311			
Sys. CUBIC S.G. O_H^5 – FM3M	1·429	9	222			
a_0 4·9506 b_0 c_0 A C	1·238	2	400			
α β γ Z 4	1·1359	10	331			
Ref. IBID.	1·1069	7	420			
	1·0105	6	422			
$\xi\alpha$ n$\omega\beta$ $\xi\gamma$ Sign	0·9526	5	511			
2V D_x 11·341 mp Color						
Ref. IBID.	0·8752	1	440			
	0·8369	9	531			
SPECTROGRAPHIC ANALYSIS SHOWS FAINT TRACES OF BI AND MG. PURITY > 99·999% AT 26°C	0·8251	4	600			
TO REPLACE 1-0995, 3-1156, 2-0811, 1-0972, 2-0799 AND 3-1153						

Figure B42. Two cards from the ASTM Index. Reproduced by courtesy of the American Society for Testing Materials

After checking that any necessary collimators or shielding are in place the x-ray beam is switched on at the recommended value (~ 40 kV and 15 mA) and the film left for about 40 minutes.

The film is then removed and developed according to the specific instructions given with the particular film used.

Using the calibrated scales which will be provided, the film is compared with each scale in turn and when a series of corresponding lines is observed the d spacings are noted.

The most prominent lines (say the first three) are then sought among the ASTM cards from which the substances (or mixture of substances) may be identified.

Discussion of Results

Care in alignment of the mounting is essential and time spent on this pays good dividends in the result obtained.

A thin even layer of finely ground powder should be spread over the circumference of the tube. 'Lumpiness' or unevenness in the specimen coating will result in a lack of sharpness and a 'grainy' quality in the lines obtained on the photograph.

An alternative method of mounting sometimes used, is to mix the finely powdered specimen material with gum tragacanth and to roll a solid cylinder of the mixture about $\frac{1}{2}$ mm in diameter and 3 to 4 cm long. This is then allowed to dry and aligned in the chuck, as in the account above.

As well as the line spacings, relative intensities of the lines are also of importance and the latter corresponding to the one single substance will be shown on the ASTM card.

However, if a mixture is chosen the lines of one substance may coincide with those of another and the addition of intensities will complicate the pattern.

This factor presents one of the serious limitations of the method for workers not experienced in photograph interpretation.

In the first instance a single compound or simple mixture should be chosen as is the case in the photographic results given in *Plate B2* (copper oxide and copper chloride mixture).

Further Work

Several different simple mixtures may be provided such that no serious confusion in identification can arise, and only when these have been investigated should more difficult examples be taken.

Reading and References

A general elementary account of x-ray analysis is given in *The History of X-Ray Analysis*, by L. Bragg (British Council—Longmans Green, 1948).

Plate B1. X-ray diffraction photograph of a tungsten specimen: 8 strands of stretched lamp filament twisted together; outside diameter 0·17 mm

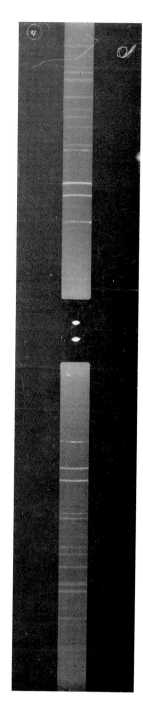

Plate B2. Plate showing half of x-ray powder diffraction photograph of a simple mixture of copper oxide and copper chloride

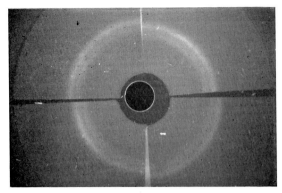

Plate B3. Determination of coefficient of expansion of a metal by an X-ray back-reflection photograph. Exposure 120 min at 6·4 cm distance to Cu K_α x-radiation, current 15 mA, e.h.t. 50 kV, 'Crystallex' film

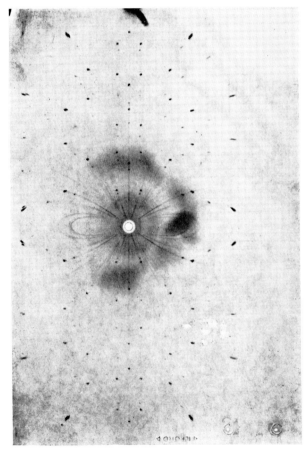

Plate B4. Single crystal x-ray diffraction photograph obtained using the rotation method

Plate B5. Photographic pattern resulting from absorption of x-rays by foils of varying thicknesses (see also *Figure B6*)

Plate B6. Photograph showing typical variations in intensities of the calibration plate

Plate C1. Diffraction pattern obtained by reflection using the neon gas laser

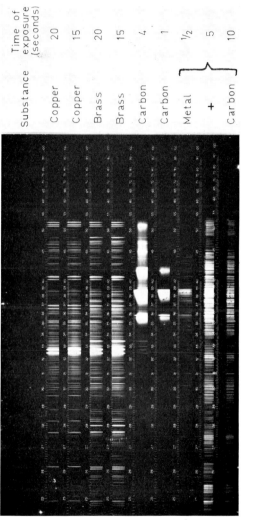

Plate C2. Typical photograph obtained with the Hilger medium quartz spectrograph

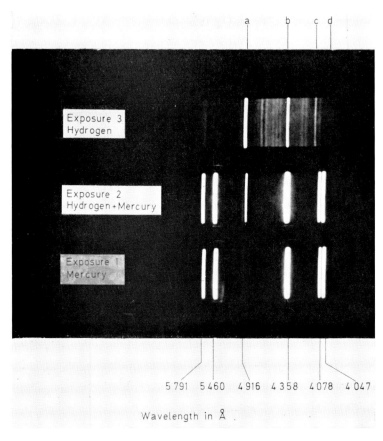

Plate C3. Determination of wavelength of hydrogen lines by photographic method in order to determine the Rydberg constant

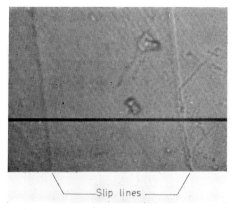

Plate D1. Slip lines of lithium fluoride single crystals ($\times 100$)

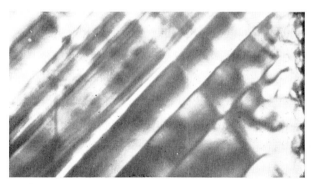

Plate D2. Birefringence pattern of stressed single crystal of lithium fluoride viewed in polarized light ($\times 300$)

Plate D3. Etch pits on the surface of a single crystal of lithium fluoride under stress ($\times 650$)

Plate D4. Zone refining of naphthalene

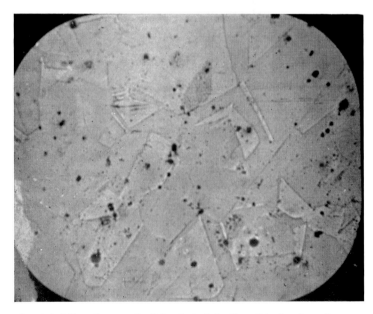

Plate D5. Microphotograph of the electrolytically polished surface of copper. Operating voltage, 5 V; polishing time, 30 seconds; magnification ($\times 100$)

LINEAR COEFFICIENT OF EXPANSION BY X-RAYS   B2.3

The theory of this experiment should be read in conjunction with Experiment Nos. B2.1 and C2.6. The latter gives details of the use of the Hilger Microdensitometer which enables relative intensities of lines to be assessed.

Experiment No. B2.3

DETERMINATION OF THE LINEAR COEFFICIENT OF EXPANSION OF A METAL BY USE OF 'BACK REFLECTION' X-RAYS

Introduction

In this experiment a collimated beam of copper K_α x radiation is allowed to strike a metal tube filled with water and ice at 0° C. The beam is reflected back at a small angle to the incident beam and the cone of diffracted x-rays is intercepted by a film so that a diffraction ring is obtained. On raising the temperature of the water in the tube to boiling point the resulting increase in interplanar spacing within the metal crystals, gives an increase in the radius of the diffraction ring.

Theory

Let r = radius of the diffraction ring
  δr = increase in radius at higher temperature
  x = distance from the metal tube to the film
  θ = Bragg angle of x-ray diffraction

then $2\theta = \pi - \phi$ where ϕ = semi-vertical angle of the cone of diffracted x-rays (*Figure B43*) and $2\delta\theta = -\delta\phi$ then $r/x = \tan \phi$ and differentiating with respect to ϕ

$$\delta r = x \sec^2 \phi \, \delta\phi$$

By the Bragg equation

$$n\lambda = 2d \sin \theta \quad \text{or} \quad d = \frac{n\lambda}{2} \operatorname{cosec} \theta$$

where n = order of diffraction
  λ = wavelength of x radiation

d = interplanar spacing (i.e. the distance between 'repeat units' of the crystal structure)
θ = angle of diffraction

then
$$\delta(d) = -\left(\frac{n\lambda}{2}\right)\frac{\cos\theta}{\sin^2\theta}\delta\theta$$

whence
$$\frac{\delta(d)}{d} = -\left\{\frac{\cos\theta}{\sin^2\theta}\bigg/\left(\frac{1}{\sin\theta}\right)\right\}\delta\theta = -\cot\theta\,\delta\theta$$

(since λ and n are constant).
Thus remembering that
$$\cot\theta = \cot\left(90 - \frac{\phi}{2}\right) = \tan\frac{\phi}{2}$$

and
$$\delta\theta = -\frac{1}{2}\delta\phi = \frac{1}{2}\frac{\cos^2\phi}{x}\delta r$$

$$\frac{\delta(d)}{d} = \frac{1}{2}\tan\frac{\phi}{2}\frac{\cos^2\phi}{x}\delta r$$

Now by definition the coefficient of linear expansion α

$$= \frac{\text{Change in length}}{\text{Original length} \times \text{temperature change}}$$

$$= \frac{\delta d}{d \times \text{temperature change}}$$

i.e.
$$\alpha = \frac{1}{2t}\left(\tan\frac{\phi}{2}\cos^2\phi\right)\delta r$$

where t = rise in temperature.

Method

First consult the x-ray operator, or safety officer, with regard to the setting up of the apparatus and the operation of the set. Ensure that any appropriate lead screens are in position before the start of the experiment itself.

At all times, and particularly during beam alignment, make certain that the beam is NOT allowed to fall upon the hands or

LINEAR COEFFICIENT OF EXPANSION BY X-RAYS B2.3

upon any other part of the body (see last paragraph of this experiment).

The film is mounted on the 'back reflection' plate (*Figure B43*) after a locating hole has been made in the centre. The lead plate is placed in position and held by screwing in the collimator.

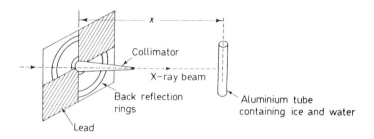

Figure B43

Note that the lead plate exposes two quadrants only to the scattered x-ray beam.

Switch on the set and align the beam using a fluorescent screen. If the safe mounting and alignment system referred to in the footnote is available then much difficulty is obviated.

Place any shields in position, note the temperature in the tube and expose the film for about 40 minutes (unless a longer period is recommended for the particular film being used). Switch off and again note the tube temperature.

Reverse the lead plate to expose the other two sectors and raise the temperature of the water to boiling point. Repeat the process as above being careful to make sure that there is sufficient water to avoid 'boiling dry'. Use of a naked flame should be avoided and some form of electrical heater should be arranged.

A photograph of the type shown in *Plate B3* is obtained and by carefully measuring the distance x the linear coefficient of expansion of the metal can be calculated as above. (An aluminium tube makes a most suitable specimen.)

Discussion of Results

If the diameters of the rings are measured carefully using a travelling microscope then the method is seen to be capable of an accuracy that is at first somewhat surprising.

Further Reading and References

Take care to read and follow any Safety Regulations which are displayed in the x-ray laboratory. A safe alignment system is described in the *Journal of Scientific Instruments*, vol. 38, No. 12, Dec. 1961, p. 493 by W. Hughes and C. P. Taylor (see footnote below). The theory behind the Bragg equation may be found in many degree level physics texts, e.g. *Physics* by S. G. Starling and A. J. Woodall (Longmans Green, 1963).

Hughes and Taylor in the reference given above describe a 'Safe Universal Mounting and Alignment System for X-ray Goniometers'. It consists of an optical system which allows alignment to be carried out in complete safety from radiation and on transferring the goniometer to the x-ray set no further adjustment is necessary. It makes a great contribution to safety particularly in teaching laboratories and those in which alignment has to be carried out by inexperienced personnel.

Experiment No. B2.4

TO DETERMINE THE LATTICE SPACING OF A SINGLE CRYSTAL BY THE X-RAY SINGLE CRYSTAL ROTATION METHOD

Introduction

In the x-ray powder method, a statistically large number of values of the Bragg diffraction angle θ were provided by using a powdered specimen consisting of an extremely large number of small crystallites.

If a single crystal is to be investigated then to provide the various values of θ the specimen may be rotated slowly with one of its major crystallographic axes aligned with the axis of rotation.

Theory

It is convenient in using this method to consider the reciprocal lattice rather than the direct lattice considered when the Bragg equation was applied to the analysis of a polycrystalline specimen.

LATTICE SPACING OF SINGLE CRYSTAL

Thus if $d^*_{h,k,l}$ is the distance from the origin to the reciprocal point in question which corresponds to a family of planes in the direct lattice having a spacing $d_{h,k,l}$ we may define

$$d^*_{h,k,l} = \frac{\text{constant}}{d_{h,k,l}}$$

The constant may be arbitrarily chosen for mathematical convenience and in our case may be taken as equal to λ the wavelength of the x-rays used.
Thus

$$d^*_{h,k,l} = \frac{\lambda}{d_{h,k,l}}$$

In the direct lattice we saw that the intercepts from the origin to the (h, k, l) plane were $(a/h), (b/k), (c/l)$.

Using the reciprocal notation with unit vectors

$$d^*_{h,k,l} = ha^* + kb^* + lc^*$$

The crystal rotates about a vertical axis lying at the centre of a circular film distance R from the axis (*Figure B44*).

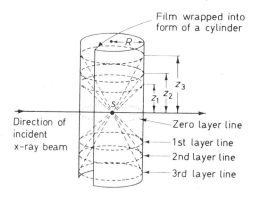

Figure B44. Diagram illustrating the mode of formation of layer lines in a single crystal rotation photograph. S represents the position occupied by the rotating crystal. z represents vertical distance between the chosen layer line and the zero layer line. Subscripts 1, 2, 3 refer to first, second and third layer lines respectively

A series of spots result (*Plate B4* and *Figure B45*) on the developed film which form a series of layer lines, the centre one being the 'zero layer line' and it is seen that in terms of the direct lattice the angle of diffraction is given by z/R, where z represents the distance of the layer line under consideration from the zero line.

Hence considering the (111) plane in terms of the reciprocal lattice

$$d_{111} = \frac{\lambda}{z}$$

so that using filtered copper K_α radiation ($\lambda = 1\cdot54$ Å) the crystal spacing may be found by measurement of z.

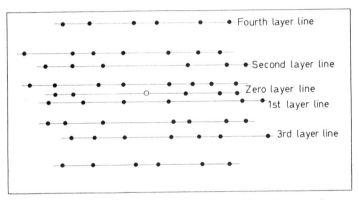

Figure B45. Diagram showing typical arrangement of layer lines on a film unwrapped from the cylinder of a single crystal rotation photograph (taken from an actual photograph: rotation about the (111) plane)

In practice measurements may be simplified by use of a Bernal chart, *Figure B46*, appropriate to the particular camera used.

Method

Using a suitable optical goniometer system (*Figure B47*) the crystal is mounted by a touch of shellac to a thin glass rod of about 1 mm diameter and placed in the detachable mount of the goniometer head. This allows for adjustment of the crystal to any desired position.

A well-defined crystallographic axis is set vertically by eye and the telescope cross wires of the goniometer are focused on the specimen. Set one 'rocker' of the goniometer parallel to the length of the telescope and adjust the one at right angles to it to bring the crystal axis parallel to the vertical cross wire. Turn the head through 90° and adjust the other rocking slide in the same way. Finally adjust the centre of the specimen to lie on the axis of rotation, using each rocker slide in turn.

Rotate the head and check that the mounted crystal does not display significant movement relative to the cross wires.

Place the mounted crystal in the x-ray set and under the operator's supervision check the crystal alignment once more.

Load the x-ray film in the camera so that it forms a circle about the specimen.

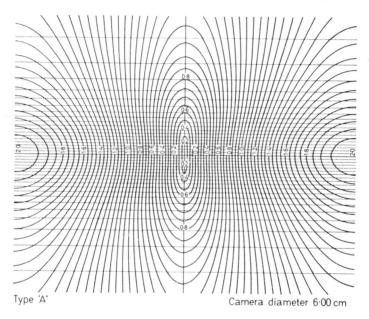

Figure B46. The Bernal chart. Reproduced by courtesy of Professor J. D. Bernal

After placing all collimators and screens in position switch on the x-ray beam and expose the film to radiation for about 1 hour at the recommended rating (~ 14 mA 50 kV).

Develop the film in accordance with the specific instructions given with it.

When dry, place flat over a suitable viewer and superimpose upon the appropriate Bernal chart (*Figure B46*).

After carefully checking the alignment of the chart read off the z values, i.e. the vertical distances between the layer lines under consideration.

Compute the average spacing appropriate to the particular mode of orientation of the crystal.

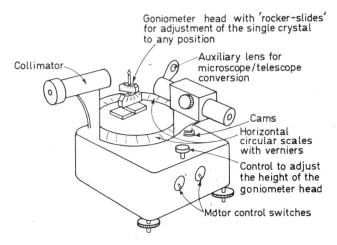

Figure B47. Typical single crystal goniometer showing the head with graduated 'rocker slides'

Discussion of Results

By far the most critical factor in the experiment is the alignment of the crystal. Failure in this respect results in lack of sharpness of the spots in the layer lines.

A crystal of well-defined form should be chosen to simplify the process of alignment.

Plate B4 shows the result obtained using a single crystal specimen of silicon, with its $[111]$ axis vertical and filtered copper K_α radiation.

Thus reading z from the Bernal chart gives d_{111} in this instance ($=9\cdot34$) and cubic symmetry is assumed (the silicon having the 'diamond structure').

If crystals other than cubic are used it is, of course, necessary to orientate the crystal with each axis in turn vertical, if the complete picture of the unit cell is to be found.

Further Work

As suggested above, a crystal of non-cubic form should be investigated where it is necessary to evaluate more than the one single dimension.

It should be noted that if a flat film is used rather than the circular arrangement above, the spots lie not on simple layer lines but on a series of hyberbolae.

The intensities of the spots might be found by plotting a calibration curve linking the density of blackening with the time of exposure (and hence the number of photons) in the x-ray beam (see Experiment No. C2.6).

Since the crystal is three-dimensional, three parameters are necessary for complete definition of the reciprocal lattice. The rotation photographs give only two parameters, no indication being given of the third, so that a number of points may satisfy particular values of the first two but may have different values of the third one. To overcome this limitation a number of oscillations of the crystal through angles of say 15° are made, until the whole 360° have been covered—resulting in oscillation photographs.

Reading and References

The background theory of the reciprocal lattice and the full proof of

$$d^*_{h,k,l} = ha^* + kb^* + lc^*$$

is given in *The Interpretation of X-Ray Diffraction Photographs* by N. F. M. Henry, H. Lipson and F. A. Wooster (Macmillan, 1961).

The rotation and oscillation methods are dealt with in *Diffraction Methods in Materials Science* by J. B. Cohen (Macmillan, 1966).

Experiment No. B2.5

TO INVESTIGATE THE ABSORPTION OF X-RADIATION BY COPPER

Introduction

Two distinct cases must be considered here:

(1) the absorption of monochromatic x-radiation, and
(2) the absorption of heterogeneous radiation.

Theory

Consider a monochromatic beam of x-rays of initial intensity I_0 incident normally upon a slice of absorber of thickness x and let the intensity after passing through this distance x be I.

The fractional loss in intensity $-(\delta I/I)$ in passing through an element of thickness δx is proportional to δx,

i.e.
$$\frac{dI}{I} = -\mu\, \delta x$$

and the constant of proportionality μ is known as the linear absorption coefficient of the absorber.

Thus
$$\int \frac{dI}{I} = -\mu \int dx$$

or
$$\log_e I = -\mu x + C,$$

where C is a constant of integration.

To evaluate the constant it should be noted that when $x = 0$, $I = I_0$ hence
$$C = \log_e I_0$$

therefore
$$\log_e \left(\frac{I}{I_0}\right) = -\mu x$$

or
$$I = I_0 \exp(-\mu x)$$

Thus for *monochromatic* x-rays a plot of $\log I$ against x gives a straight line (*Figure B48*) but if other wavelengths are present the curve is non-linear (*Figure B49*) since the absorption coefficient μ can no longer be considered linear but is itself a function of wavelength. Absorption of the 'softer' x-rays, i.e. those of longer wavelengths, is stronger than for those of shorter wavelengths.

For a monochromatic beam of radiation the half value thickness of the absorber is of importance since it provides a convenient measure of the quality of the radiation.

It may be defined as that thickness of absorber which reduces the intensity of the x-ray beam to half its original value, i.e.

$$I = \frac{I_0}{2} = I_0 \exp(-\mu l)$$

where l now represents the half value thickness.

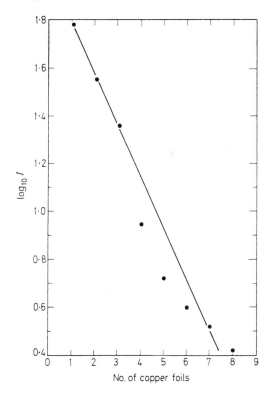

Figure B48. Graph to obtain the 'half value' thickness of copper, using filtered x-radiation

Thus
$$\log_e \tfrac{1}{2} = -\mu l$$
or
$$\log_e 2 (= 0\cdot 69) = +\mu l$$
therefore
$$l = \frac{0\cdot 69}{\mu}$$

Hence having plotted a graph of log I against x for monochromatic x radiation the slope gives μ and the half value thickness may also be calculated. Alternatively the intensity I may be plotted directly against x and the half value thickness read off straight away.

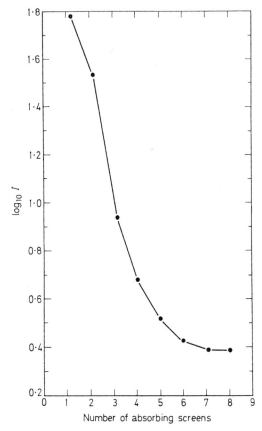

Figure B49. Graph of log intensity against absorber thickness for the absorption of heterogeneous x radiation by copper

It should be noticed that although the above theoretical formula cannot strictly be applied to a beam of heterogeneous x-rays and thus no single value of absorption coefficient obtained, the half value thickness read off directly from the curve still provides a convenient 'yard-stick' for measurement of the quality of the radiation.

Method

A series of eight or so copper foils each of the order of 0·1 mm thick are cut to give a rectangular aperture of increasing area and mounted in sequence in a film carrier over an x-ray film so that parallel layers of increasing thickness of copper cover the paper which enfolds the sensitive surface (*Figure B50*).

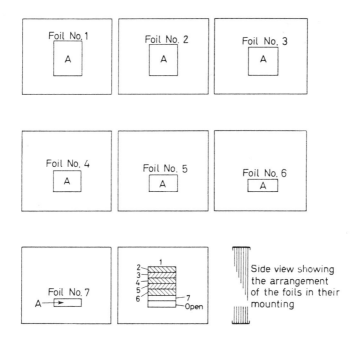

Figure B50. Method of using copper foils for absorption measurements using x-rays. A represents the extent of the aperture cut out in each case

The complete carriage is mounted in the x-ray set (under the guidance of the operator) making sure that the safety regulations appertaining to the use of the particular set are fully obeyed.

The screens are placed in position and the film is exposed to the radiation corresponding to 2 mA or so at 40 kV, for about 3 seconds.

The set is switched off and the film removed and developed in accordance with instructions given with the film being used (Ilford

Industrial G is suitable for this purpose). The result will be somewhat as shown in *Plate B5*.

A calibration plate should now be made by mounting in position in the set a similar film (without screens of course) before which is a rotating sector disc (*Figure B51*). The cutaway sectors in the disc

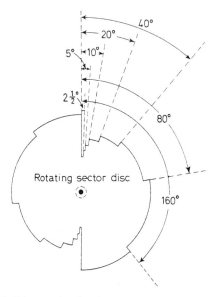

Figure B51. Diagram showing the arrangement of the angles in the sector disc that is used to obtain a photographic density calibration curve

are of increasing angle and the number of x-ray quanta passing through each aperture is proportional to the angle of the sector, hence as the angle increases the resulting blackening of the film increases (*Plate B6*).

This blackening may be measured using a densitometer such as the Hilger model and a calibration graph of the blackening against the sector angle may be plotted (*Figure B52*).

Corresponding to a particular density of blackening on the first film the angle is read off on the x axis of the calibration graph and this value (proportional to the intensity of the transmitted x-rays) is then plotted against the number of absorbing foils.

If the thickness of the foils is then accurately determined the half value thickness of copper may now be found. A second graph of the logarithm of the previous ordinate against the number of absorbing screens should also be plotted.

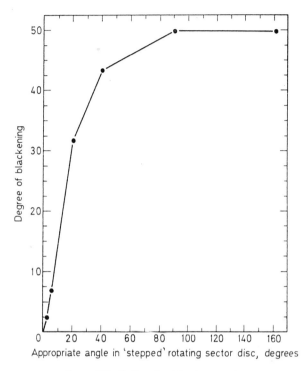

Figure B52. Calibration 'densitometer' curve

The experiment should now be repeated after a nickel foil filter has been placed before the x-ray tube window and the same quantities plotted as before.

Discussion of Results

The most striking feature of the log I against thickness graph is its lack of linearity shown in *Figure B49* (and therefore the inapplicability of the simple theory).

Notice that introducing the filter does give some improvement and a 'best fit' straight line may be drawn, but truly monochromatic radiation is still not achieved, some obvious scatter and

deviation from strict linearity does remain in the resulting graph (*Figure B48*).
Compare the half value thickness obtained in each case.

Further Work

A convenient alternative method of measurement is to use an x-ray dose-rate meter, when dose rate may be plotted as ordinate against thickness as abscissa.

A better straight line graph of log I/x may be plotted if an x-ray spectrometer is available. A more nearly monochromatic beam can then be obtained using a rock salt crystal and copper K_α radiation although care must be taken to restrict the operating voltage on the tube in order not to excite the 'harmonics' of the copper K_α radiation.

Reading and References

Details concerning the use of the Hilger densitometer may be found in Experiment No. C2.6.

Section C. Applied Optics, Spectroscopy, Photometry

Experiment No. C1.1

TO DETERMINE THE SPECIFIC ROTATION OF A SOLUTION OF SUGAR

Introduction

Many substances have the ability to rotate the plane of polarization of light as it passes through them, and sugars in solution show this effect (known as optical activity) to a marked degree. The rotation depends upon the number of solute molecules present.

Theory

Since the rotation depends upon the number of molecules of optically active materials present, the important factors are:

(1) the length l of the path which the light travels through the optically active medium, and

(2) in the case of a solution, the concentration C of the solution.

Thus rotation $\theta = SCl$ where S is known as the specific rotation of the optically active substance and is a characteristic of the substance.

Method

Set up the polarimeter before a monochromatic source of light, as shown in *Figure C1*. Most of these instruments carry a Lippich auxiliary Nicol prism system (*Figure C1*) which bears a small arm moving over a scale and this arm is set at an angle predetermined by the manufacturers (usually about 6°). This auxiliary device enables more accurate observations to be made and careful study of the detail of the instrument before making this initial setting is essential. Tubes of length, 10, 20 and 30 cm will be provided and taking the largest tube it should be filled with water and placed in the instrument chamber which is then closed. Set the vernier scale at approximately 0° and 180°. Make the final adjustment by tightening the screws H and rotating the fine screw G until exact alignment of the 0°–180° is seen on the vernier. The Nicol-adjusting screw J is then turned until equal illumination of the field of view occurs.

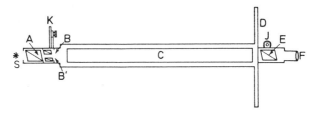

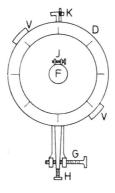

Figure C1 Optical activity—the polarimeter.
A, Main polarizing Nicol prism
B and B', Lippich auxiliary Nicol prism
C, Tube holding optically active solution
D, Analyser scale with vernier adjustment
E, Analyser   F, Eyepiece   V, Verniers
G, Vernier adjustment screw
H, Main scale locking screw
J, Fine screw adjustment for analysing Nicol
K, Angular setting lever for the auxiliary Nicol
S, Monochromatic light source

Remove the tube of water and replace this by an N/10 solution of sugar, then replace the tube in the chamber. It will be seen that the field of view now has one half darker than the other, and the vernier should be rotated (releasing the clamping screw if necessary) to give equal illumination again. This gives the angle of rotation of the plane of polarized light. Repeat the experiment using tubes of different lengths but keeping the strength of the solution constant. Plot a graph (*Figure C2*) of θ against l. The slope of the graph gives SC. Now take the longest tube and vary the concentration (N/20,

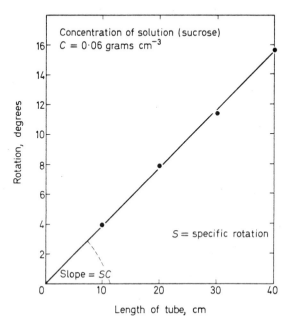

Figure C2. Graph to determine the specific rotation of a solution of sucrose

N/40, etc.) so that a graph can be plotted of θ against concentration C (*Figure C3*). The slope of this graph is Sl. Hence it is possible to determine the specific rotation S.

Discussion of Results

The accuracy of this experiment is very high with an instrument carrying Nicol prisms and the auxiliary Lippich system, but it falls off drastically for the simple instrument carrying no auxiliary prism and using 'Polaroid' analyser and polarizer. An estimate of the accuracy of the experiment can again be deduced from the deviation of particular points of the graph from the 'best fit' straight lines obtained.

Further Work

The term 'sugar' has been used above to denote sucrose which has a constant value of specific rotation.

If glucose is used rather than sucrose it will be found that the rotation is a function of time, i.e. that a freshly prepared solution has a higher value of specific rotation than has an older solution.

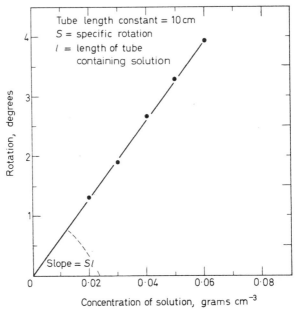

Figure C3. Graph showing rotation of plane of polarized light by optically active solution of increasing concentration

Experiment might also be performed to show that the rotation varies with temperature, but stabilization of temperature must be very strict otherwise it becomes very difficult to 'balance' the half shadow device of the polarimeter.

Optical activity within turpentine and/or quartz is also of interest.

Reading and References

Greater detail concerning polarimeters and optical activity may be found in *Optical Rotatory Power* by T. M. Lowry (Dover/Constable, 1964).

Experiment No. C1.2

TO INVESTIGATE THE RELATIONSHIP BETWEEN OPTICAL ACTIVITY AND WAVELENGTH

Introduction

In the experiment to determine the specific rotation of sugar solution a sodium lamp (or mercury lamp and filter) was used in order to obtain an effectively monochromatic source of light of known wavelength.

Here it will be assumed that the rotation θ is an unknown function of λ and the law governing the variation of specific rotation with wavelength λ investigated.

Theory

Assuming $\theta = f(\lambda)$, in particular that the function is of the form $\theta = a\lambda^n$, one wishes to find the value of n. Taking logarithms on both sides of the equation

$$\log \theta = n \log \lambda + \log a.$$

This is the equation of a straight line whose slope is n and whose y intercept gives $\log a$.

Method

Set up the polarimeter in the usual way as was described in the previous experiment. Using a tube of fixed length (say 20 or 30 cm) and a solution of known concentration (say N/10) find the rotation produced in the case of approximately monochromatic light from each of several sources in turn, including cadmium, sodium and mercury lamps used in conjunction with suitable red, green and blue filters. A series of well-defined predominant wavelengths may be obtained and these can be found from the list of Wratten filters in the Kodak data book. Tabulate the wavelength and rotation values, compute the logarithms and plot the graph of $\log \theta$ against $\log \lambda$ to find the value of n (*Figure C4*).

Discussion of Results

It may be found that the lines, particularly towards the low wavelength end of the spectrum, are very faint and the polarimeter should be used in a dark room with shields to ensure that extraneous light is eliminated.

The slope of the resulting graph of log θ against log λ is negative and of the order of -2. It should be remembered that the filters

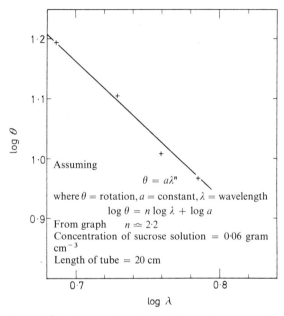

Figure C4. Variation of rotation of plane of polarized light with wavelength of light used

transmit a band of wavelengths and not just a single wavelength of light so that it is not surprising that wide scatter of points about the 'best fit' line of the graph may well occur (*Figure C4*).

Notice that the value of a in the equation will depend upon the units chosen for λ and θ (i.e. whether Ångström units and degrees, or centimetres and radians are used).

An alternative method of treatment is to plot θ against $1/\lambda^2$ when again an approximate straight line should be obtained (*Figure C5*).

Further Work

If the facilities are available the specimen tube should be filled with nitrobenzene and placed in a uniform magnetic field. It is then

OPTICAL ACTIVITY AND WAVELENGTH   C1.2

possible to measure Verdet's constant for the liquid since the rotation $\theta = klH$, where H = magnetic field, l is the length of the column of liquid and k is Verdet's constant. The variation of the rotation with wavelength can again be investigated and found to be

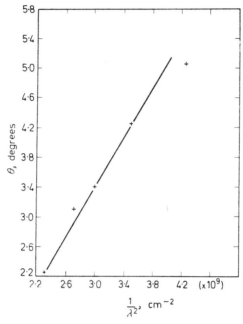

Figure C5. Graph of rotation θ of the plane of polarized light against $1/(\text{wavelength } \lambda)^2$

a law similar to the one above, but the accuracy to which measurements can be made is seriously reduced, and once again it is essential to maintain a constant temperature since the rotation is itself a function of temperature.

Reading and References

Information both on general theory and instrumentation may be found in *Optical Rotatory Power* by T. M. Lowry (Dover/Constable, 1964).

APPLIED OPTICS, SPECTROSCOPY, PHOTOMETRY

Experiment No. C1.3

TO DETERMINE THE LINEAR COEFFICIENT OF EXPANSION OF BRASS BY FIZEAU'S INTERFERENCE METHOD

Introduction

The method makes use of the familiar Newton's rings technique to enable the linear coefficient of expansion of brass or other material to be found. Change in length of metal supports carrying the lens varies the air gap between lens and plate (*Figure C6*) and this variation is measured in terms of optical fringe shift.

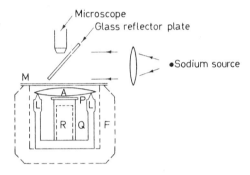

Figure C6. Measurement of linear coefficient of expansion of a metal by Fizeau's interference method.
A, Lens ⎱ between which the interference
P, Quartz plate ⎰ fringes are formed
Q, Quartz tube located over brass stud R
L, Adjustable screws at head of each support
M, Mica cover to minimize convection
F, Cylindrical lagged furnace
N.B. Thermocouple to measure temperature near the point where fringes are formed is omitted for clarity

Theory

Using the apparatus as shown in *Figure C6*, Newton's rings are observed in the usual way. If the furnace is now gently heated, the brass limbs supporting the lens increase in length but the position of the optical flat (a small piece of microscope slide will do) may be assumed to remain constant if we consider the linear coefficient of

LINEAR COEFFICIENT OF EXPANSION OF BRASS C1.3

expansion of the quartz, upon which the slide is supported, to be negligible.

This effectively changes the length of the air gap between the slide and the lens surface, and for an increase of one wavelength of light in the optical path length a movement of one fringe is observed. This corresponds to a geometrical increase in air gap of half a wavelength λ of the monochromatic light used.

For a movement of n fringes therefore the change in length = $n\lambda/2$, but the change in length

$$(l_\theta - l_0) = l_0 \alpha \theta$$

where l_θ = length at temperature $\theta°$ C
  l_0 = length at temperature $0°$ C
  α = coefficient of linear expansion of the metal
  θ = change in temperature

from the definition of the coefficient of linear expansion, hence

$$\frac{n\lambda}{2} = l_0 \alpha \theta \quad \text{or} \quad \alpha = \frac{n\lambda}{2 l_0 \theta}$$

so that if the number n of fringes moved is plotted against θ a straight line graph will result of slope = $2l_0 \alpha / \lambda$. Thus knowing l_0 and λ, α may be found.

Method

Set up the apparatus within the furnace chamber, as shown in *Figure C6*, taking great care in levelling the lens by adjusting the screws of the actual supports. This is most important if a good result is to be obtained.

Carefully position the 45° reflecting glass plate to give maximum illumination over the central area of the optical flat, and observe first *by eye* to find the optimum conditions for viewing the fringes, which appear as a series of concentric rings about a millimetre or so in diameter.

Then gently adjust the supports and the lens to centralize the rings in the field of view.

Only when satisfied that Newton's rings are clearly centred in the arrangement should the vernier microscope be placed in position.

A small ink mark on the optical flat will enable the microscope to be focused more easily and the telescope should be adjusted so that the cross wires coincide with the centre of the Newton's rings system.

Having done so, a thermometer (or preferably a thermocouple) should be placed as near to the lens–flat interface as is conveniently possible, and a mica cover placed over the furnace to minimize convection effects. (The focus will have to be readjusted slightly.)

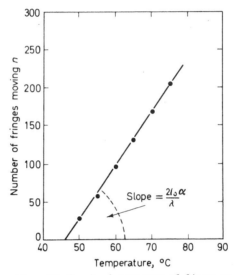

Figure C7. Graph of movement of fringes as furnace cooled against temperature to determine the linear coefficient of expansion of a metal

The furnace temperature is then gradually increased and as the fringes move the temperature should be noted at convenient points between 20° C and 80° C.

Readings should also be taken as the furnace cools.

Plot a graph of n against θ and from its slope determine α in the manner indicated above (*Figure C7*).

Discussion of Results

Unless great care has been taken in initial alignment and adjustment of the optical system it may be found that the visibility of the fringes decreases so rapidly at higher temperatures that effective readings can no longer be taken, hence the stress which was laid upon obtaining a sharp, well-centralized fringe system initially.

The resulting straight line should be quite a good one, although it will not in general pass through the origin as indicated in the theory, since an arbitrary point will have been chosen at which to commence counting fringes.

It will probably be found that the points cluster more closely about the straight line with readings taken as the furnace cools, than as the temperature rises. This is explained remembering that temperature stabilization will not be achieved until some time after the commencement of the experiment.

The results obtained may vary from the accepted one to a far greater extent than is indicated by the inaccuracies in the graph, but it should be remembered that convection about the furnace is high, and the temperature recorded will certainly not be exactly that in the air gap in which the interference fringes are formed. Remember, too, that the expansion of the glass slide has been neglected. (If a quartz plate is available this will reduce the error on this score.)

Further Work

One very useful extension of this method is to assume the linear coefficient of expansion of brass and to use the system to measure the coefficient of expansion of very small specimens or crystals placed beneath the lens.

Reading and References

The detailed theory of Newton's rings may be found in almost any standard degree textbook on physical optics, e.g. *Fundamentals of Optics* by F. A. Jenkins and H. E. White (McGraw-Hill, 3rd edition, 1957).

Experiment No. C1.4

TO VERIFY CAUCHY'S EQUATION USING THE LINE SPECTRA OF THE ELEMENTS NEON AND MERCURY

Introduction

Experiment shows that the refractive index of a medium is not truly constant, but in fact varies with frequency. Cauchy developed an empirical equation of the form

$$\mu = A + \frac{B}{\lambda^2} + \frac{C}{\lambda^4}$$

where μ = refractive index of the medium
  λ = wavelength of light
  A, B and C are constants

to account for this dispersion. The aim of this experiment is to verify this equation neglecting terms of higher order than the second.

Theory

Although the equation was originally propounded empirically it can be shown to have a sound basis in electromagnetic theory even though it is an approximation and in a sense, incomplete.

If we consider that the mechanism of polarization in a polarizable medium involves movement of electrons and that the electrons are set into forced vibration by electromagnetic radiation energy then each electron will possess its own natural frequency and the motion may be represented by an equation of the form

$$m\frac{d^2y}{dt^2} + \omega_0^2 y = Ee \cos \omega t \tag{1}$$

where m = mass of the electron
  e = charge on the electron
  $E \cos \omega t$ = the force producing the oscillation
  y = displacement of the electron.

The solution of this equation is of the form

$$y = \frac{Ee}{m(\omega_0^2 - \omega^2)} \tag{2}$$

By Gauss' equation well known in electrostatics the total normal induction = q/ε_0 where q represents the charge and ε_0 is the permittivity of free space and hence

$$E = \frac{\sigma}{\varepsilon_0}$$

where σ charge density = q/A and A = area over which charge is spread. Therefore

$$\frac{dE}{dt} = \frac{i}{\varepsilon_0} \quad \text{since } \sigma = \int i \, dt$$

where i represents current density.

VERIFYING CAUCHY'S EQUATION   C1.4

Therefore *in vacuo*

$$i = \varepsilon_0 \left(\frac{dE}{dt}\right).$$

The flow of charge (across unit area) due to the movement of the electrons $= Nev$, where $v = $ velocity and $N = $ number of electrons, i.e.

$$\text{current density} = Ne\left(\frac{dy}{dt}\right).$$

Therefore

$$\text{total current density} = \varepsilon_0 \frac{dE}{dt} + Ne\frac{dy}{dt} = \varepsilon_0\left(\frac{dE}{dt} + \varepsilon_0 Ne\frac{dy}{dt}\right)$$

but

$$Ne\frac{dy}{dt} = \frac{Ne^2}{m(\omega_0^2 - \omega^2)}\frac{dE}{dt} \quad \text{from (2)}$$

hence

$$i_{\text{total}} = \varepsilon_0 \left\{\frac{dE}{dt} + \frac{Ne^2}{\varepsilon_0 m(\omega_0^2 - \omega^2)}\frac{dE}{dt}\right\}$$

but

$$i_{\text{total}} = \varepsilon\left(\frac{dE}{dt}\right) = \varepsilon_0 \frac{dE}{dt}\left\{1 + \frac{Ne^2}{\varepsilon_0 m(\omega_0^2 - \omega^2)}\right\}$$

where ε is the total effective permittivity and therefore the effective permittivity

$$\varepsilon = \varepsilon_0\left\{1 + \frac{Ne^2}{\varepsilon_0 m(\omega_0^2 - \omega^2)}\right\} \quad \text{or} \quad \frac{\varepsilon}{\varepsilon_0} = \left\{1 + \frac{Ne^2}{\varepsilon_0 m(\omega_0^2 - \omega^2)}\right\}$$

but

$$\frac{\varepsilon}{\varepsilon_0} = \mu^2 \quad \text{(by Maxwell's theory)}$$

where μ is the refractive index of the dielectric medium (N.B. it must not be confused with permeability μ in this context)

therefore

$$\mu^2 = 1 + \frac{Ne^2}{\varepsilon_0 m(\omega_0^2 - \omega^2)}$$

and remembering that $\omega = 2\pi f = 2\pi c/\lambda$

where c = velocity of electromagnetic radiation
f = frequency of the radiation
λ = wavelength of the radiation

$$\mu^2 = \left\{1 + \frac{Ne^2}{\varepsilon_0 m\left(\frac{4\pi^2 c^2}{\lambda_0^2} - \frac{4\pi^2 c^2}{\lambda^2}\right)}\right\} = 1 + \frac{Ne^2}{\varepsilon_0 m 4\pi^2 c^2\left(\frac{1}{\lambda_0^2} - \frac{1}{\lambda^2}\right)}$$

$$= 1 + \frac{Ne^2 \lambda_0^2}{4\pi^2 \varepsilon_0 mc^2(1 - \lambda_0^2/\lambda^2)}$$

and by the binomial expansion

$$1 \bigg/ \left(1 - \frac{\lambda_0^2}{\lambda^2}\right) = 1 + \frac{\lambda_0^2}{\lambda^2} + \frac{\lambda_0^4}{\lambda^4}$$

hence

$$\mu^2 \simeq 1 + \frac{Ne^2 \lambda_0^2}{\varepsilon_0 m 4\pi^2 c^2}\left(1 + \frac{\lambda_0^2}{\lambda^2}\right)$$

$$= \left(1 + \frac{Ne^2 \lambda_0^2}{\varepsilon_0 m 4\pi^2 c^2}\right) + \left(\frac{Ne^2 \lambda_0^4}{\varepsilon_0 m 4\pi^2 c^2}\right)\frac{1}{\lambda^2}$$

which is Cauchy's equation, viz. $\mu^2 = A + B/\lambda^2$ where A and B are constants.

The refractive index μ is determined from the known formula

$$\mu = \sin\left(\frac{A + D}{2}\right) \bigg/ \sin\left(\frac{A}{2}\right)$$

familiar from earlier work using the spectrometer. (A = refracting angle of the prism used, D = angle of minimum deviation.)

The determination of refractive index corresponding to the spectral lines of different wavelengths needs careful setting up and use of the spectrometer and this experiment should be carried out only after one has experience of using the laboratory spectrometer with success.

Method

The spectrometer is carefully levelled and focused using Schuster's method, with the mercury lamp before the spectrometer slit.

The refracting angle A of the prism is first carefully measured in the usual way by observing the light reflected from either face of the prism about the refracting angle itself.

Minimum deviation is then found for each line of the mercury

VERIFYING CAUCHY'S EQUATION   C1.4

spectrum in turn. (It corresponds to the point at which movement of the spectrum over the field of view of the eyepiece is reversed, although direction of rotation of the telescope continues in the same sense.)

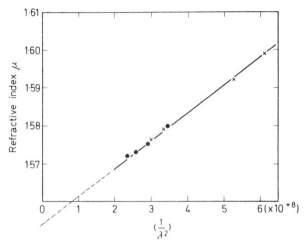

Figure C8. Verification of Cauchy's dispersion formula.

$$\mu \simeq A + \frac{B}{\lambda^2}$$

μ = Refractive index; λ = wavelength of line, cm; A and B are constants. A is given by the y intercept of the graph, B by the slope of the graph
× represents wavelengths associated with lines of the mercury spectrum.
• represents wavelengths associated with lines of the neon spectrum.

The refractive index is then calculated for each line in turn using the formula

$$\mu = \sin\left(\frac{A + D}{2}\right) \bigg/ \sin\left(\frac{A}{2}\right)$$

and $1/\lambda^2$ plotted against μ, the wavelengths for each line being obtained from standard tables. Repeat the procedure using neon and sodium lamps in turn. *Figure C8* shows a typical resulting graph obtained.

Discussion of Results

The accuracy obtained depends critically upon the care taken in levelling and focusing the spectrometer. The use of a dark room and the shielding of extraneous light from the lamp can add significantly to the ease of conducting the experiment, and it is of considerable help to allow one's eyes to grow used to the darkness before taking measurements.

Further Work

It is worth while to plot a graph showing the dispersion directly in terms of refractive index against wavelength (*Figure C9*).

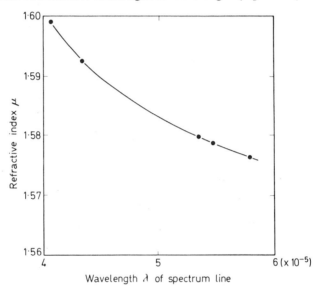

Figure C9. Verification of Cauchy's dispersion formula. The dispersion curve for a glass prism

This should be compared with the curves sketched in theoretical texts to note the limited experimental range.

Reading and References

A detailed treatment of dispersion formulae may be found in *Modern Optics* by E. B. Brown (Reinhold, 1965).

Schuster's method for the focusing of a spectrometer is dealt with in some detail in *Physics* by S. G. Starling and A. J. Woodall (Longmans Green, 1964).

Experiment No. C1.5

TO INVESTIGATE THE FARADAY EFFECT AND TO DETERMINE VERDET'S CONSTANT

Introduction

Faraday discovered that if a transparent isotropic medium is placed in a strong magnetic field the substance becomes optically active and therefore on passing polarized light through the medium the plane of polarization suffers rotation (see Experiment No. C1.1).

Theory

The Faraday effect is one of several phenomena which show the interaction between light and matter under the influence of a magnetic field. It is due to the distortion produced in the orbital movement of the electron within the atom by the application of the magnetic field.

Plane polarized light suffers a rotation θ when the applied field is in the same direction as the light beam such that

$$\theta = klH$$

where k = a constant known as Verdet's constant
l = length of the specimen through which the polarized light passes
H = applied magnetic field.

By convention k is considered positive if the direction of rotation of the plane of polarized light is the same as that of the current necessary to produce the magnetic field.

Method

The apparatus is set up as shown in *Figure C10*, the pole tips being removed and polarizer and analyser being fitted, as shown.

The specimen tube is filled with nitrobenzene and placed in position in the pole gap of the magnet.

The 'half shadow' device in the analyser is set to give even illumination in the field of view.

A current of several amperes is passed through the magnet coils when a distinct disturbance is noted in the field of view.

The polarizer is then rotated until even illumination is restored

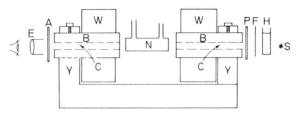

Figure C10. Apparatus used to determine Verdet's constant.
N, Specimen tube containing nitrobenzene
C, Optical channel through pole pieces of magnet left by removing 'bolt on' type pole tips
W, Magnet windings   Y, Magnet yoke
B, Magnet pole pieces   S, High intensity source
H, Water cell acting as heat filter   E, Eyepiece
F, Chosen optical filter   P, Polarizer
A, Analyser

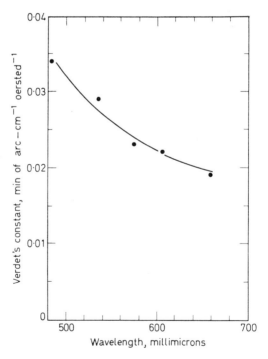

Figure C11. Graph showing the variation of Verdet's constant with wavelength of light used, for nitrobenzene

in the eyepiece, the angle of rotation being noted from the vernier scale attached to the polarizer.

Hence using the formula given in the theory, Verdet's constant $k(=\theta/lH)$ can be calculated once the electromagnetic field calibration for known currents has been made.

Discussion of Results

The experiment is best carried out in a darkened laboratory and it is wise to allow time for the eyes to adapt themselves to the dark-

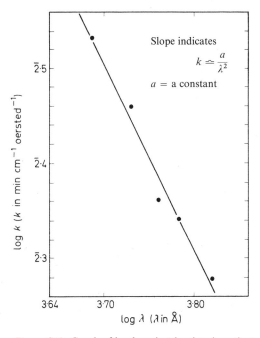

Figure C12. Graph of log k against log λ to investigate the form of the law governing the variation of Verdet's constant with the wavelength of the light used

ness before actually taking readings. Extraneous light from the source should be shielded from the viewer as far as possible.

It should be noted that θ is in fact also a function of wavelength λ and it is of advantage to use an optical filter tending towards the blue end of the visible spectrum rather than the red, since the lower the transmitted wavelength the greater the rotation. However, this is counterbalanced by the fact that it grows more difficult to obtain a good even illumination of the field of view with the lower wavelength filters and one has to make a compromise.

Further Work

A number of liquids other than nitrobenzene should be tried (see third reference) and also one should investigate the variation of the rotation with wavelength (*Figure C11* and *C12*) which shows that Verdet's constant is not strictly a constant but is a function of the wavelength of the light used.

Reading and References

The Faraday effect is dealt with in many standard degree texts on optics, e.g. *Fundamentals of Optics* by F. E. Jenkins and H. E. White (McGraw-Hill, 1957), or *Optics* by C. J. Smith (Arnold, 1960).

Values of Verdet's constant for a number of substances may be found in the *American Institute of Physics Handbook* (McGraw-Hill, 1963).

Experiment No. C1.6

TO CALIBRATE THE BABINET COMPENSATOR

Introduction

The Babinet compensator consists essentially of two thin quartz wedges whose optic axes are mutually at right angles (*Figure C13*). One wedge can be moved to vary the effective thickness of the pair so that the degree of retardation through the quartz plates may be varied. The instrument is thus used for the investigation of elliptically polarized light, but must first be calibrated so that a measured shift of the wedge can be expressed in terms of phase differences introduced into light passing through the instrument.

The aim of this experiment is to carry out such a calibration.

Theory

Plane polarized light entering normally the first plate of the compensator is resolved into two beams (the vibrations parallel to the optic axis being retarded if the axis lies in the plane of the paper).

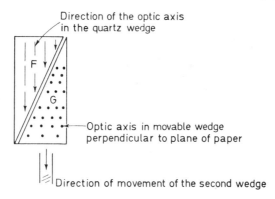

Figure C13. The Babinet compensator

On entering the second prism the vibrations perpendicular to the paper are retarded so that the path difference δ is given by

$$\delta = (\mu_e - \mu_0)(t_1 - t_2)$$

where μ_e = refractive index of the medium appropriate to the extraordinary ray
μ_0 = refractive index appropriate to the ordinary ray.

Then the phase difference

$$\phi = \frac{2\pi\delta}{\lambda}$$

and from one dark fringe to the next corresponds to $\delta = \lambda$, i.e.

$$\phi = 2\pi = 360°.$$

A graph of micrometer reading against fringe number thus provides calibration for the instrument.

Method

The compensator B is placed between crossed polarizers (P and A, *Figure C14*), the polarizer being set at approximately 45° to the optic axis of the wedges. Using a white light source the compensator

APPLIED OPTICS, SPECTROSCOPY, PHOTOMETRY

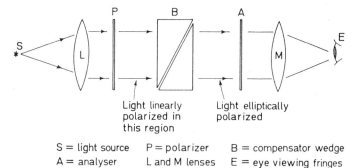

S = light source   P = polarizer   B = compensator wedge
A = analyser   L and M lenses   E = eye viewing fringes

Figure C14. Arrangement of apparatus with the Babinet compensator

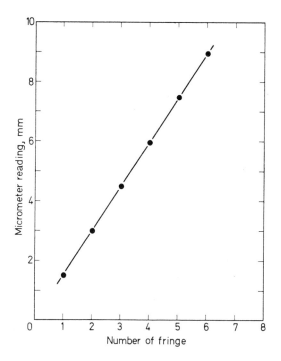

Figure C15. The Babinet compensator—a graph of fringe location against micrometer scale reading. N.B. Graph may not necessarily pass through origin. Readings given are actually differences from a chosen origin

CALIBRATING THE BABINET COMPENSATOR   C1.6

is rotated until the position is reached when the fringes which cover the field of view disappear. On rotating through a further 45° equally spaced fringes of maximum contrast are obtained, the central fringes being black and white with a series of coloured fringes either side.

The micrometer adjustment on the compensator is then set to align the cross wires on this central fringe. The white light source is now replaced by a monochromatic source (sodium lamp and filter may be used) and the micrometer movement, to traverse from the one fringe to the next and subsequent fringes is noted. A graph is plotted of the fringe number against the micrometer movement, when a straight line results (*Figure C15*), from which it is possible to determine the micrometer movement corresponding to one fringe shift or a phase difference of 360°.

Discussion of Results

As the resulting straight line graph should show (*Figure C15*). the instrument is capable of a high order of accuracy in the hands of a careful experimenter.

In assessing the position of maximum fringe visibility it is of considerable help to work in a dark room and to reduce extraneous light to a minimum. Adjustments of the micrometer and of rotation of the compensator should be made always turning in one direction to obviate the 'backlash' errors.

Further Work

The corollary to the above calibration is to use the instrument for the analysis of elliptically polarized light. To do this first align a dark (monochromatic) band on the cross wires using plane polarized light. Then turn the micrometer through a distance corresponding to one quarter of the calibration movement, i.e. $\pi/2$ or 90°. Introduce a thin sheet of mica between the polarizer and the compensator.

A movement of the fringe is noted and the analyser and compensator are independently rotated to once more re-align the fringe on the cross wire under conditions of optimum visibility. The light leaving the compensator must, once again, be plane polarized.

The compensator has annulled the $\pi/2$ phase difference introduced by the mica and the axes of the ellipse corresponding to the polarized light entering the compensator must be parallel and perpendicular to the length of the wedges.

The ratio of the axes = $\tan \theta$, where θ = angle between the vibration direction of the analyser and the principal direction of the compensator.

Reading and References

Background reading upon the Babinet compensator and its use in the analysis of elliptically polarized light is given in *Degree Physics—Optics* by C. J. Smith (Arnold, 1960) or in many standard degree texts.

Experiment No. C1.7
THE USE OF THE RAYLEIGH REFRACTOMETER TO DETERMINE THE REFRACTIVE INDEX OF AIR AT N.T.P.

Introduction

The refractometer is used here to confirm that the refractive index of air increases linearly with pressure and by making use of this, to obtain an accurate value of the refractive index at n.t.p.

Theory

The refractometer is shown diagramatically in *Figure C16*. Parallel light beams pass through the slits S_1 and S_2 and are brought

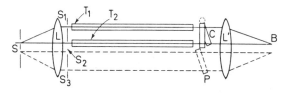

Figure C16. The Rayleigh refractometer diagrammatically. The reference system shown in broken lines is displaced downward for clarity.
S, Main slit
S_1 S_2 S_3, Slits associated with interference
T_1 T_2, Refractometer tubes to contain gas
C, Compensating plates
P, Plate used to place reference fringes in juxtaposition with experimental fringes

to a focus at B to produce an interference pattern. If one tube is evacuated and the other contains gas of refractive index μ, a path difference $(\mu - 1)l$ is created, where l is the geometrical length of the tube (the refractive index corresponding to the vacuum being assumed to be 1).

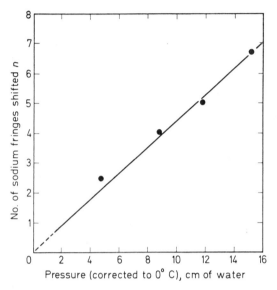

Figure C17. Graph of interference fringe shift in the Rayleigh refractometer corresponding to changes in pressure in the gas (measured by water manometer)

Such a path difference leads to a displacement of n fringes and hence

$$(\mu - 1)l = n\lambda \quad \text{or} \quad \mu = 1 + \left(\frac{n\lambda}{l}\right).$$

A graph of fringe movement against pressure difference is found by experiment to be linear (*Figure C17*), i.e.

Slope $\delta n/\delta p$ = constant = n/p or $n = p$ (Slope $\delta n/\delta p$)

Hence

$$\mu = 1 + \left\{\frac{(\text{Slope})\lambda}{l}\right\}p$$

where

$$\left\{\frac{\text{Slope}}{l}\right\} = \text{constant.}$$

Ideally from the above it is seen that monochromatic light should be used if λ is to be constant, but it is not then possible to determine n the number of fringes moved. To do this white light must be employed, the central white fringe being used as the fiducial mark. In order to produce a fixed fringe reference system, the tubes only cover half the field of view, the lower half acting as the reference and being brought into juxtaposition just below the experimental fringe system by means of a tilted glass plate P (*Figure C16*).

In addition a compensating arrangement of two similar glass plates one in the path of each experimental beam is provided, so that rotation of one introduces a calibrated compensating path, in order that the two sets of fringes may be brought into coincidence one with the other and a 'null displacement' method employed. Calibration of the micrometer rotation required, in terms of a standard wavelength, then gives direct measure of the path retardation.

This is done as shown in the graph of *Figure C18*, where fringe shift n is plotted against micrometer reading r.

Then the slope of this graph gives (dn/dr) so that when the pressure is varied, the appropriate fringe-shift is given by multiplying the micrometer reading by this slope, $n = r(dn/dr)$. This value is then plotted against the pressure (*Figure C17*) to give the graph whose slope is (dn/dp).

Method

It is convenient to employ small changes in pressure on a water manometer to determine (dn/dp) rather than to exhaust the air from the tube and to commence with both tubes at atmospheric pressure.

Illuminate the main slit of the instrument with white light and focus the eyepiece to give a clear bright image of the system of coloured fringes. In general the upper set will not coincide with the lower reference fringes and the micrometer of the compensating device is adjusted to bring the white central fringe opposite the fiducial reference, the micrometer reading being carefully noted.

It is first necessary to calibrate the system using monochromatic light and for this purpose the white light is replaced by a sodium source.

The micrometer readings corresponding to shifts of 10, 20, 30 fringes and so on are noted, and a graph of fringe-number against micrometer reading is drawn (*Figure C18*).

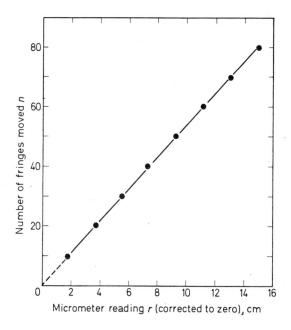

Figure C18. Calibration graph for the Rayleigh refractometer giving the micrometer movement in terms of the shift of sodium light fringes

One of the chambers should be connected through a T-junction via a water manometer to a suitable reservoir which may be raised or lowered to provide means of varying the pressure of the gas in the refractometer tube. The tap previously open to the atmosphere is closed and by raising the reservoir the pressure slightly increased. At the same time the white light source is once more placed before the instrument and the micrometer adjusted to bring the zero order fringe opposite the fiducial mark of the reference system. The pressure is again increased in small steps (say 5 cm of water pressure) and the process repeated. The temperature is also carefully noted.

The number of fringes shift (of sodium, for which the micrometer is calibrated) corresponding to the observed micrometer movements is noted from the calibration graph and hence a graph of fringe movement against pressure change may be drawn. Since this is linear (*Figure C17*), the slope of the graph (dn/dp) is readily obtained and hence μ the refractive index of air at n.t.p. may be calculated in the manner indicated in the theory.

Discussion of Results

It is important to note the way in which the points in the calibration graph closely fit the line. In the second graph however, the points show some scatter about the 'best-fit' line due in the main to inaccuracies in measuring the pressure.

In making the final calculations do not forget that the pressures are in terms of the head of water and must be converted to appropriate absolute units.

Further Work

Compare the above apparatus with the modified form used for the measurement of the refractive indices of liquids (see references) and actually make these measurements if the apparatus is available.

The instrument as described here has been used for measurement of changes in refractive index of gases due to chemical reaction.

Reading and References

Excellent background reading on the Rayleigh interferometers and upon other types may be found in *Introduction to Interferometry* by S. Tolansky (Longmans Green, 1955).

An alternative description of this experiment may be found in *Advanced Practical Physics for Students* by B. L. Worsnop and H. T. Flint (Methuen, 1957).

Experiment No. C1.8

TO DETERMINE THE THICKNESS OF THIN FILMS USING FRAUNHÖFER DIFFRACTION FRINGES

Introduction

Methods of measuring the thickness of thin films have become of great importance during recent years in (among other applications)

THICKNESS OF THIN FILMS   C1.8

the blooming or 'silvering' of optical surfaces and in research involving for example electron microscope techniques and properties of magnetic thin films.

This experiment uses the thin film of gelatine on an exposed photographic plate in order to illustrate how Fraunhöfer diffraction fringes may be used to measure its thickness once its refractive index is known.

Theory

If a pair of slits (~ 0.05 mm apart) are cut *not* in an opaque screen but in the 'semi' transparent portion of an exposed photographic film, the interference pattern corresponding to the double slit is obtained and in addition the diffraction pattern due to the large single slit that is effectively produced by that portion of the film between the two slits which lies in the parallel beam of monochromatic light. *Figure C20* represents the interference pattern due to the double slit and *Figure C19* the diffraction pattern due to the single slit. The amplitude of the central maximum here will vary with changes in phase between rays passing through the slits and those passing through the film, thus by varying the optical path length through the film we can obtain a condition where the central maximum of the diffraction pattern is annulled leaving only two fine sharp lines about the dark centre, the remainder of the interference pattern being little affected (*Figures C21* and *C22*).

The variation in optical path length is achieved by mounting the double slit arrangement on a spectrometer table in the same way as one would mount the usual diffraction grating. Then considering *Figure C23*:

Let δ = path difference between rays through the slits and rays through the film
μ = refractive index
i = angle of incidence
r = angle of refraction
and t = thickness of the film,
then

$$\delta = \mu \text{NM} - \text{NO} = \frac{\mu t}{\cos r} - \text{NM} \cos (i - r)$$

$$= \frac{\mu t}{\cos r} - \frac{t \cos (i - r)}{\cos r} = \frac{t}{\cos r} \{\mu - (\sin i \sin r + \cos i \cos r)\}$$

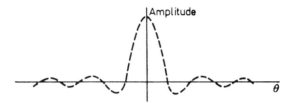

Figure C19. Single slit amplitude diffraction pattern

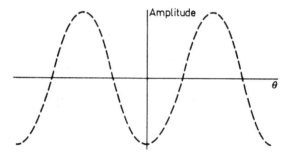

Figure C20. Double slit interference pattern

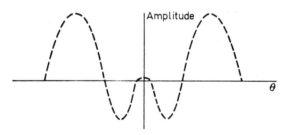

Figure C21. Sum of amplitude patterns

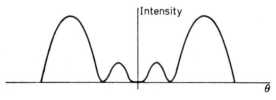

Figure C22. Resultant intensity pattern

i.e.
$$\delta = \frac{t}{\cos r}\left(\mu - \frac{\sin i}{\sin r}\sin^2 r - \cos i \cos r\right)$$

and substituting
$$\mu = \frac{\sin i}{\sin r}$$

we have
$$\delta = \frac{t}{\cos r}(\mu - \mu \sin^2 r - \cos i \cos r)$$

i.e.
$$\delta = t(\mu \cos r - \cos i) \quad \text{but} \quad \cos r = (1 - \sin^2 r)^{\frac{1}{2}}$$
$$\therefore \delta = t\{(\mu^2 - \mu^2 \sin^2 r)^{\frac{1}{2}} - \cos i\}$$

or
$$t\{(\mu^2 - \sin^2 i)^{\frac{1}{2}} - \cos i\} = \delta.$$

For minimum intensity of the central fringe
$$\delta = (2n + 1)\frac{\lambda}{2}$$

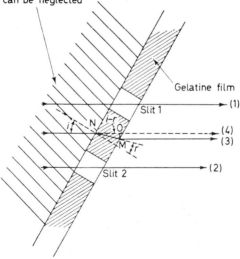

Figure C23

where n = an integer, λ = wavelength of the light used, hence

$$t\{(\mu^2 - \sin^2 i_n)^{\frac{1}{2}} - \cos i_n\} = (2n + 1)\frac{\lambda}{2} \tag{1}$$

Method

Two fine parallel lines should be made upon an exposed piece of photographic plate using a razor blade. They should be about 0·05 mm apart and to help to do this mount the film on a calibrated lathe cross slide and use the micrometer scale adjustment.

```
DETERMINATION OF LAMBDA/T FOR THIN FILMS
FREE STORE= 4630- 6265
MU =   1.45
```

ORDER N	ANGLE OF INCIDENCE DEG. MIN.		SIN I	$SIN^2 I$	COS I	$(MU^2-SIN^2 I)^{1/2}$ -COS I	LAMBDA/T
1	15	12	0·26219	0·06874	0·96502	0·49157	
2	25	55	0·43706	0·19102	0·89943	0·51456	0·02299
3	32	37	0·53901	0·29054	0·84230	0·53606	0·02150
4	35	31	0·58274	0·38781	0·76243	0·56018	0·02412
5	43	30	0·68835	0·47385	0·72558	0·58480	0·02462
6	47	45	0·74022	0·54792	0·67237	0·60922	0·02442
7	51	50	0·78621	0·61813	0·61795	0·63595	0·02672

```
END OF PROGRAM
```

Figure C24. Computer print-out of the value of λ/t for thin films

Set up the spectrometer for parallel light by Schuster's method (see first reference) and place the specimen in a diffraction grating holder upon the table. Observe the pattern and by gently rotating the table note the positions either side of the 'straight through' position where the faint double lines at the centre appear.

Discussion of Results

A computer 'print-out' of results from equation (1) is shown (*Figure C24*) but an alternative method is to plot

$$\{(\mu^2 - \sin^2 i)^{\frac{1}{2}} - \cos i\}$$

against the fringe order when we have

$$\{(\mu^2 - \sin^2 i)^{\frac{1}{2}} - \cos i\} = \left(\frac{\lambda}{t}\right)n + \frac{\lambda}{2}$$

and the slope of the graph *Figure C25* should give (λ/t). The value of n is not known absolutely, but since we are concerned here only with the slope this does not prevent us obtaining (λ/t).

THICKNESS OF THIN FILMS C1.8

Note that under these circumstances μ must be found using Brewster's method.

Careful adjustment of the spectrometer is essential if a satisfactory result is to be obtained by this method. A minority of students find great difficulty in recognizing the critical condition concerned and for these a standard mounted pair of slits previously produced by the demonstrator have been found to provide the easiest solution. Other students may profit from cutting slits of different spacings in plates of varying density and choosing the best by trial and error.

Further Work

Values of μ and n can be found using the computer method indicated. This involves taking differences between pairs of a series of

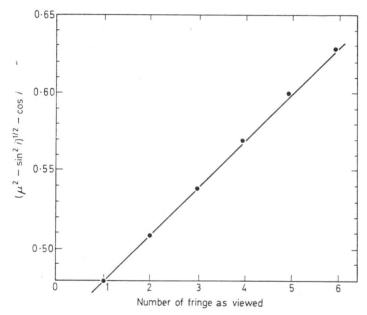

Figure C25. Typical graphical results for determination of thickness of a thin film when its refractive index is known.

Fringe No.	Angle of incidence i	$(\mu^2 - \sin^2 i)^{\frac{1}{2}} - \cos i$
1	4° 08′	0·48
2	24° 01′	0·509
3	33° 23′	0·54
4	40° 20′	0·570
5	45° 55′	0·601
6	50° 38′	0·629

245

consecutive values of $(\mu^2 - \sin^2 i)^{\frac{1}{2}} - \cos i$ for a single value of t and finding μ or m where $m = (n - 1), (n - 2), (n - 3)$, etc.

Try making your own films by coating cellulose acetate (or formvar) onto a microscope slide (see *Figure C26*).

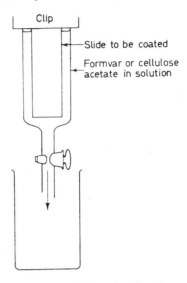

Figure C26. Use of dropping funnel to produce a thin film of even thickness upon a microscope slide

An alternative treatment is to scrape or dissolve the film from the glass leaving an untouched ridge of gelatine (*Figure C27*) and to use Newton's interference wedge method to estimate the thickness of the film. The method then allows us to measure the refractive index of the film.

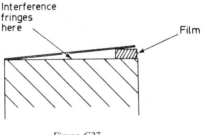

Figure C27

Further Reading and References

Schuster's method for the setting up of the spectrometer is described in *Advanced Practical Physics for Students* by B. L. Worsnop and H. T. Flint (Methuen, 1957) and the background theory of Fraunhöfer diffraction in *Geometrical and Physical Optics* by R. S. Longhurst (Longmans Green, 1957). The method itself is described by Krishna Rao in *The American Journal of Physics*, vol. 28, No. 5, p. 447, 1960.

Experiment No. C1.9

THE USE OF A GAS LASER AND A STEEL RULE TO MEASURE THE WAVELENGTH OF THE MONOCHROMATIC LIGHT FROM THE LASER

Introduction

The light beam from a gas discharge laser has such a high degree of spatial and temporal coherence that if it is allowed to impinge at almost grazing incidence upon a steel rule a Fraunhöfer diffraction pattern may be projected onto a vertical photographic plate and by measuring the spacings between the diffraction images the wavelength of the light emitted may be calculated.

Theory

The diffraction grating formula may be written

$$m\lambda = d(\sin i + \sin \theta_m)$$

where m = order of diffraction image
 λ = wavelength of light used
 i = angle of incidence
 θ_m = angle of mth order diffraction and
 d = grating spacing.

Note that for the zero order $i = \theta_0$ (*Figure C28*) corresponding to specular reflection. Considering the mth and zero order reflections $m\lambda = d(\sin \theta_0 - \sin \theta_m)$ and by Pythagoras

$$\sin \theta_m = \frac{x}{(x^2 + y_m^2)^{\frac{1}{2}}}$$

Using the binomial expansion

$$m\lambda = d\left\{\left(1 - \frac{1}{2}\frac{y_0^2}{x^2}\right) - \left(1 - \frac{1}{2}\frac{y_m^2}{x^2}\right)\right\}$$

or

$$(y_m^2 - y_0^2) = \left(\frac{2x^2\lambda}{d}\right)m.$$

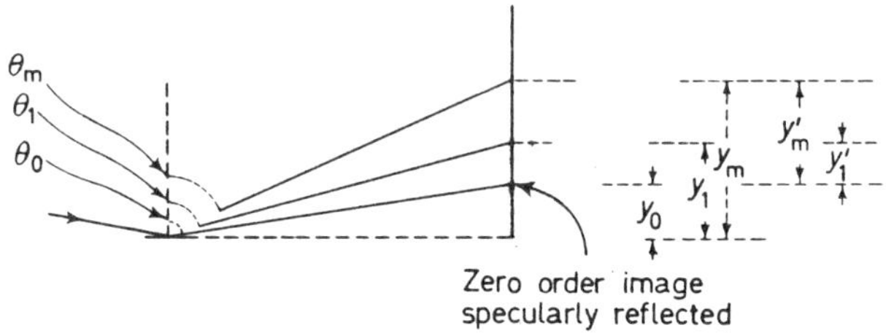

Figure C28. Arrangement for measuring the wavelength of light from a gas laser. N.B. $y'_m = (y_m - y_0)$ is the actual distance measured from the film

Plotting y_m^2 as ordinates and the order m as abscissae gives $(2x^2\lambda/d)$ as the slope of the graph and hence λ may be found since both x (the distance from the grating to the screen) and d (the grating spacing) are known.

Method

Before setting up the apparatus carefully note any laser safety regulations that are provided in the laboratory. The laser chosen here will be of low energy output and one can use it with confidence but should at all times avoid looking into the incident or specularly reflected beams (see first reference).

Switch on the laser in accordance with instructions given with the particular model being used.

Arrange that the beam strikes a metal rule at almost grazing incidence and by careful adjustment of the small angle involved throw the diffraction pattern onto a non-reflective screen or wall at some distance from the grating (say > 2 metres). It is more convenient to place the rule on an adjustable tilting platform rather than to tilt the laser (inferred from *Figure C28*) as some models only operate satisfactorily in the horizontal position.

By loading a long strip of film in a suitable carriage which can be mounted on the wall one can arrange to obtain a photograph of the type shown in *Plate C1*.

Taking measurements of y'_1, y'_2, y'_3, etc. from the negative plot y_m^2 against m, the fringe order, to give a straight line of slope

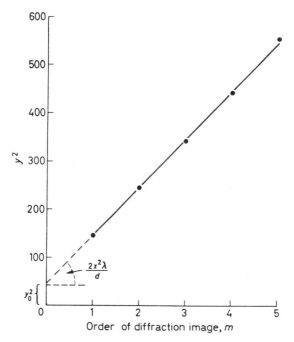

Figure C29. Graph from slope of which wavelength of monochromatic laser light is obtained

$(2x^2\lambda/d)$ (*Figure C29*). The intercept then gives a check upon the value y_0 and knowing x, the distance from the centre of the illuminated region of the diffraction grating ruler to the film, one can now obtain a value of the wavelength of the laser beam.

Repeat the procedure for graduations of various different magnitudes noting the effect on the diffraction pattern.

Discussion of Results

In considering the accuracy of the experiment recall that you have:

(*a*) employed the binomial theorem,

(b) invoked the 'small angle' approximation,
(c) had to allow for quite a large area of the rule being covered by the laser beam,
(d) to take into account the accuracy to which the distances involved may be measured with the measuring instruments chosen.

Despite this the experimental graph shows that the method is capable of a good order of accuracy.

Further Reading and References

An article on laser safety giving a typical set of 'General safety rules for a laser laboratory' will be found in *Scientific Research*, vol. 1, No. 7, p. 26, July 1966.

Background reading upon lasers may be found in Masers and Lasers by R. A. Smith, *Endeavour*, vol. 21, No. 82, p. 108, April 1962.

An alternative, non-graphical treatment of this experiment is given in the *American Journal of Physics*, vol. 33, p. 922, November 1965.

Experiment No. C1.10

TO DETERMINE THE WAVELENGTH OF THE LIGHT EMITTED BY THE NEON GAS LASER AND TO VERIFY THE LAW GOVERNING INTERFERENCE FROM A YOUNG'S DOUBLE SLIT

Introduction

The highly coherent light beam from the laser is projected onto a suitable scale and from direct measurement of the fringe separation the wavelength of the emitted laser light is calculated.

Theory

The distance y_n of the nth order interference fringe (from the optical axis) of the system is given by

$$y_n = \frac{n\lambda D}{b} \quad \text{(see *Figure C30*)}$$

where λ = wavelength of the light used
 D = the distance between the double slit and the screen, and
 b = the separation between the centres of the slits.

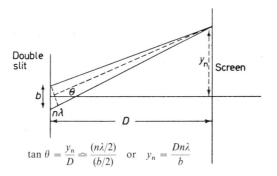

$$\tan \theta = \frac{y_n}{D} \simeq \frac{(n\lambda/2)}{(b/2)} \quad \text{or} \quad y_n = \frac{Dn\lambda}{b}$$

Figure C30. The geometry of Young's double-slit interference arrangement

Hence fringe separation

$$y_1 = \frac{\lambda D}{b}.$$

The aim here is to confirm the form of the above law by:

(1) showing y_1 is proportional to D when b is constant, and
(2) $y_1 \propto (1/b)$ for a fixed value of D.

Method

Set up the laser in the darkened laboratory in accordance with the instructions appropriate to the particular instrument used. Cut two parallel lines about 0·2 mm apart on an exposed photographic plate and mount it before the laser. Project the beam onto a vertical scale (*Figure C31*) when the characteristic interference pattern will be seen. A photograph of the pattern may be taken in a similar way to that of the previous experiment so that by direct measurement on the film the separation corresponding to say ten fringes may be measured.

The procedure is repeated for several different values of 'slits-to-screen' distance D and a graph of the fringe separation y_1 against D plotted. This results in a straight line (*Figure C32*) the slope of

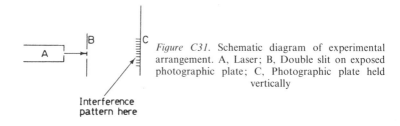

Figure C31. Schematic diagram of experimental arrangement. A, Laser; B, Double slit on exposed photographic plate; C, Photographic plate held vertically

which is (λ/b); hence λ, the wavelength of the light emitted by the gas laser, may be calculated if b, the slit separation, is measured by vernier microscope.

The experiment is then carried out for double slits of different separations so that $y_1 \propto (1/b)$ may be confirmed.

Discussion of Results

A freshly cut slit arrangement whose edges are sharp and clean helps in obtaining a good result and care in making them is worthwhile.

The use of a photograph, from which measurements are easily taken, is to be preferred rather than simply marking the positions of the fringes on a white screen (although it is possible to obtain an answer in this way).

Further Work

The laser should be used to observe the single slit diffraction pattern and also that for a diffraction grating (whose dispersive power may be calculated).

The razor blade used to cut the slits in this experiment may be taken to investigate diffraction at a straight edge and it is possible using the densitometer to investigate the fringe intensities at the edge directly.

Brewster's law ($\tan i_B = \mu$) can be verified (but don't forget that the laser beam itself is polarized) and by magnifying the area of the laser beam, using a telescope arrangement, Airey's disc may be

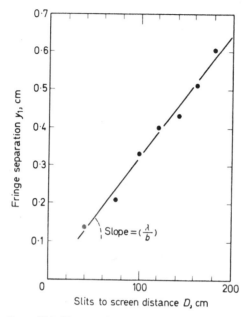

Figure C32. Young's double slit method used to determine the wavelength of the monochromatic light from a neon gas laser and to verify that $y_1 \propto D$

observed for a lens whose aperture is less than that of the resulting beam.

Reading and References

The theory of Young's double slit interference method may be found in standard texts, such as *Fundamentals of Optics* by F. A. Jenkins and H. E. White (McGraw-Hill, 1957).

An outline of this experiment and also suggestions for a number of others, including those mentioned in further work may be found in *Simple Optical Experiments with Ferranti GP Series Laser* (published by Ferranti Ltd., Kings Cross Road, Dundee, 1967).

Experiment No. C2.1

THE QUALITATIVE ANALYSIS OF A SAMPLE OF BRASS USING THE HILGER MEDIUM QUARTZ SPECTROGRAPH

Introduction

The Hilger medium quartz spectrograph is described here as being the instrument most likely to be encountered by students studying spectrographic techniques. Other models are available however, and the relevant manufacturers' handbooks should be consulted for their operation.

Theory

The medium quartz instrument employs a range of wavelengths from 2000 to 1000 Å and gives this spectrum over a single plate. The instrument (*Figure C33*) makes use of a Cornu prism and the

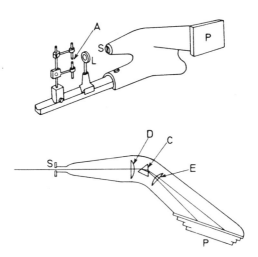

Figure C33. Schematic diagram for the quartz spectrograph

A, Region between electrodes where the arc is fired
L, Lens to focus light from the arc onto the slit S
P, Photographic plate holder   C, Quartz prism
D, Collimating lens   E, Camera lens

adjustable slit carries a Hartmann diaphragm which enables spectra of three different substances to be placed side by side on the one plate without removal of the plate holder.

Method

First check that the camera bellows attachment is set in the recommended position (4). Remove the plate holder P from its housing by pushing the locking device upwards.

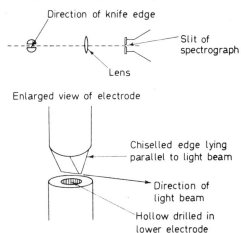

Figure C34. Diagram showing the arrangement of the 'knife edge' of the upper electrode relative to the direction of the light beam

Load with the recommended panchromatic plate (R40 may be used) in total darkness. (The emulsion side of the plate can be detected by its smoothness and should be placed facing upwards.)

Replace the mounted film in the housing, locking it firmly in position. Push the lever on the right of the plate holder upwards to put the scale pointer out of action.

Using the wheel on the left of the holder set the plate at the correct height within the carriage (position 1). Place the quartz lens (f.958) on the bar, 28 cm from the slit, and 10 cm from the position at which the arc will be fixed. Clean two 6·5 mm diameter copper rods and screw them into position in the clamp with a gap of 3 mm between their tips. The tips should be filed to the form of a wedge and the ridge edge of each wedge placed so that it is parallel to the central axis of the optical system of the instrument (*Figure C34*).

Switch on the 110 V d.c. supply to the arc and adjust the regulating resistance to give a current of 3·5–4 A. Strike the arc by touching both electrodes with an insulated carbon rod to bring the firing electrodes into contact. (Do NOT do this or look directly at the arc without wearing the protective dark goggles which are supplied.)

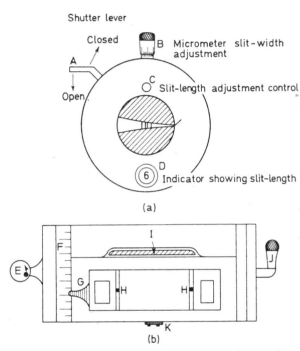

Figure C35. (a) Shutter and slit adjustments on the Hilger medium quartz spectrograph. (b) Photo plate carriage and adjustments on the spectrograph

E, Plate height control H, Back clips of plate holder
G, Plate height pointer J, Wavelength scale lever
I, Plate slide cover K, Locking device
F, Plate height positioning scale

Set the slit width on the instrument to 0·001 mm using the micrometer B (*Figure C35*), and also the length of the slit to 6 mm by control C.

Open the dark slide on the plate holder and expose the plate by depressing the shutter lever. Close the shutter after 10 seconds and pull down the wavelength scale by means of the lever on the right

ANALYSIS OF A SAMPLE OF BRASS   C2.1

of the plate holder. Switch on the lamp at the rear of the camera for 2 seconds.

Push the wavelength scale lever upwards once again and reset the plate height to position 2 to align a fresh area of the plate. Repeat the above procedure using exposure times of 15 to 20 seconds for the copper arc (but always 2 seconds only when photographing the scale) ensuring that each exposure to the arc is made on a fresh strip of the plate.

Hence the optimum exposure time for the copper arc may be determined upon developing the film.

The same procedure is carried out using zinc rods to produce the arc. Here it may be that the arc splutters and extinguishes itself, but the full exposure of 10 seconds (or other chosen time) should be given.

Finally an arc is struck across brass rods but a shorter exposure time should be chosen, starting at 5 seconds.

Discussion of Results

The use of the wedge on the electrode allows the arc to 'run' along the wedge during filming. If a point is used the 'burn up' rate of the tip of the electrode tends to be too rapid.

As the main constituents of brass are zinc and copper, it should be fairly easy to pick out corresponding appropriate lines in the spectra of the elements and brass, the alloy composed of these elements (*Plate C2*).

On the photograph will be seen also a number of impurities such as lead, nickel and iron which are common to commercial brass.

Further Work

A powder should also be analysed held in a hollow between pure graphite rods. If the rods have had previous use they should be broken off (about 25 mm will do) at the used end, and filed to a wedge shape as before, the other hollowed out to a depth of about 4 mm by means of a small twist drill. This becomes the lower electrode, the anode.

First 'fire' the arc between the electrodes alone using a current of 4 A, and an exposure time of 10 seconds.

Introduce the powder into the hole and repeat the procedure.

APPLIED OPTICS, SPECTROSCOPY, PHOTOMETRY

Reading and References

A chapter on 'Emission spectrographic analysis' may be found in *Quantitative Inorganic Analysis* by A. Vogel (Longmans Green, 1964). This also gives greater detail of the instrument and of suitable photographic plates and their treatment.

Experiment No. C2.2

TO DETERMINE THE RYDBERG CONSTANT

Introduction

Ritz and Rydberg showed that the lines of the hydrogen spectrum could be expressed by a general equation of the form

$$\frac{1}{\lambda} = R\left(\frac{1}{n_1^2} - \frac{1}{n_2^2}\right)$$

where λ represents the wavelength of the appropriate line
        R represents the Rydberg constant, and
        n_1 and n_2 are integers.

If $n_1 = 1$, and $n_2 = 2, 3, 4$, etc. then the Lyman series of spectral lines corresponding to the ultra-violet region of the spectrum is obtained.

If $n_1 = 2$, and $n_2 = 3, 4, 5$, etc. then we have the Balmer series in the visible region, whilst if $n_1 = 3$ and $n_2 = 4, 5, 6$, etc. the resulting spectrum lines lie in the infra-red.

The portion of the spectrum investigated here lies in the visible where $n_1 = 2$, and n_2 has values 3, 4 and 5.

Theory

According to the above note the Balmer spectroscopic formula for hydrogen may be written

$$v = \frac{1}{\lambda} = R\left(\frac{1}{2^2} - \frac{1}{n^2}\right)$$

where v = the wave number = (1/wavelength) of the line concerned,
and $n = n_2$ above has values 3, 4 and 5.

In order to find R it is necessary to obtain a value for λ in the

THE RYDBERG CONSTANT

above expression and to do this, use is made of the diffraction grating, the formula for the use of which is

$$m\lambda = (a + b)\sin\theta = \frac{\sin\theta}{N}$$

where m is the order of the line
$(a + b)(= 1/N)$ is the grating spacing, and
θ is the angle of diffraction.

Method

First focus the spectrometer by Schuster's method using a glass prism. Remove the prism and replace it by the grating in its mount (*Figure C36*). Position the cross wires on the central image of the

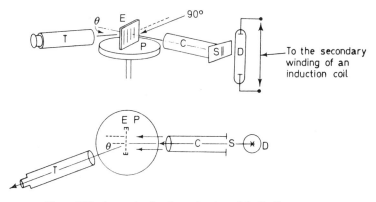

Figure C36. Apparatus for determination of the Rydberg constant
D, Hydrogen discharge lamp
S, Adjustable slit of spectrometer
C, Collimator adjusted to give parallel light
E, Diffraction grating at right angles to the incident beam
θ, Angle of diffraction   P, Spectrometer table
T, Telescope adjusted to focus parallel light

slit and note the reading on the spectrometer vernier. Rotate the telescope 90° from this reading and clamp it. After adjusting the table screws so that the image is vertical rotate the table until the reflected image of the slit (the same colour as the original monochromatic light) is centred on the cross wires. Read the table position. Now rotate the table through 45° (or 135°) until the ruled face (which should *not* be fingered) is turned to the collimator.

Parallel light now falls normal to the grating. Replace the original monochromatic source (sodium is suitable) by a hydrogen discharge tube which may be activated by the high voltage from an induction coil as shown in *Figure C36*. Locate the central maximum

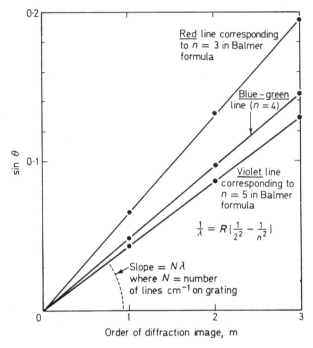

Figure C37. Graph of sine of angle of diffraction θ against order m of diffraction image used to determine the wavelength λ of the hydrogen lines

first, and then locate the first diffraction image of the brightest line (probably red) in the spectrum and note the telescope reading when the cross-wire is centred on the image. The angle between the first diffraction image positions of the one line on either side the centre is twice the diffraction angle for the line, hence tabulate this angle. Repeat the procedure for several lines in the spectrum for first order images.

Now do the same for second and third order spectra. The value of $\sin \theta$ can be found and this is plotted against m the order of the diffracted image as shown in *Figure C37*. The slope of each line is equal to $N\lambda$ and hence the wavelength λ for the red, blue, green and

THE RYDBERG CONSTANT C2.2

violet lines can be determined. From this value $1/\lambda$ is computed and tabulated and is itself plotted (*Figure C38*) against $1/n^2$ where n has the values previously noted (3 for the red line, 4 for blue-green line and 5 for the violet line). The slope of the single line obtained then

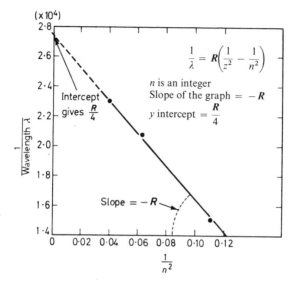

Figure C38. Graph of (1/wavelength λ) against ($1/n^2$) used to determine the value of the Rydberg constant R from the Balmer formula

gives R the Rydberg constant and the intercept on the y axis gives a confirmatory value for R, since it should be equal to $R/4$.

Discussion of Results

The experiment should be carried out in a dark room and extraneous light from the lamp shielded from the rest of the apparatus. Under these conditions a result of high accuracy is possible.

Try a number of different gratings, for although the greater the number of lines, the greater the resolution, it may be, that with too many lines per centimetre of grating, sufficient orders of diffraction image cannot be obtained to employ a graphical method and it may be found too, that the line intensity falls off very rapidly and that the violet lines do not remain accurately measurable. (The graphs shown were obtained using a grating of 10^3 lines cm^{-1}.)

Further Work

If a constant deviation spectrograph is available it provides an alternative method of determination of the Rydberg constant and a photograph of the spectrum may be readily obtained (*Plate C3*).

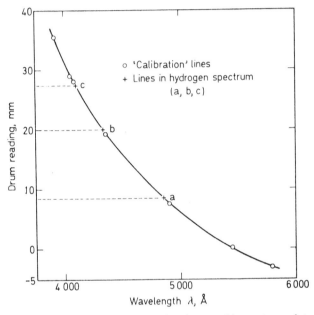

Figure C39. Calibration curve for the photographic spectrum plate used to determine the wavelengths of the hydrogen lines in the Balmer series

By superimposing the hydrogen spectrum over that of mercury which is well known (and can be found in tables), the wavelength of up to six Balmer lines may be computed (*Figure C39*).

Reading and References

Background reading upon the series of spectral lines may be found in *Introduction to Atomic Spectra* by H. E. White (McGraw-Hill, 1934).

Details of Schuster's method for the focusing of the spectrometer are given in *Advanced Practical Physics for Students* by B. L. Worsnop and H. T. Flint (Methuen, 1957).

Experiment No. C2.3

USE OF THE SPECTROPHOTOMETER TO VERIFY BEER'S LAW GOVERNING ABSORPTION OF LIGHT BY COLOURED MEDIA

Introduction

Selective absorption of light by coloured media has found many uses such as, for example, in chemical inorganic analysis, in physiological identification of chlorophyll, carotene and derivatives of blood, and in purity checks of foods and drugs.

In this experiment the wavelength corresponding to maximum absorption is first found and the law governing absorption verified.

Theory

The fundamental law of absorption of light shows that the fractional loss of intensity is proportional to the geometrical thickness of the absorbing medium and thus, in the case of a solution, to the number of solute molecules in the solution, i.e.

$$\frac{\delta I}{I} = -\mu l$$

where μ = linear absorption coefficient and l represents thickness, when integration gives

$${}_{I_0}^{I}[\log_e I] = -\mu[l]_0^l$$

and

$$I = I_0 \exp(-\mu l).$$

Since logarithms to the base 10 are often more convenient than logarithms to the base e an extinction coefficient k is defined in such a way that it is the thickness of medium that will absorb one-tenth of the incident light, then

$$\exp(-\mu/k) = \tfrac{1}{10}$$

i.e.

$$\exp(+\mu/k) = 10 \quad \text{or} \quad \frac{\mu}{k} = \log_e 10 = 2\cdot 303$$

$$\therefore k = \frac{\mu}{2\cdot 303}.$$

APPLIED OPTICS, SPECTROSCOPY, PHOTOMETRY

In the case of a solution the optical density of the solution ρ is defined as

$$\rho = \log_{10} \frac{I_0}{I}$$

whence linking the two equations

$$\rho = \log_{10} \frac{I_0}{I} = \frac{\mu l}{2 \cdot 303} = kl.$$

For an absorbing substance dissolved in a transparent solvent the decrease in intensity is, as pointed out above, proportional to the number of solute molecules in solution, and thus we may write

$$\rho = \log \frac{I_0}{I} = \varepsilon' cl$$

where ε' is known as the molar extinction coefficient, and
c is the concentration in moles litre^{-1}.

Method

The type of spectrophotometer described here is by Unicam. Other models are available and whilst the fundamentals are the same, the instructions appropriate to the particular model used should be studied carefully.

The spectrophotometer produces a beam of monochromatic light whose chosen wavelength may be varied by movement of the diffraction grating (or prism) with respect to the incident beam of white light.

Using a coloured solution the wavelength corresponding to the maximum absorption of light is characteristic of the medium and it is first necessary to check the wavelength calibration of the instrument.

Plug the instrument into the mains and switch on the heater. After allowing about 5 minutes to warm up, adjust the galvanometer to zero on the scale and the wavelength control to '583' corresponding to 5 830 Å or 583 millimicrons (mµ).

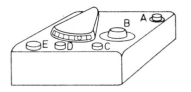

Figure C40. Typical small type visible range spectrophotometer (Unicam SP 400)
A, 'Heater/test' switch
B, Wavelength control and dial
C, Solution tube chamber
D, 'Increase light' control
E, 'Dark current' adjustment

Place the didymium glass filter in position in the holder (at right angles to the pin slots) when all light is cut off.

Switch from 'heater' to 'test' (*Figure C40*), and adjust the 'dark current control' to bring the galvanometer spot back to zero.

Remove the filter holder and darken the filter chamber by putting the cover into place. Adjust the 'increase light' control until the galvanometer reads 100% transmission then replace the filter in position 2 with the pins in the slots so that the didymium filter lies in the beam. The wavelength control is then rotated to pass from 580 mµ upwards through 583 mµ. Maximum absorption should occur at 583 mµ and if this is not so the demonstrator should be consulted.

Now pipette 5, 10, 15, 20 and 35 ml respectively of bench copper sulphate solution into 100 ml measuring flasks. Add 10 ml of bench ammonia to each flask when a deep blue colouration will result. Make up each solution in turn to 100 ml and shake thoroughly.

Using first the weakest solution and adjusting the 'increase light' control to give maximum galvanometer scale deflection, then each other solution in turn, in the tube holder of the spectrophotometer, plot a graph of galvanometer reading (i.e. transmission) against

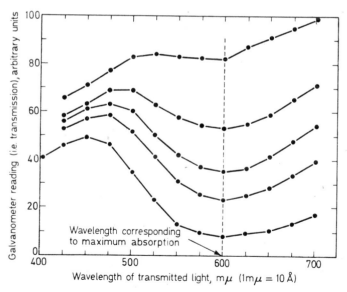

Figure C41. Absorption spectrum of ammoniacal copper sulphate solution

APPLIED OPTICS, SPECTROSCOPY, PHOTOMETRY

wavelength and determine the wavelength at which maximum absorption (i.e. minimum transmission) takes place (*Figure C41*).

Then taking the logarithm of the galvanometer reading, plot this against the concentration of the solution in ml of the original

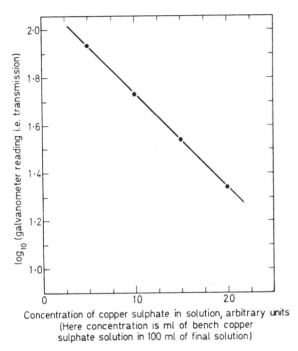

Figure C42. Graph to verify Beer's law

copper sulphate solution per 100 ml of final solution (or better still in grams of copper sulphate per litre of final solution). A straight line results (*Figure C42*), verifying Beer's law.

Discussion of Results

Unless the concentration of the solutions is known absolutely and the transmission can be expressed in absolute terms it is not possible to determine the extinction coefficient k, but only to verify the form of Beer's law.

If possible, however, the molar extinction coefficient should be calculated by plotting the concentration in moles per litre along the x axis of the graph.

Further Work

From the results obtained above, the experiment may be extended to determine the concentration of a solution of copper sulphate of unknown strength.

Some solutions show marked deviations from Beer's law and this may be demonstrated using aqueous solutions of cobalt chloride.

Reading and References

Further details of a number of different models of visible and ultra-violet spectrophotometers may be found in *A Textbook of Quantitative Inorganic Analysis* by A. I. Vogel (Longmans Green, 3rd edition, 1964). Also in this book is a chapter on colorimetric analysis and an outline of the theory behind this experiment.

Experiment No. C2.4

TO DETERMINE THE MAGNITUDE OF THE ENERGY GAP IN THE BAND STRUCTURE OF INDIUM ANTIMONIDE FROM THE INFRA-RED SPECTRUM OF THE SEMI-CONDUCTOR

Introduction

Many semiconductors of which germanium and indium-antimonide are examples are 'transparent' to infra-red radiation. *Figure C43* shows the way in which the absorption varies with the infra-red wavelength and it is seen that from longer wavelength (i.e. smaller energy) values the absorption coefficient becomes less, showing that more infra-red radiation is being transmitted. There comes a point, however, when the absorption rapidly rises again (the 'cut-off' value) and this may be interpreted in terms of the forbidden zone which exists between the valence band and the conduction band of the semiconductor.

Theory

It may be supposed that the forbidden zone of the semiconductor represents an energy gap E_g.

If the energy of the infra-red radiation is given by

$$E = hc/\lambda = E_g$$

where h = Planck's constant
 c = the velocity of infra-red radiation
 λ = the wavelength of the infra-red radiation

then the photon of radiation can (theoretically) excite an electron to an extent sufficient to raise the electron from the top of the

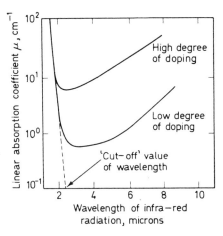

Figure C43. The absorption of infra-red radiation with variation in the wavelength of the radiation

valence band to the bottom of the conduction band. In doing so, however, the energy of the photon is completely absorbed and hence the crystal appears to be 'opaque' to the radiation, which results in the abrupt rise of the absorption coefficient at one particular value (the cut-off value) of the wavelength.

Using Planck's formula one may then write

$$E_g = hc/\lambda_c$$

where λ_c represents the cut-off value of wavelength.

This cut-off wavelength may be found from the infra-red spectrum and *Figures C44* and *C45* show typical i.r. spectra for indium antimonide and indium arsenide.

ENERGY GAP IN INDIUM ANTIMONIDE C2.4

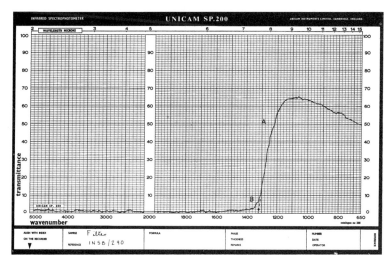

Figure C44. Infra-red spectrum of indium antimonide obtained on the Unicam SP. 200. Curve obtained by the author on paper supplied by, and reproduced by permission of Pye Unicam Ltd.

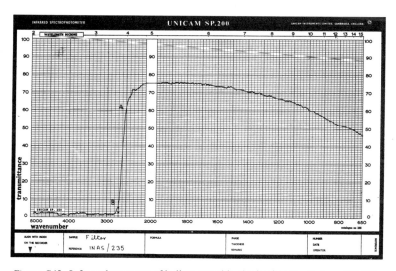

Figure C45. Infra-red spectrum of indium arsenide obtained on the Unicam SP. 200. Curve obtained by the author on paper supplied by, and reproduced by permission of Pye, Unicam Ltd.

Method

The results shown here were taken on the Unicam SP.200 infrared spectrometer. Students should take care to follow implicitly any specific instructions given with the particular instrument they use. Those given here are appropriate only to the SP.200.

First place the chart paper in position with the fiducial arrows adjacent to one another, making sure that it lies flat. Using the transmittance control set the pen to register 100% transmittance before the specimen is placed in position. Make sure that the scanning mode switch is in the appropriate position (in general No. 1 on the SP.200).

Now simply place the indium antimonide in the form of a thin film with protective mounting in the infra-red beam and switch to 'scan'.

Discussion of Results

In assessing the accuracy of the result, note that the fall of the transmission curve is not instantaneous and that it is necessary to extrapolate the region AB to zero, in order to obtain the 'cut-off' value of wavelength. This typical fall is to be expected since we are concerned not with a single photon but with a large number of them, which will not all have exactly the same energy.

It has also been assumed that the excitation is from the very top of the valence band to the lowest conduction level, when in fact jumps greater than this will occur.

Ensure that the paper is correctly fitted and that it lies flat otherwise serious errors in wavelength reading can occur, particularly at the shorter wavelength end of the scale.

A wavelength check can be made using a standard specimen such as polystyrene.

Further Work

The spectra of other semiconductors, e.g. indium arsenide, should be plotted and the energy gap associated with these computed.

Reading and References

A simple treatment of the optical properties of materials in terms of zone theory is given in *Electronic Processes in Materials* by L. V. Azaroff and J. J. Brophy (McGraw-Hill, 1963).

A deeper treatment is to be found in *Elements of Infra Red Technology* by P. W. Kruse, L. D. M. Glauchlin and R. B. McQuistan (Wiley, 1962).

The wider applications of infra-red spectroscopy are treated in *Ultra-Violet and Infra-Red Engineering* by W. Summers (Pitman, 1962).

Details of a number of i.r. spectrometers including the Unicam SP.200 are given in *An Introduction to Infra-Red Spectroscopy* by W. Brugel (Methuen, 1962).

An Introduction to Practical Infra-Red Spectroscopy by A. D. Cross (Butterworths, 3rd edition, 1969), gives essential elementary theory and deals with the uses of the method as well as giving details of instruments.

Experiment No. C2.5

TO VERIFY LAMBERT'S LAW BY THE USE OF THE SPEKKER ABSORPTIOMETER

Introduction

The absorption of light and the verification of Beer's law has been dealt with in Experiment No. C2.3 where the Unicam spectrophotometer was used. This instrument enabled measurements to be made throughout the spectral range from 250 to 1 000 mμ (i.e. 2 500 Å to 10 000 Å).

The Spekker absorptiometer uses white light in conjunction with a chosen range of optical filters (i.e. it does not scan the visible spectral range fully) but the use of a selected filter, or filters, is often sufficient for routine and rapid colorimetric measurements.

For that reason the use of the instrument is described here. In the previous absorption experiment Beer's law was investigated, i.e. the effect of variation of concentration of absorbing solution upon intensity. In this experiment verification of Lambert's law governing the variation of intensity with thickness of absorbing medium is chosen.

The theory of absorption is precisely the same as that of Experiment No. C2.3 and a glance at this will show that to verify Lambert's law a graph of log I against absorber thickness t should be a straight line.

APPLIED OPTICS, SPECTROSCOPY, PHOTOMETRY

Theory

The instrument makes use of two photocells P and P'. The first receives light from the central tungsten lamp T (*Figure C46*), after

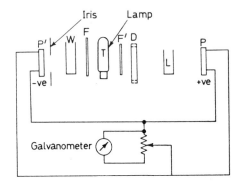

Figure C46. Optical arrangement of the Spekker absorptiometer

it has passed through the liquid L whose light absorbing properties are to be investigated and also through an appropriate wavelength filter. The second photocell P' receives light through a variable aperture which makes adjustment possible and the two cells are connected in such a way that when the photoelectric currents are equal the galvanometer connected across the cells reads zero. In addition heat absorbing filters F and F' are introduced into each light path and the focusing onto the cells is achieved by introduction of lenses L' and L" (not shown on diagram). A rheostat enables the sensitivity of the instrument to be adjusted in use.

The drum of the Spekker is calibrated so that when the iris aperture is partly open the drum reading

$$R_1 = \log \frac{A_0}{A_1}$$

where A_0 = area of the aperture when fully open
 A_1 = area of aperture in the partly opened position,
i.e. when fully open

$$R = \log \frac{A_0}{A_0} = \log 1 = 0.$$

Now suppose the galvanometer of the instrument reads zero with

VERIFYING LAMBERT'S LAW

the thickest absorber cell filled and in position. Then with the iris aperture fully open

$$I = I_0 \exp(-kt)$$

where I_0 = incident light intensity upon the cell
 I = intensity of light emerging from the cell, and
 k = appropriate absorption coefficient (compare μ in Experiment No. C2.3).

If a narrower cell of thickness t_1 is now introduced

$$I_1 = I_0 \exp(-kt_1) \quad (I_1 > I).$$

To reduce the intensity of light on the right-hand photocell from I_1 to its previous value I, the aperture area is reduced by adjustment of the iris control

then

$$\frac{I}{I_1} = \frac{A}{A_1} = \frac{I_0 \exp(-kt)}{I_0 \exp(-kt_1)}$$

and taking logarithms

$$\log_e \frac{A}{A_1} = k(t_1 - t).$$

Since the drum reading is calibrated so that $R = \log(A/A_1)$, $R = -kt + kt_1$.

Method

The coloured solution is used to fill the 'thickest' cell and is placed in position between the Spekker lamp and the photocell.

A matching cell is filled with distilled water (or the appropriate solvent used in the coloured solution) and placed in the other arm of the instrument (*Figure C47*).

The heat filters F and F' are then inserted before the lamp and the appropriate optical filters (enabling measurements to be made at one narrow wavelength band) between the liquid cells and the photocells. The calibrated drum is set to the position corresponding to the largest aperture and then the next 'thickest' solution-cell is used to replace the first one. Balance is restored by reducing the aperture using the drum control and the drum reading is noted. This is repeated for the other cells and the results plotted (*Figure C48*).

Discussion of Results

With careful use the instrument is an accurate one as may be inferred from the graphical results such as obtained here. In the graph shown, the colour was that due to the presence of iron in

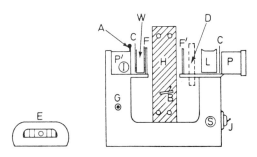

Figure C47. The Spekker absorptiometer
H, Lamp housing
F and F', Heat filters
C and C', Selected optical filters
A, Main iris diaphragm control
I, Fine control of diaphragm
B, Lamphouse shutter
W, Water cell
D, Calibrated drum
L, Cell containing liquid
P and P', Photocells
G, Galvanometer switch
S, Sensitivity control
E, Galvanometer
J, 'On/off' switch

solution and a deep green colour was developed by the addition of a small quantity of reagent to the original solution. The optical filters chosen were Ilford deep red. (In fact there were 0·1 grams of iron per litre of the solution. The reagent consisted of 0·1 g of nitro-resorcinol mono-ether in 25 ml of glacial acetic acid, and 1 ml of the reagent was added to 20 ml of the solution containing iron.)

Other and simpler colorimetric solutions (such as the copper sulphate one described in Experiment No. C2.3) are of course quite suitable for this experiment—other optical filters may then be necessary.

Further Work

It is worthwhile to make up a series of standard solutions and to construct a calibration chart for the instrument if routine tests are to be made with it.

Reading and References

The instrument described is by Hilger and Watts, Ltd., and full detail of it is given in their operation-manual.

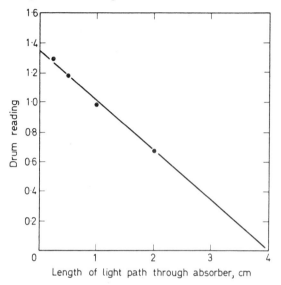

Figure C48. Verification of Lambert's law using the Spekker absorptiometer

The instrument is also described in *A Text-Book of Quantitative Inorganic Analysis* by A. I. Vogel (Longmans Green, 3rd edition, 1964) together with further suggestions for the ways in which the Spekker may be used and notes upon its mode of operation.

Experiment No. C2.6

TO DETERMINE THE CHARACTERISTIC DENSITY–EXPOSURE CURVE OF A PHOTOGRAPHIC FILM USING THE HILGER MICRO-DENSITOMETER

Introduction

It is well known that the longer a photographic film is exposed to light (or other appropriate electromagnetic radiation) the more dense will be the resultant blackening. The variation of density (or blackening) with exposure gives a curve characteristic of the particular film used, and the aim of this experiment is to plot the density–exposure graph in order that qualitative measurements may be made.

Theory

The density of blackening B of the film may be defined by the equation

$$B = \log_{10} \frac{I_0}{I}$$

where I_0 represents the intensity of light through an exposed portion of the film, and I the intensity through the exposed portion considered.

Method

The selected film should be exposed so that narrow strips, say 6 or 7 mm wide, of differing density are obtained. A commercial ortho fine-grain film may be used and exposed to the light through a yellow darkroom filter (Kodak OB) in increasing steps of 5 seconds or so at a distance of approximately 6 feet. A strip at either edge of the film should remain covered at all times so that an estimate of the intensity through the clear part of the film may be made.

After developing and fixing under the conditions recommended for the particular film used, it should be placed, emulsion side upwards, on the carrier plate K of the Hilger densitometer (*Figure C49*).

Switch on the densitometer lamp and the 'galvoscale' lamp and check that the shutter lever A on the front of the instrument is in the 'closed' position. Align a clear portion of the film in the beam (approximately) then open the shutter lever and check that the image falls symmetrically on the slit. Then focus the image by setting the sensitivity control, on the right-hand side of the shutter lever, to give 25 cm deflection on the galvanometer scale and move the lower objective to give maximum deflection.

Close the shutter and set the galvanometer to zero using the knob F on the lower left of the viewing screen. Open the shutter once more and by means of the sensitivity control adjust for maximum deflection of approximately 50 cm. Close the shutter and move the region of film to be investigated into alignment with the beam. Check that the galvanometer reading is zero then open the shutter and note the new scale reading, closing the shutter straight away. Move to the next region of darkness and repeat the procedure. Plot the film-density against a measure of the time of exposure (*Figure C50*).

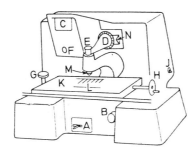

Figure C49. Diagram showing the essential arrangement of the micro-densitometer
A, Lever to move the viewing lens in or out of the path of the light
B, Galvanometer 'throw' adjustment
C, Sensitive galvanometer
D, Knurled ring to adjust slit width
E, Eyepiece for viewing
F, Zero adjustment for the galvanometer scale
G, Up and down movement control for carriage
H, Lengthwise adjustment control for plate carrier
J, On/off switch   K, Photo plate carriage
L, Portion of plate under observation
M, Objective lens
N, White adjustable vee-jaws before photocell

Adjustment of the slit length can be made if necessary using the sliding wedge diaphragm N on the instrument and of the slit-width by rotating the ring D (with the knurled divisions) which lies behind the wedge, but this should have been done and the demonstrator should be consulted in case of doubt. (For ordinary purposes of spectrum line observation a suitable slit width is of the order of one-tenth the width of the spectrographic slit used, with a slit length of 2 mm.)

Discussion of Results

The accuracy to which the curve can be plotted will depend upon the intervals of exposure time chosen. Notice that white light has not been used here since it is possible using filtered light to maintain more delicate control of the blackening. A number of trials with the particular film being used under different conditions of exposure are needed before a satisfactory curve can be plotted.

Make sure that the photocell shutter is firmly closed, immediately after taking each reading. Fatigue in the photocell can seriously affect the results.

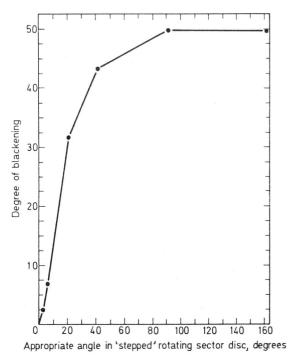

Figure C50. Calibration densitometer curve

The area of each strip should be large enough for several readings to be taken corresponding to one exposure time. The variation that occurs in these data enables a good estimate to be made of the experimental accuracy.

Further Work

If a micro-densitometer is not available it is still possible to plot the characteristic curve of bromide paper by the reflection of light from it under standard conditions (for detail see Experiment No. C3.3).

Further Reading and References

Further details of the particular instrument used will be found in the appropriate operations manual.

The Hilger instrument referred to above is also described in *A Text-Book of Quantitative Inorganic Analysis* by A. I. Vogel (Longmans Green, 3rd edition, 1964), under the Instrumental Analysis.

The theory given here is purposely brief and basically may be considered in conjunction with that given in Experiment No. C3.3.

Experiment No. C3.1
TO PLOT THE POLAR CURVE OF A FILAMENT LAMP AND TO DETERMINE ITS MEAN SPHERICAL INTENSITY

Introduction

In general if a filament lamp is rotated about a fixed position the intensity of illumination varies.

If the intensity in a large number of directions is measured and marked off to scale along the appropriate direction a 'polar diagram' is obtained.

Theory

From the polar diagram the mean spherical intensity may be found. Perhaps the most convenient way to do this is to read off the intensities at the Russell angles (*Tables C1* and *C2*).

These are predetermined angles chosen so that they are equally spaced ordinates on the Rousseau diagram. (The Rousseau diagram being a graphical method used to determine the value of the total luminous flux I involved.)

$$I = 2\pi \int_0^\pi I_\theta \sin \theta \, d\theta$$

where I_θ = the intensity over the chosen range θ_1 to θ_2, θ being the chosen direction with respect to an arbitrary axis (see *Figure C53*).

Method

Place a standard (or known sub-standard) lamp over the fixed end mark (0′) (*Figure C51*) of the optical bench, and the bare lamp

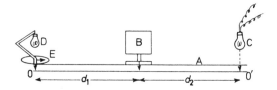

Figure C51. Schematic diagram of apparatus used in the plotting of a polar curve
A, Optical bench scale
B, Lummer-Brodhun or other suitable photometer head
C, Standard lamp, distance d_2 from photometer
D, Lamp under investigation
E, Horizontal scale and pointer

to be tested (in its special holder) over the mark 0 at the other end. (If no actual standard is available an ordinary coiled filament 60 W bulb may be used and its candle power assumed to be 45 candelas.)

Align the chosen photometer (a Lummer-Brodhun is eminently suitable), so that it lies at the same height as the filaments of the lamps and at the centre mark of the optical bench.

Check that on rotating the arm of the special holder of the test lamp the filament remains vertically above the mark 0 on the bench.

Switch on the supply to both lamps and move the photometer to give equal illumination on each side of the instrument. Note the position of the photometer, then from the inverse square law

$$\frac{\text{Power of the test lamp}}{\text{Power of the standard lamp}} = \frac{d_2{}^2}{d_1{}^2}$$

where d_1 = distance of the test lamp from the photometer
d_2 = distance of the standard lamp from the photometer.

Now rotate the test lamp through 10° as noted on the base scale of its holder. Check once again that the lamp lies vertically over its fiducial point 0 and repeat the above procedure.

Finally tabulate the results corresponding to photometric balance for every 10° rotation through 360° (and also at those angles, the Russell angles, which are given in *Table C1*).

Join the points to obtain the polar curve (*Figure C52*).

CHARACTERISTICS OF A FILAMENT LAMP C3.1

Table C1. The Russell angles appropriate to ten subdivisions at which luminous intensities should be read off when the '6 o'clock' position of the hour hand corresponds to 0° and the '12 o'clock' position to 180°

25·8°	95·7°
45·6°	107·5°
60·0°	120·0°
72·5°	134·4°
84·3°	154·2°

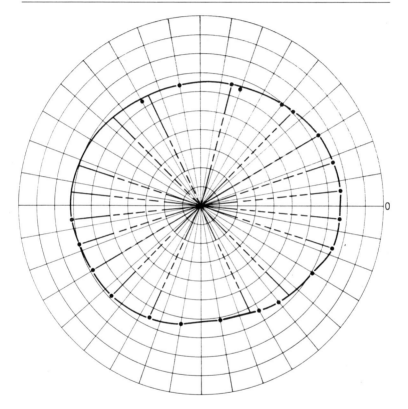

Figure C52. Polar curve of a lamp showing intensities at the Russell angles to be used to determine the mean spherical intensity

Read off from the resulting curve the appropriate intensities at the Russell angles, add them and divide by the number of readings to obtain the mean spherical intensity of the lamp (*Figure C53*).

Discussion of Results

The alternative set of Russell angles quoted in *Table C2* are those corresponding to the situation when the positive x axis is the 0° reference position, the negative one the 180° reference, and the positive y axis 90°.

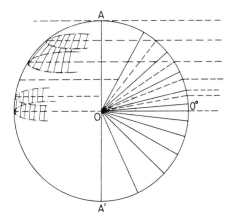

Figure C53. Diagram to illustrate the use of predetermined Russell angles to obtain the mean spherical intensity of a light source. The dotted lines divide the circle into ten zones such that the surfaces of the zones of the sphere swept out by revolution about AA' (and indicated by the shaded regions) are each equal

Table C2. Russell angles corresponding to the eight subdivisions at which luminous intensities should be read off when the 0 and 180° positions correspond to the usual Cartesian system

7·2°	119·0°
22·0°	141·3°
38·7°	158·0°
61·0°	182·8°

Details of these and the Russell angles appropriate to other subdivisions may be found in Glazebrook's *Dictionary of Applied Physics* (see references).

If the y axis is used as zero a different set of Russell angles is of course needed (*Table C1*).

The intensities should be read off at the Russell angles on both sides of the zero so that the calculated mean spherical intensity is the mean of 20 not 10 readings. (N.B. Subdivision into 10 zones has been assumed here.) Russell angles appropriate to other chosen subdivisions are to be found in the reference below. The theory of mean spherical intensity has assumed a certain symmetry which in general does not exist. The initial drawing of the polar curve to obtain some idea of the axis most nearly corresponding to symmetry, does enable a more accurate result to be obtained from the Russell angle calculation (see Glazebrook).

Further Work

After obtaining the polar curve of the bare lamp a suitable fitting should be used to obtain the curve due to the flux emitted with the fitting in position.

The efficiency ε of the complete unit is then the ratio of the mean spherical intensity with the fitting in position to that of the bare lamp.

Reading and References

The Lummer-Brodhun Photometer is described in standard degree texts on light. *Principles of Illumination* by H. Cotton (Chapman and Hall, 1960) also deals with different types of photometers and their use as well as with polar diagrams and Rousseau diagrams.

The theory of Russell angles and tables of them corresponding to various zonal subdivisions are given in *Dictionary of Applied Physics* by R. Glazebrook (Macmillan, 1923), p. 433, and a number of polar diagrams are shown.

Photometry by J. W. T. Walsh (Constable, 1958) gives valuable background reading.

Experiment No. C3.2

TO COMPARE THE LUMINOUS INTENSITIES OF TWO LAMPS EMITTING LIGHT OF DIFFERENT COLOURS USING A FLICKER PHOTOMETER

Introduction

In viewing the two comparison surfaces of a photometer in order to compare their brightness the eye may sense flicker in the field of view due to two causes:

(1) difference in screen luminosity
(2) difference in colour of the light emitted by the two sources.

In the flicker photometer the field of view alternates rapidly to present that due to each source in turn and using different coloured sources the flicker due to colour difference is found to be distinct from that due to luminosity difference. Moreover, if the speed of the flicker produced by the instrument is slowly raised the 'colour flicker' is found to disappear before the luminosity flicker.

Although, therefore, the flicker photometer may be used to compare intensities of sources of the same colour its particular value lies in its suitability for heterochromatic photometry.

To do this it is first necessary to adjust the photometer motor speed to minimize colour difference and then to adjust for minimum luminosity flicker.

Theory

The instrument used here is the Simmance-Abady flicker photometer (*Figure C54*) which consists of a plaster disc in the form of two

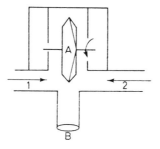

Figure C54. The Simmance-Abady flicker photometer
A, Rotating disc of magnesium carbonate which reflects light from each source into the eyepiece (the speed of rotation is variable)
B, Viewing eyepiece
1, Light beam from first source
2, Light beam from second source

shallow truncated cones in combination (*Figure C55*). As the disc rotates the cones present surfaces such as are indicated in *Figures C55c* and *d* thus giving alternate fields of view first from one source and then from the other.

When photometric balance is finally achieved

$$\left\{\frac{\text{Intensity due to the 1st source}}{\text{Intensity due to 2nd source}}\right\} = \frac{d_2^{\,2}}{d_1^{\,2}}$$

where d_1 = distance of 1st source from photometer head
       d_2 = distance of 2nd source from photometer head.

Hence if a number of values of d_1 and d_2 are taken a graph of d_2^2 against d_1^2 may be drawn (*Figure C56*) which should be a straight

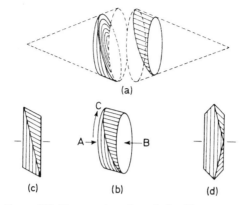

Figure C55. The rotating disc of the Simmance-Abady flicker photometer
(a) Shows the two portions of the photometer disc as sections of truncated cones
(b) Shows the portions together as a composite disc
(c) Shows the disc in position corresponding to (b) as seen in elevation
(d) Shows the composite disc in elevation when the edge corresponding to AB has rotated to C

line whose slope gives the ratio of the intensities of illumination due to the two sources.

Method

Set up both lamps, each giving a different colour, at convenient distances either side of the photometer on a suitable optical bench.

Set the photometer head in motion and note that in general flicker is observed which will be due to differences in both colour and brightness. Slowly increase the speed of rotation of the photometer head until 'colour flicker' is eliminated. It should then be possible to adjust the position of the test lamp to reduce 'brightness flicker' to a small minimum if not to zero. Make several repeat observations with the first lamp in the one fixed position to confirm the position of photometric balance of the second source.

Do this for a number of different positions of the first lamp and plot a graph of d_2^2 against d_1^2.

APPLIED OPTICS, SPECTROSCOPY, PHOTOMETRY

Discussion of Results

Care must be taken not to run the photometer motor too fast since the sensitivity of the eye to brightness-flicker is then seriously impaired.

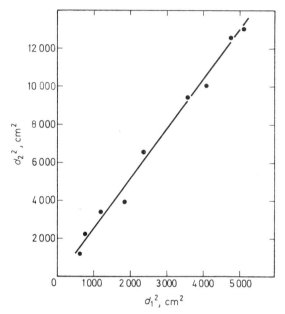

Figure C56. Comparison of intensities of lamps of different colours using the flicker photometer. By reference to the inverse square law the comparison is made in terms of the squares of the distances of the lamps from the photometer
d_1, Distance of the first lamp from the source
d_2, Distance of the second lamp from the source

The sensitivity also varies with the luminosity of the field of view (optimum condition corresponds to about 2·5 lumens ft^{-2}).

A Guild flicker photometer may be used to advantage—it gives a sharp edge to a field of view of optimum 'angular' size ($\sim 2°$) and an illuminated 'surround' of slightly lower intensity from that of the source provides for accurate photometric balance.

Using the Guild photometer it is advisable to obtain balance between the lamps to be compared, say a white source and the sodium lamp, and then to substitute a standard lamp for the latter and readjust for balance.

Whilst the results given here are for 40 and 60 W bulbs used in conjunction with red and green filters it should be borne in mind that the greater the colour difference in the sources employed the less satisfactory will be the photometric balance obtained and the use of a white light as standard and a sodium lamp may be preferable.

If lamps are used with filters, the lamps themselves should be suitably housed so that extraneous light of different colour from that under investigation does not affect the result. The use of irises enabling adjustment of the aperture presented to the photometer to be made is also to be recommended.

To ensure that the speed of rotation is not too high for good sensitivity one may run the motor too rapidly and adjust the test lamp distance to minimize the sensation of brightness flicker, then reduce the speed of rotation until flicker reappears. The test lamp distance should then be adjusted again until the flicker is *minimal*.

Further Work

It is of interest to use the flicker photometer to compare sources of the same colour and to compare its performance with that of a standard white light photometer such as the Lummer–Brodhun.

Reading and References

A number of photometers are described in *Dictionary of Applied Physics* by R. Glazebrook, vol. 4 (Macmillan, 1923). Graphs are given of the effect of speed on sensitivity of flicker.

The Guild flicker photometer and heterochromatic photometry are dealt with in *Principles of Illumination* by H. Cotton (Chapman and Hall, 1960).

Contrast sensitivity, visual acuity and the effect of surrounding glare as well as flicker are dealt with in *The Scientific Basis of Illuminating Engineering* by P. Moon (Dover, 1961).

Experiment No. C3.3

TO INVESTIGATE THE CHARACTERISTIC EXPOSURE CURVE OF PHOTOGRAPHIC PAPER

Introduction

In a previous experiment (No. C2.6) using the Hilger densitometer, variation of density with exposure has been investigated.

APPLIED OPTICS, SPECTROSCOPY, PHOTOMETRY

In this experiment use is made of bromide paper and the characteristic exposure curve is plotted using reflected light under standardized conditions. This enables the 'blackening' curve to be obtained without the use of a micro-densitometer.

Theory

In this experiment to estimate the density we make use of the relation

$$\text{Density } D = \log_{10}(1/R)$$

where R is the reflectance of the treated paper (usually expressed as a percentage) and may be defined as

$$R = \left(\frac{\text{light energy reflected by exposed paper}}{\text{light energy reflected by unexposed (white) paper}}\right).$$

Since both specular and diffuse reflection takes place from the paper surface it is necessary to standardize the conditions under

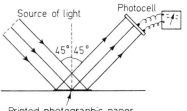

Figure C57. Reflection of light from the exposed photographic bromide paper under standard conditions. Angle of incidence and angle of reflection both 45°

which measurements are made and it is usually convenient to illuminate the paper by parallel light incident at 45° and to measure the reflected light also at 45° as shown in *Figure C57*.

Method

Sheets of bromide paper are cut into squares about 25 mm along each side, and in the photographic darkroom they are exposed in turn to the light from a 6 V bulb with thin white paper before it, placed at a convenient and constant distance from the bromide paper, the exposure varying from 0 to 1 minute in increasing steps of 5 seconds. The exposed sheets are placed in a light-tight box until they are ready to be developed together for the recommended time, according to the instructions given with the particular paper used and then fixed, washed and dried in the usual way.

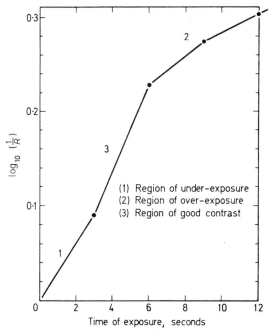

$$\frac{1}{R} = \frac{\text{Amount of light reflected by unexposed paper}}{\text{Amount of light reflected by specimen of exposed paper}}$$

Figure C58. Characteristic exposure curve of bromide paper

Parallel light is arranged to fall onto the sheets in turn at 45°, the reflected light being allowed to fall onto a suitable photocell also at 45° to the paper. The photocell current reading is tabulated and the density calculated. A graph (*Figure C58*) is then plotted of density against exposure time.

Discussion of Results

Remember that in fact relative density has been plotted and that it has been assumed that the unexposed paper is truly white and has a reflection coefficient of 100%. The more dense papers will be difficult to assess in blackness and it is unlikely that readings below 1% or 2% ($R = 1\%$ gives $D = 2$) can be obtained accurately.

Care in the alignment of the light-reflecting system pays dividends in the final result and a mirror in the place later to be occupied by the sheets of photographic paper helps in alignment.

Tube mountings for the lamp and photocell prevent light other than that reflected from the paper from reaching the photocell and it may be helpful to make use of a spectrometer arrangement with a wide slit and the normal eyepiece, replaced by a suitable photocell. Note that the resulting curve shows clearly the regions of (1) under-exposure, (2) over-exposure and (3) the region of good contrast.

Further Work

A recommended developing time will be given for the paper but the effect upon the resulting curve of varying the developing time should be investigated.

Reading and References

Typical characteristic exposure curves for a number of different types of films and papers will be found in the *Kodak Data Book*.

Background reading upon the relation between exposure and optical density, as well as upon the physics of the photographic process will be found in *The Theory of the Photographic Process* by C. E. K. Mees and T. H. James (Macmillan, 3rd edition, 1966).

Section D. The Physics of Engineering Materials

Experiment No. D1.1

TO OBSERVE THE PLASTIC DEFORMATION PROCESSES IN LITHIUM FLUORIDE CRYSTALS

Introduction

The theory of dislocations relates the atomic structure of crystals to their elastic (or more strictly their plastic) properties. Many crystals are brittle and display little or no plasticity, but others, in particular metals and alloys, are ductile and show plasticity to a marked degree. It is, in fact, this property that allows them to be used in many engineering projects without fracture. Lithium fluoride is an optically transparent material of simple structure which is readily cleaved into convenient rectangular form and which lends itself to investigation of internal defects within it. It is also an example of a crystal that can undergo plastic deformation and is therefore valuable for studying defect mechanisms.

The aim of this experiment is to examine the deformation processes which occur in the lithium fluoride.

Theory

One of the most important mechanisms involved in plastic deformation is that where one section of the crystal slips or shears across another as shown in *Figure D1*, giving a 'step' on the crystal surface which on microscope examination appears as a slip line (*Plate D1*).

The slip is really the movement of an imperfection in the crystal and this particular defect is known as a dislocation. It is these dislocations and their movements which to a large extent determine the elastic properties and mechanical strength of the material as a whole.

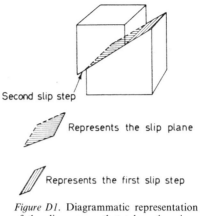

Figure D1. Diagrammatic representation of the slip process that takes place in a crystal of lithium fluoride under stress

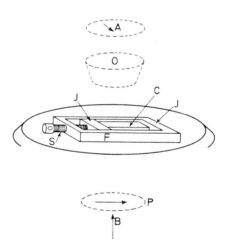

Figure D2. Diagram of stressed lithium fluoride crystal on the stage of the polarizing microscope

A, Microscope analyser
P, Microscope polarizer
(Note that the axes of polarization lie parallel to the edges of the crystal)
B, Transmitted light beam
F, Frame of the jig
C, Lithium fluoride crystal
O, Microscope objective
J, Jaws compressing the specimen
S, Screw providing means of stressing the specimen

Method

The presence of strain within the crystal may be detected by placing the lithium fluoride on the stage of a polarizing microscope and observing the crystal in polarized light with the polarizer and analyser 'crossed'. If completely free from strain the specimen appears dark whatever its orientation on the stage. The presence of strain results in a coloured pattern since under these conditions the

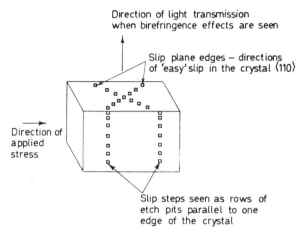

Figure D3. Diagram showing etch pits on intersecting planes and on adjacent faces of the lithium fluoride crystal

material becomes birefringent (see also Experiment No. E2.3) and this is usually apparent near the edges of the crystal where cleavage has taken place.

If the specimen is now stressed slightly in a jig such as shown in *Figure D2* then the birefringence pattern becomes more marked (*Plate D2*) and the slip system becomes apparent. Rotation of the crystal will soon give the optimum orientation for viewing.

To observe the step dislocation the specimen is now etched by immersion in a dilute solution of hydrogen peroxide (20 vols) for several minutes after which it is thoroughly cleaned by immersion in fresh industrial methylated spirits and then ether. (Beware—no naked flames.) Preferential attack takes place at the defect, yielding etch pits which mark the site of the dislocation (*Plate D3*). Note the presence of intersecting slip planes on adjacent faces of the crystal (*Figure D3*).

Discussion of Results

Great care and delicacy should be exercised in handling the crystals. Philatelists' broad-ended tweezers should be used, the tips having short lengths of soft rubber tubing pushed over them. This prevents the introduction of extraneous dislocations into the crystal.

Various dilutions of the etching solution should be tried as well as different etching times in order to discover the optimum conditions to develop clear 'etch pit' patterns.

Further Work

The dislocation patterns created by producing micro-indentations upon the surfaces of the crystals by gentle pressure from a sharp needle should also be studied.

Reading and References

Elasticity, plasticity and dislocation kinetics are treated in *Deformation and Strength of Materials* by P. Feltham (Butterworths, 1966).

The Theory of Crystal Dislocations is treated in some detail by A. H. Cottrell in a monograph of that title (Documents of Modern Physics Series, Blackie, 1964).

Dislocations in Lithium Fluoride by C. W. A. Newey and R. W. Davidge (Metallurgical Services, 1965) gives a more comprehensive account of dislocation theory than is given here; also details of other experiments and pictures of indentation patterns among others. Screw dislocations and mixed defects are discussed.

Deeper treatment of all aspects of the subject may be found in a series of papers by Gilman J. J. and Johnston W. G. as well as in their work *Dislocations and the Mechanical Properties of Crystals* (Wiley, 1957).

Metallurgical Services will also supply cleaved lithium fluoride crystals.

Experiment No. D1.2

TO INVESTIGATE THE CREEP OF METALS

Introduction

Andrade showed that the conventional stress–strain diagram did not fully represent the behaviour of common engineering materials under tension and that deformation can continue to take place at a decreasing rate over long periods, after the application of stress. The phenomenon is referred to as creep and a steady state condition may finally be achieved, after which further deformation is negligible.

In this experiment the materials chosen to illustrate the phenomenon are copper and aluminium in the form of wire. Using these materials only comparatively short time intervals are required to produce curves displaying characteristic features of creep and apparatus of the simplest nature is utilized.

Theory

Andrade represented his results in the form

$$l = l_0 (1 + \beta t^{\frac{1}{3}}) \exp(kt)$$

where l = length of the specimen at time t
l_0 = length of the specimen when the initial sudden extension at loading has ceased
k and β = constants.

The results may thus be divided into two distinct parts, one identified by the β term when $k = 0$

i.e.

$$l = l_0(1 + \beta t^{\frac{1}{3}}) \quad \text{or} \quad \frac{dl}{dt} = \tfrac{1}{3}\beta l_0 t^{-\frac{2}{3}}.$$

This is a condition of transient creep (sometimes referred to as β flow) and is shown as OA in *Figure D5*. The second part is identified by the k term when $\beta = 0$

i.e.

$$l = l_0 \exp(kt)$$

whence

$$\frac{dl}{dt} = kl_0 \exp(kt) = kl$$

and

$$\frac{(dl/l)}{dt} = \text{a constant } k$$

(dl/l) is of course the strain and hence the equation represents a constant rate of strain (region AB in *Figure D5*).

Method

One end of the wire (copper or aluminium is suitable) is clamped in a sonometer type arrangement (*Figure D4*), the other end being passed over a pulley and attached to a weight carrier. A convenient mark is made on the wire (at say 100 cm) and the cross wires of a

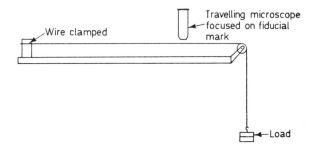

Figure D4. Investigation of creep in metal wires

travelling microscope focused on the mark. A 4 kg weight is placed on the carrier and, using the microscope, the extension noted every half minute for several minutes. A graph of the type shown in *Figure D5* is obtained.

Discussion of Results

Notice that the curve obtained shows the two distinct regions of transient and steady state creep quite clearly. The detailed graph of the β flow region (*Figure D6*) shows only approximate verification of the transient law suggested by Andrade. This is due to the fact that the creep term is 'cubed' and any small errors in measurement are thus highly magnified. This is aggravated by the fact that rapidity of measurement is required (particularly if 15 second intervals are to be used as in the case quoted).

Further Work

Copper is used in the experiment illustrated, but as suggested both copper and aluminium should be investigated.

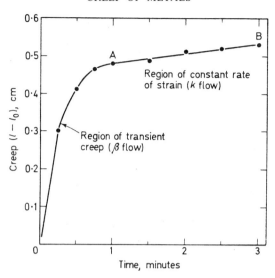

Figure D5. The creep curve for a sample of copper wire

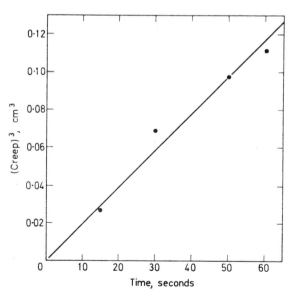

Figure D6. Graph of (creep)^3 against time to investigate the region of transient creep

The effect upon the stress–strain curves of first loading and then removing the weights at intervals of one minute should be noted as

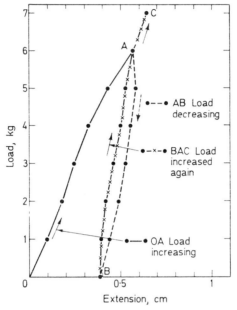

Figure D7. Stress–strain curve for aluminium wire showing the effect of loading (OA region), decreasing load (AB region) and re-loading (BAC region)

well as the effect of finally loading the wires until they snap. Compare and contrast the resulting curves, and note particularly how the amount of creep rapidly decreases due to 'work hardening'. Such a curve for aluminium is shown in *Figure D7*.

Reading and References

An excellent monograph on *Creep of Metals* by L. A. Rotherham is published by the Institute of Physics (1950). It gives an extensive bibliography and should provide ample ideas for further work as well as sound background reading.

EQUILIBRIUM DIAGRAMS OF LEAD–ANTIMONY ALLOYS D1.3

Experiment No. D1.3

TO CONSTRUCT THE EQUILIBRIUM DIAGRAM APPROPRIATE TO A SERIES OF LEAD–ANTIMONY ALLOYS OF VARYING COMPOSITION

Introduction

The equilibrium diagram of an alloy, as its name suggests, gives a pictorial representation of the condition of the particular alloy when in a state of equilibrium. It is often known as a phase diagram since the mathematical foundations of its interpretation are based on the phase rule. For our purpose here it is sufficient to say that in the phase diagram of a binary alloy the two descending portions of the liquidus curve (see also Experiment No. D1.5) meet at a point known as the eutectic point.

Theory

Figure D8 shows what happens in general when an alloy cools from the liquid state. The initial cooling may be represented by a

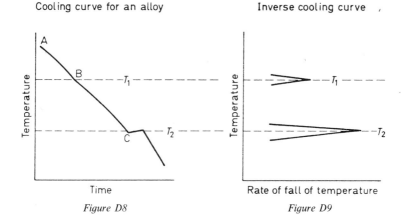

Cooling curve for an alloy   Inverse cooling curve

Figure D8   Figure D9

smooth curve (region AB) until a particular temperature (T_1 on the graph) is reached, when a discontinuity occurs corresponding to the fact that one component of the alloy begins to separate out from the alloy. The second smooth curve region BC occurs when the

299

separation of the one particular component is complete and the 'eutectic' point is reached. The 'eutectic' substance then solidifies giving up its latent heat of fusion and yielding another discontinuity on the curve. If sufficient alloys, composed of (say) the same two elements in different proportions, are taken the equilibrium diagram may be drawn as shown in *Figure D15*. Lead–antimony alloys are chosen here.

Method

A standard form of potentiometer is used and the circuit set up as shown in *Figure D10* with the cold junction of a chromel–alumel

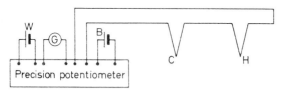

Figure D10. The equilibrium diagram of lead–antimony alloys potentiometer circuit
W, Weston standard cell   G, Galvanometer
C, Cold junction of thermocouple in ice and water
H, Hot junction of thermocouple in the alloy
B, Two volt accumulator

thermocouple in a mixture of ice and salt at $0°$ C and the hot junction in the liquid lead or lead–antimony alloy, contained in a crucible. The crucible itself may be conveniently housed in a fireclay furnace brick above a Meker burner.

First standardize the potentiometer in accordance with the operating instructions of the particular model being used.

Meanwhile, heat the crucible containing pure lead stirring the metal when it melts (the sheath of the hot-junction thermocouple itself may be used).

It is more convenient to read thermocouple voltages rather than actual temperatures and the thermocouple dial should be set to a predetermined value and a stop-clock started when the detector denotes that 'balance' has been achieved.

Immediately, the thermocouple setting is set at the next lower conveniently small interval (e.g. 1 mV on the graph of *Figure D11*) and the time is noted as the new balance point is passed.

This procedure is repeated until solidification is complete and for several points beyond this.

It is now found more convenient to plot the inverse cooling curve, i.e. the temperature against the rate of fall of temperature (*Figure D9*). An actual graph for lead is shown in *Figure D11*.

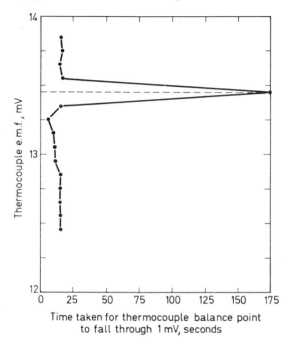

Figure D11. Inverse cooling curve for unalloyed lead

The rate of fall of temperature (or more strictly the time taken for a fall of 1 mV on the thermocouple) is obtained simply by subtracting the first reading of time from the second, the second from the third and so on.

The readings are repeated for alloys of lead with antimony and *Figures D12, D13* and *D14* show results obtained with

100% lead– 0% antimony
95% lead– 5% antimony
90% lead–10% antimony
60% lead–40% antimony

The various values of electromotive force at temperatures corresponding to the sharp discontinuities are tabulated and from the thermocouple calibration the actual temperatures obtained.

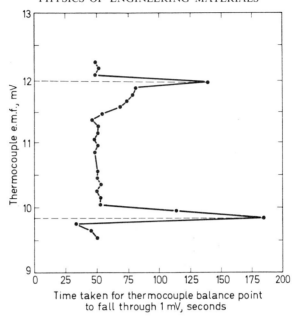

Figure D12. Inverse cooling curve for 95% lead–5% antimony alloy

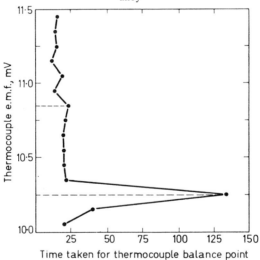

Figure D13. Inverse cooling curve for 90% lead–10% antimony alloy

EQUILIBRIUM DIAGRAMS OF LEAD-ANTIMONY ALLOYS D1.3

These are plotted as abscissae against percentage of antimony in the alloy. The melting point of antimony (corresponding to 100%) may be taken from tables.

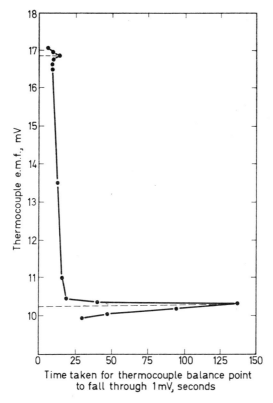

Figure D14. Inverse cooling curve for 60% lead–40% antimony alloy

The features of the equilibrium diagram then become apparent (*Figure D15*).

Discussion of Results

This elementary method provides a reasonably accurate means of determination of the form of the equilibrium diagram for simple alloy systems of the lead–antimony (or lead–tin) type. For complex alloy systems more elaborate laboratory procedures are required.

Lack of homogeneity throughout the alloy and a variation in temperature in different parts of it gives rise to experimental errors here as do supercooling in the liquid phase and the existence of a time lag between the temperature change occurring within the alloy and that registered by the thermocouple.

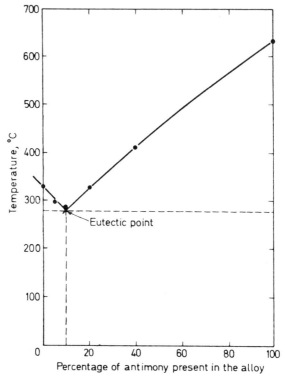

Figure D15. The equilibrium diagram for a lead–antimony alloy constructed from the inverse cooling curves

Further Work

Students should observe under the microscope mounted specimens of lead–antimony alloys of the compositions used above.

Reading and References

Deeper background information may be found in *The Phase Rule* by T. K. Jones and F. D. Ferguson (Butterworths, 1966).

The eutectic and metallurgical equilibrium diagrams of a number of alloy systems are shown and discussed in the *Encyclopaedic Dictionary of Physics* (Pergamon, 1961).

Experiment No. D1.4

TO INVESTIGATE THE EFFECT OF VARIATION IN CARBON CONTENT UPON THE TENSILE PROPERTIES OF STEEL

Introduction

It is found that variation of the carbon content of steel within the range from 0·1 to 1% has significant effect upon the tensile properties of the steel and the aim of this experiment is to investigate how the breaking stress, ultimate tensile stress and the yield stress are affected.

Theory

We may define (*a*) the ultimate tensile stress as the ratio of the maximum load applied to the original cross sectional area of the specimen; (*b*) the true breaking stress as the ratio of final value of the load applied to the final resulting cross sectional area; (*c*) the yield stress as the ratio of the load at the lower yield point to the original area.

The stress in this experiment may be applied using the Hounsfield tensometer and some models may differ in slight detail from the one described here. However, it will operate on the following general principles.

The specimen is held by pins between the load screw and the beam (*Figure D16*). Load is applied by rotation of the straining handle shown. The drum upon which graph paper is wrapped is geared to the 'load' screw so that the rotation of the drum is proportional to the movement of the screw.

Application of load causes bending of the beam and resultant movement of the piston connected to it. This forces mercury into the capillary tube beside the scale. Since the deflection of the beam is proportional to the load applied, the length of the mercury column in the tube is a direct measure of the magnitude of the load and may be read off in appropriate units, after first adjusting the 'zero' of the scale to be in line with the mercury meniscus when no load is applied.

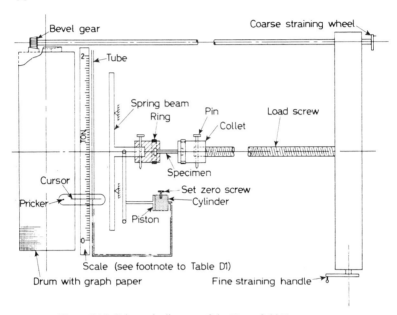

Figure D16. Schematic diagram of the Hounsfield Tensometer

To take account of the deflection of the beam itself a thick bar is first placed in the tensometer. It is assumed that the bar is so thick that no appreciable strain is produced within it and an extension/load graph corresponding to the beam deflection is plotted (*Figure D17*).

Method

Insert the graph paper in the clips of the recording drum and insert the 'no strain' bar between the beam and the load screw. Depress the stylus control and record the line due to the bending of the beam alone as load is applied.

Remove the thick bar and insert the steel specimen of low carbon content. Take measurements of the diameter of the specimen before applying the load. (If a standard form of specimen is used then the tensometer usually has an attachment for measuring area and reduction in area directly.)

Apply the load and record the graph of extension against load for the specimen, noting the changes that take place in the specimen until it finally breaks (*Figure D18*).

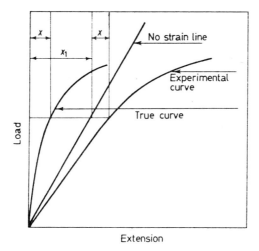

Figure D17

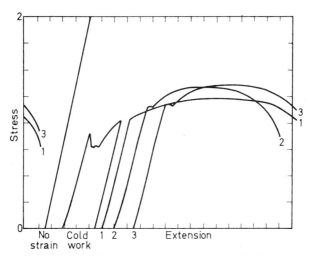

(Actual values depend upon chosen gear-ratio between the drum and driving mechanism of the tensometer)

Figure D18

Plot the 'corrected' graph to make allowance for the bending of the beam as in *Figure D19*.
Examine the fracture carefully noting its nature (*Figure D20*).
Repeat for several other specimens of increasing carbon content. Calculate the yield stress, the ultimate tensile stress and the true breaking stress and tabulate the results as in *Table D1*. Plot graphs of variation of yield stress, ultimate tensile stress and true breaking stress with carbon content (*Figure D21*).

Table D1

Specimen No.	Carbon content %	Percentage elongation	Percentage reduction in area	Yield stress* tons inch^{-2}	MN m^{-2}
1	0·1	42	70	15·2	234·7488
2					
3	0·18	32	57	21	324·324
4	0·34	30	53	20·8	321·2352
5	0·54	28	47	21·2	327·4128
6	0·81	22	40	28	432·432
7	1·10	12	18	—	—

Specimen No.	Ultimate tensile strength* tons inch^{-2}	MN m^{-2}	True breaking stress* tons inch^{-2}	MN m^{-2}
1	25·2	389·1888	57·3	884·9412
2	—	—	—	—
3	36	555·984	56·7	875·6748
4	36	555·984	63	972·972
5	41·3	637·8372	50	772·200
6	46·4	716·6016	69	1065·636
7	56·8	877·2192	65·5	1011·582

* The figures for yield stress, ultimate tensile strength and true breaking stress are shown converted to metric units (basis: 1 ton inch^{-2} = 15·444 MN m^{-2}), but in practice it may not be possible to measure the stress to such accuracy as four or even three decimal places. Probably the greatest accuracy attainable will be to the nearest whole number.

Discussion of Results

Examination of the fractures should show that with the lower percentages of carbon content 'necking' occurs in the specimen and a jagged 'cup and cone' fracture results (*Figure D20*). As the carbon content grows higher, however, the break becomes cleaner and exhibits stronger evidence of crystalline properties.

The graph showing yield stress against carbon content should show that the stress increases at first with increase in carbon content,

TENSILE PROPERTIES OF STEEL D1.4

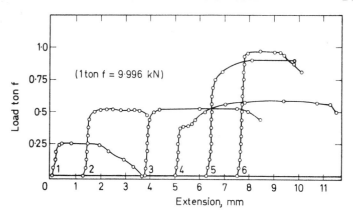

Figure D19. True load-extension curves

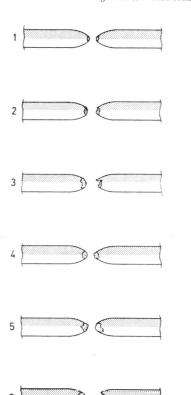

Figure D20. Different types of fracture in test pieces (see *Table D1*)

PHYSICS OF ENGINEERING MATERIALS

but beyond a certain point further increase in carbon content produces but little increase in yield stress.

The graph of ultimate tensile stress shows an increase with carbon content and indicates that addition of further carbon beyond 1·1% would produce further increase in ultimate tensile stress.

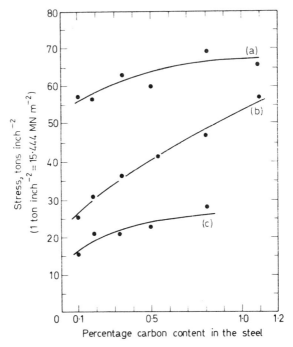

Figure D21. Graphs showing variation of (a) breaking stress, (b) ultimate tensile stress and (c) yield stress with change in carbon content

The graph of breaking stress against percentage carbon content indicates again that the breaking stress tends to increase with increase in carbon content and that little further increase would result from further increase in the carbon content of the steel.

It may be found that the points of the graph of breaking stress against the percentage content of carbon, show considerable scatter. This may be due to one or more of several factors since the breaking stress is critically dependent upon the rate at which the load is applied and upon the presence of impurities within the specimen.

One conclusion which may be drawn from the results is that the strength of brittle material is best defined in terms of the ultimate tensile stress rather than the other two parameters.

Further Work

Could include measure of hardness and of impact properties of the steels with variation in carbon content.

Reading and References

An elementary account of yield stress, ultimate tensile stress and breaking stress may be found in *Metallurgy for Engineers* by R. C. Rollason (Arnold, 3rd edition, 1961).

Typical data for steels of the above type can be found in Kempe's *Engineers Year Book, 1959* and later issues (published by Morgan Bros.).

Considerable variation may be found in the results obtained from the published values owing to unknown factors in the previous history of the specimens.

Experiment No. D1.5

TO ILLUSTRATE THE ZONE REFINING PROCESS USING NAPHTHALENE

Introduction

The zone refining technique has achieved primary importance in the production of germanium and silicon of extremely high purity for use in semiconductor devices and research.

The aim of this experiment is to illustrate the process by purifying naphthalene which has been contaminated with a small quantity of blue dye (Duranol brilliant).

Theory

If one substance is found to be soluble in another, both in the liquid and the solid states, then the system may be represented in the manner of *Figure D22* and in the region above the upper line (the liquidus ALB) all is liquid; below the lower line all is solid solution, whilst between the two lines we have liquid and solid existing in equilibrium.

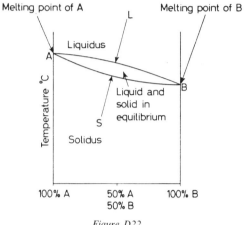

Figure D22

Considering a mixture containing $x\%$ of substance A and $(100 - x)\%$ of substance B cooling from above the liquidus to the liquidus at $T_1\,°C$ (point x) we see that a corresponding point y lies on the solidus at T_1 and thus the solid separating out at this temperature will have a higher percentage of substance A in the mixture (*Figure D23*).

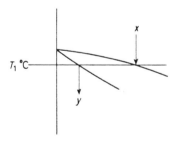

Figure D23

This sort of effect occurs when a molten zone is caused to traverse a nearly pure substance, the separating solid being very pure and the impurities remaining in the traversing zone of liquid.

Method

Plate D4 shows the apparatus used which has a low voltage geared motor allowing the tube A to pass *slowly* through the loop (B) of nickel chrome wire (see also *Figure D24*). A rheostat in series with the wire allows the current to be adjusted until the molten zone occupies about 5 mm as it slowly traverses the tube.

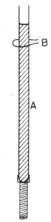

Figure D24

Take a small quantity of naphthalene and mix it with the recommended blue dye to give a fairly strong coloration. Melt this into a larger quantity of pure naphthalene in a test tube to give a slight but even coloration. Fill the specimen tube taking care to eliminate air bubbles. Place the tube in position with the heater coil about it and adjust the motor speed and the heater current to obtain a 5 mm molten zone. Traverse the full length of the tube. Repeat if necessary.

Observe the tube contents carefully (*a*) in daylight (*b*) in ultra-violet light.

Discussion of Results

A typical result is shown in *Figure D25*.

A region of refined naphthalene which appears white in daylight and brilliantly white under the ultra-violet lamp exists after passage through the heated zone. Immediately below this is a small region of dark blue and then a sharp ring of black, whilst any untraversed region will display the original light blue.

The dark blue region represents the concentration of the contaminant dye by the zone refining process and the black region the tar which is often removed from laboratory naphthalene.

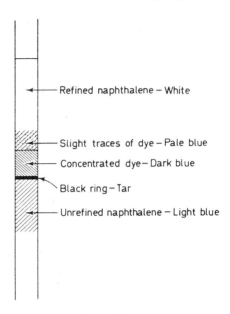

Figure D25

Conclusions

Before summarizing your conclusions you should

(*a*) try several 'runs' on the same specimen,
(*b*) vary the quantity of dye to help assess the efficiency of the process.

Further Reading and References

There is a variation upon this experiment (using a radioactive tracer as the indicator) given by S. Z. Mikhail in the *American Journal of Physics*, vol. 33, No. 5, p. 399, May 1965.

This is a more sophisticated version and graphs are plotted showing distribution of impurities for passage of many zones and zone lengths.

Experiment No. D1.6

TO INVESTIGATE THE CHARACTERISTICS OF AN ELECTROLYTIC POLISHING CELL USING A COPPER OR BRASS SPECIMEN

Introduction

Mechanical methods of polishing metal surfaces suffer from the disadvantage that they produce strain at the surface of the polished specimen (particularly in soft metals).

For many applications this is undesirable and may be obviated by use of the electrolytic polishing method.

This technique has seen wide application in the study of magnetic domains where it is important to obtain a pattern corresponding to the character of the *bulk* material (see *Plate D5*) and not to that of the Beilby layer upon the surface, which displays high strain induced anisotropy. It has also been used in polishing the copper substrates used in magnetic thin-film research.

Theory

Any minute projections or irregularities, such as scratches due to grinding and mechanical polishing, suffer preferential attack by the specimen being made the anode of a suitable electrolytic bath and a

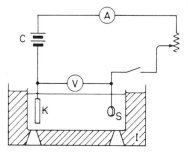

Figure D26. Circuit used for the electrolytic polishing of copper
A, Ammeter (0–1 A) V, Voltmeter (0–25 V)
S, Specimen acting as anode K, Cathode of stainless steel
I, Iced water bath for cooling C, 20 V d.c. supply

typical arrangement is shown in *Figure D26* where the specimen is made the anode of an electrolytic bath after being first polished mechanically using a suitable series of fine emery papers.

Method

The circuit is assembled as shown in *Figure D26*, the stainless steel cathode being of much greater size than the anode specimen to be polished. The electrolyte in this case consists of 1 part of concentrated nitric acid with 2 parts by volume of absolute methyl alcohol.

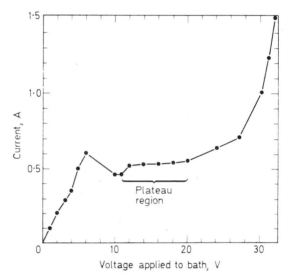

Figure D27. Voltage-current characteristic obtained using the electrolytic polishing technique with a copper specimen

A cold water bath should surround the cell and the surface to be polished should be horizontal about 12 mm below the cathode. In some commercial forms of electrolytic polishing equipment (e.g. the Shandon) the specimen is held in position by a spring-loaded pointer and a pump is used to force the liquid up to the face of the specimen.

Voltage is supplied across the bath from a d.c. supply and is increased in SMALL steps from zero to about 20 V. Note the current at each point and plot a graph of the current against the voltage as in *Figure D27*. A plateau region should be apparent in the graph and the beginning of the plateau represents the correct point to give the ideal electrolytic polishing conditions of the particular cell. The specimen used to obtain the characteristic will be heavily pitted and a similar surface is now used to investigate the polishing conditions further. Using voltages of the order indicated by the graph (i.e. 11 V) and this particular electrolyte, the polishing time will be very short and times of the order of 10–60 seconds should be tried.

Before switching off the current the polished specimen is removed and washed immediately in water, then in industrial methylated spirit and finally dried in the warm draught from a hair dryer.

Discussion of Results

A copper specimen is perhaps the easiest specimen to try first, since it is homogeneous. If brass is chosen 60/40 brass should be used, since this is of homogeneous nature, whereas 70/30 contains two phases and the complication of preferential areas of attack in the specimen arises.

Further Work

A paper by G. J. Williams in the *Journal of Scientific Instruments* vol. 42, 1965, p. 170, describes a simply constructed electrolytic polishing bath for use with transformer-type steel specimens.

Reading and References

Further details of the electrolytic polishing method may be found in *Photomicrography with the Vickers Projection Microscope* published by Cooke Troughton and Simms, Ltd. (1956). It contains a full table of data (including an alternative electrolyte for brass giving much longer polishing time) relevant to most of the commonly used metals and alloys and a very useful appendix of references.

It cannot be too strongly emphasized that if other specimens are chosen which require perchloric acid in the electrolyte, the experiment should be done only under strict supervision since danger of explosion exists, particularly in the presence of organic substances.

PHYSICS OF ENGINEERING MATERIALS

Experiment No. D1.7

PREPARATION OF MAGNETIC THIN FILMS BY ELECTRODEPOSITION AND INVESTIGATION OF THE PERMEABILITY OF THE THIN FILM

Introduction

Magnetic thin films have become of primary importance in recent years particularly with regard to their use as memory storage elements in computers.

Fundamental research upon them has shown that magnetization within thin films often differs markedly from that of the bulk material and the aim here is to prepare a thin film and to obtain a measure of its permeability.

Theory

The film chosen is Permalloy since its permeability is greater than that of iron.

The substrate is of copper in the form of a ring about 25 mm diameter, 6·5 mm deep and 1·6 mm thick. These rings may be cut conveniently from standard 1 inch copper piping.

They must first be electrolytically polished (see Experiment No. D1.6) in order that the magnetic thin film shall properly adhere to them.

The permalloy film is then deposited upon the ring to the desired thickness. Its permeability is found using the familiar 'anchor ring' method, but with alternating current applied (*Figure D30*).

Method

The preliminary electrolytic polishing is done using an electrolyte of methanol and nitric acid in the proportions of 70 to 30 by volume.

Great care must be taken in making up the electrolyte and it is important that the ACID is slowly added to the methanol (NOT the other way about) whilst the latter is surrounded by a *cooling bath of ice*.

About 200 ml of the electrolyte are poured into a beaker and a large copper cathode is immersed in it. The specimen to be polished is suspended in the liquid from a suitable 'pinch' clip and again before passing current into the cell it is surrounded by an ice bath (*Figure D28*). A thermometer should be used in the bath and at all times throughout the experiment the temperature should be kept well below 30° C.

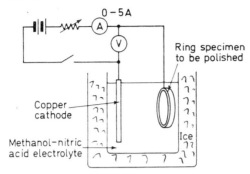

Figure D28. Apparatus for the electrolytic polishing of copper. Optimum current-density for polishing $\sim 0{\cdot}8$ A cm^{-2} of specimen surface area

Only by trial will the optimum condition for polishing be established (see also Experiment No. D1.6) but the oxidized surface film will clearly be seen to fall away from the ring when this condition is reached. If small areas of oxidation still remain on the specimen on the inner side beneath the clip, rotate the ring through the clip and continue polishing for a short time. (Typical polishing currents for

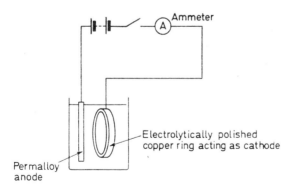

Figure D29. Circuit for the plating of a thin ferromagnetic film of permalloy on copper. Current passed $\simeq 40$ mA cm^{-2} of specimen surface area. Time of passing 15 seconds, for plating 10^{-7} m thick

such specimens in the conditions outlined above are of 1·8 A for 2 minutes with 5 V across the bath.)

After polishing, the specimen is rinsed in distilled water and immersed in methanol until ready for plating.

To plate it with Permalloy the ring becomes the anode of a second bath of electrolyte (*Figure D29*).

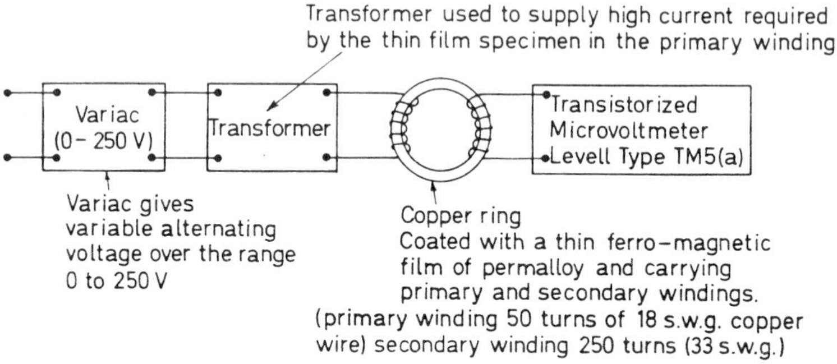

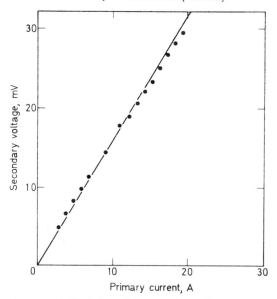

Figure D30. Block diagram of the apparatus used in the measurement of the permeability of a thin film of permalloy

Figure D31. Variation of the secondary voltage induced in the winding (over a 5000 Å thick permalloy thin film) against the current in the primary winding

This electrolyte solution is made up of 12 g of sodium chloride, 6 g of ferrous ammonium sulphate, 218 g of nickel sulphate and 25 g of boric acid dissolved in distilled water and made up to 1 litre.

The anode is a large plate of Permalloy or mu-metal and a current of 40 mA is passed for 5 minutes.

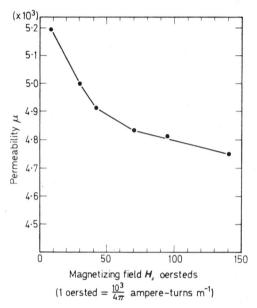

Figure D32. Variation in the permeability of a thin film of permalloy with change in magnetizing field

The mass of permalloy deposited is calculated using Faraday's law ($m = izt$) where m = mass, i = current, t = time and z = electrochemical equivalent, and knowing the surface area of the ring and the density of the permalloy ($\simeq 8$ g cm^{-3}) the thickness of the deposit may be calculated.

The specimen is removed from the bath, washed in distilled water and then in industrial methylated spirits and finally thoroughly dried in a current of warm air. It is then dipped into shellac and allowed to dry. A second coat of shellac is applied and the ring is wound with about 50 turns of heavy gauge double cotton covered wire (18 s.w.g.) in an even winding over the toroid. Dip in shellac once again and when dry, wind on 150 to 200 turns of fine insulated copper wire as the secondary winding (33 s.w.g.).

Then using a suitable a.c. source and circuit as shown in *Figure D30*, secondary voltage readings corresponding to 0 to 30 A, flowing in the primary are tabulated (*Figure D31*). From these **B** and **H** may be calculated and $\mu = B/H$ can be plotted for various values of **H** giving a graph such as shown in *Figure D32*.

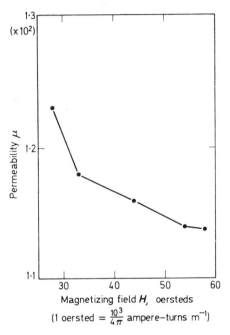

Figure D33. Graph showing the variation of the permeability of a thin film of iron with change in magnetizing field

Discussion of Results

Unless sufficient turns are wound on the ring it will be difficult to obtain accurate readings at low values of magnetic field **H**.

A grave error may be introduced in estimating the thickness of the film deposited since the system has been assumed to be 100% efficient.

One factor neglected here is that of the strain within the film as deposited. This can have a marked effect upon the value of permeability obtained.

Further Work

Films of iron may also be used for investigation and in fact *Figure D33* shows the graph obtained with such a film. Owing to the fact that it has a much lower permeability than permalloy, many more turns will be required on the secondary winding in order to obtain values of μ at low magnetic field. Otherwise interesting facets of the graph are omitted.

Reading and References

A detailed account of all aspects of electrolytic polishing will be found in *The Electrolytic and Chemical Polishing of Metals in Research and Industry* by J. J. McG. Tegart (Pergamon, 2nd edition, 1959), and *Electrodeposition of Alloys* by A. Brenner (Academic Press, 1963).

A paper by G. A. Jones, D. P. Oxley and R. S. Tebble, *Philosophical Magazine*, vol. 11, p. 993, May 1965, gives detail of the more theoretical aspects of the electrodeposition of thin films.

The distinction between permeability as measured using conventional d.c. methods and the 'effective' a.c. permeability is dealt with in *The Physics of Electricity and Magnetism* by W. T. Scott (Wiley, 2nd edition, 1966).

Experiment No. D1.8

TO PREPARE A SERIES OF NICKEL–COPPER ALLOYS AND TO VERIFY THAT THE CURIE POINT OF THE ALLOYS VARIES LINEARLY WITH PERCENTAGE NICKEL CONTENT

Introduction

Since copper and nickel are adjacent in the periodic table of elements, are electrochemically similar and have atoms of nearly the same size, they form a continuous series of solid solutions. The aim here is to produce a series of such solutions and to verify that a linear relationship exists between the Curie point of the alloy and the increasing percentage of nickel in the alloy. In order to determine the Curie temperature the oscillation method of Experiment No. D2.3 is used.

Theory

It is important to obtain homogeneous alloys and to prevent oxidation occurring in the specimens whilst at the same time it is necessary to control the percentage composition.

To do this finely powdered copper and nickel should be well mixed in appropriate proportions and then sintered.

Method

Carefully weighed proportions of finely powdered nickel and copper are well mixed together in a mortar, to produce a series of alloy specimens containing 70, 75, 80, 85, 90 and 95% by weight of nickel.

The mixtures are placed in a combustion boat with suitable ceramic spacers so that the resulting specimens will be about 25 mm long and 3 mm in diameter. (They may not necessarily be truly cylindrical in shape of course.)

Each specimen is labelled by a pencilled number beside it and the boat placed in the furnace tube (*Figure D34*).

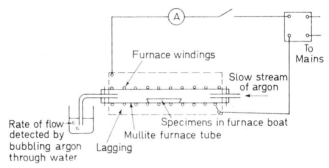

Figure D34. Diagram showing apparatus used in the preparation of nickel–copper alloys

Argon is passed in a slow steady stream over the boat so that oxidation during heating of the specimen (up to about $1\,000°$ C) may be prevented.

The rate of passage of argon may be adjusted by allowing the argon to bubble, before exit, through a wash bottle of water and once air has been flushed from the apparatus the slowest rate possible may be used. It is advisable to 'preheat' the argon by passing it through a heated silica or mullite tube before entering the furnace so that the surface temperature of the alloy is not reduced.

The specimens should remain in the furnace for about 30 hours and then be allowed to cool very slowly, after which time they may be used to determine the appropriate Curie point using the oscillation method of Experiment No. D2.3.

Discussion of Results

As the Curie temperature/composition graph of *Figure D35* shows individual points can deviate markedly from the 'best fit' straight line. This is mainly due to the fact that errors of the order of plus or minus 20° C can occur in the determination of the Curie point by extrapolation, particularly for the alloys having high nickel content.

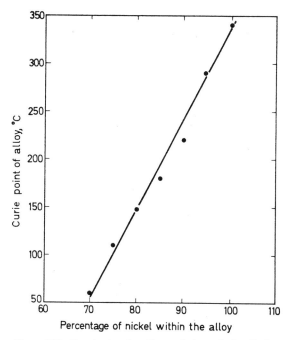

Figure D35. Graph showing the variation of the Curie point of a strongly magnetic series of nickel-copper alloys with variation in the proportion of nickel in the alloy

Other contributory factors are lack of homogeneity in the specimen and deviation in the temperatures at the point where the specimen is suspended, from those given by the thermocouple (see Experiment No. D2.3).

Care should be taken to ensure that the specimen has been 'fired' for sufficient length of time, otherwise the Curie points tend to be higher than they should be.

Further Work

It is interesting to investigate the effect of different firing times upon the Curie point of the specimens. If they are heated in the furnace for periods from 2 to 30 hours the Curie temperature is found to decrease at first rapidly and then more slowly until a virtually constant value of Curie temperature is achieved (*Figure D36*).

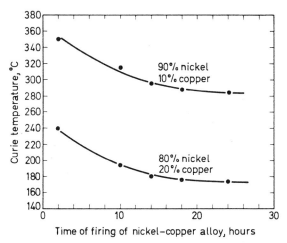

Figure D36. Graph showing the way in which the Curie point of a nickel–copper alloy varies with the 'firing' time

The important Heusler alloys may be prepared and investigated in similar fashion. With such a specimen (composed of 8 grams of manganese and 2 grams of bismuth) it is further possible, by mounting in perspex and grinding and polishing the exposed surface, to observe domain structure within the alloy.

Reading and References

The nickel–copper system of alloys, as well as many others, are discussed in relation to Curie point variation in *Magnetochemistry* by P. W. Selwood (Interscience, 1956).

Data on nickel–copper alloys are given in *Constitution of Binary Alloys* by M. Hansen (McGraw-Hill, 1958).

Experiment No. D1.9

TO GROW SINGLE CRYSTALS OF CADMIUM FROM THE MELT AND TO CONFIRM THE ORIENTATION OF THE MODE OF GROWTH

Introduction

The technique of growing single crystals from the vapour phase is dealt with in Experiment No. D2.1.

The aim of this experiment is to illustrate another mode of growth, viz. from the molten polycrystalline material and to investigate the orientation of the mode of growth.

Theory

The method used here is that of zone melting (see first reference) and can be compared with the zone refining technique of Experiment No. D1.5.

A furnace (*Figure D37*) is allowed to traverse the polycrystalline specimen contained in a tube or boat.

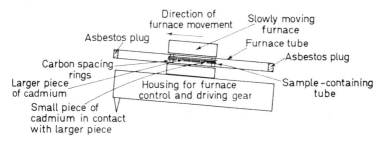

Figure D37. Preparation of cadmium single crystals from the melt

The single crystal obtained will have one of two types of orientation:

(1) where the specimen is very ductile, and
(2) where the specimen is difficult to strain along its length.

The first type is often referred to as the 'soft' or ductile mode of orientation and the second as the 'hard' or brittle mode.

Method

Cadmium in the form of polycrystalline rod is chosen for the starting material because of its low melting point (approximately 320° C). Two pieces, one about 25 mm long, and the other 75 mm long, are placed in a fused silica tube of uniform 3 mm bore and mounted in the furnace tube (*Figure D37*).

The furnace is allowed to reach equilibrium at a temperature of about 60° C above that of the melting point of the material, whilst in position about both pieces of metal.

Make sure that the short piece makes good contact with the larger one and that both pieces of metal lie just within the furnace.

When the larger piece has melted (it should take about 15 to 30 min) switch on the motor, which should allow the furnace to slowly traverse the specimen at a rate of just under 1 mm per minute.

Switch off the motor when the larger piece of cadmium has emerged clear of the furnace, then switch off the furnace and allow the whole apparatus to cool to room temperature.

Remove the cadmium from the furnace tube and gently tap the containing tube to remove the larger piece of material. Handle the

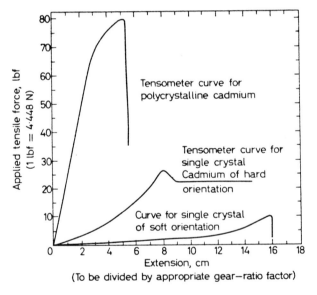

Figure D38. Tensometer curves for polycrystalline and single crystal cadmium

specimen as little as is necessary being most careful not to deform it at this stage in the experiment.

Immerse the crystal in concentrated nitric acid and allow the liquid to boil in order to remove the brown oxide scale which has been formed on the crystal. Pour off the acid and wash the crystal copiously with water. Then dip in ethyl alcohol and dry it in the warm draught from a dryer.

Check the mode of crystal growth by cutting off a small portion using tin snips and pulling gently on either end. If the crystal is ductile and extends easily 'soft' orientation has been obtained, but if it displays audible clicks and resists extension, hard orientation has been achieved (*Figure D38*).

Discussion of Results

Etch the crystal in a dilute solution of hydrochloric acid to determine whether the resulting material is a single crystal. For comparison a piece of the polycrystalline material should be etched and observed under the microscope. The polycrystalline material will display well defined grain boundaries which will be absent in the case of a single crystal specimen.

Further Work

Zinc and aluminium are other suitable materials for use with this technique of single crystal growth.

Higher temperatures are required and it is advisable to use fused silica tubes to contain the sample and a high temperature refractory material for the furnace tube.

It is advantageous too, to pass an inert gas (argon) over the specimen in the furnace in order to prevent extensive oxidation.

The 'hot' region of the furnace should be kept as narrow as is conveniently possible. Hounsfield type tests on the material can be made to investigate the tensile strength of the single crystal and the direction of the slip planes. The orientation might be checked by x-ray technique.

Other work including the process of changing the orientation of crystal growth is dealt with in clear detail in a little instruction booklet provided with the 'Metallurgical Services Single Crystal Unit' which can be obtained commercially.

Other types of furnaces are described in the second and third references.

Reading and References

Preparation of Single Crystals by W. D. Lawson and S. Nielsen, p. 18 (Butterworths, 1958).
'A simple furnace for growing metal crystals' by G. T. Clayton and C. Hendrickson, *American Journal of Physics*, vol. 32, No. 9, Sept. 1964, 679.
Modern Physical Metallurgy by R. E. Smallman (Butterworths, 2nd edition, 1963).
An Introduction to X-ray Metallography by A. Taylor, p. 253, 'Single crystals, their growth and properties' (Chapman and Hall, 1956).

Experiment No. D1.10

TO INVESTIGATE THE STRAIN-ANNEAL METHOD OF SINGLE CRYSTAL GROWTH USING ALUMINIUM

Introduction

This method depends upon the growth of recrystallized nuclei at the expense of other nuclei within a slightly deformed matrix, the impetus for growth being provided by the strain energy released within the specimen on recrystallization.

The method cannot be used for all types of specimens, for example it is not applicable to those metals that form annealing twins, such as copper and brass, but it has advantages in that the problems of separating out the single crystal material (as is necessary with the technique of growing from the melt, see Experiment No. D1.9) is avoided and control of conditions to minimize contamination can yield a final specimen largely free from defects.

Theory

The initial specimens should have as fine and uniform a grain size as possible (Region A, *Figure D39* and *Plate D6*). If necessary the specimen should first be 'cold worked' and then annealed for a short time to produce the desired recrystallization. The aluminium is then subjected to strains of the order of $\frac{1}{2}$ to 4% and after this, annealed slowly to reduce to a minimum the number of nuclei which form on recrystallization. The furnace annealing temperature

STRAIN-ANNEAL CRYSTAL GROWTH D1.10

should be raised very slowly at first but at higher temperatures the rate of rise can be increased more rapidly concluding with a short spell of heating *just below* the melting point of the specimen, when the small grains present, tend to be 'swallowed' in the process of larger grain growth.

Method

The aluminium, of 99·9% purity, should be provided in thin rolled sheet cut to the form of *Figure D39* using a steel stencil and

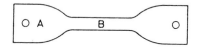

Figure D39. Original form of specimen as cut using the steel template. Note: The region in *Plate D6* corresponding to region A in the above diagram displays the fine grain structure which originally covers the whole specimen

should have as small and uniform a grain size as possible. If doubt exists about these criteria the aluminium should first be cold worked for some time before being heated for about three hours at 300° C and allowed to cool slowly to room temperature, as noted above.

Taking one of eight specimens it is subjected to strain of 0·5% using the Hounsfield tensometer (see Experiment No. D1.4), a second to 1% strain and the others to strains increasing in steps of 0·5% up to 4%.

All the specimens are then carefully placed in a suitable furnace and slowly heated to 300° C at which temperature they are held for 30 minutes. The temperature is raised to 400° C and held again for 30 minutes, subsequently to 450, 500 and 600° C at which steps the same period of temperature stabilization is allowed to elapse. Finally the whole is allowed to cool slowly to room temperature.

The specimens should then be etched using a solution made up of the following constituents by volume:

10% hydrofluoric acid
15% nitric acid
25% hydrochloric acid
50% distilled water.

Exercise great care with the etching reagent which should be already prepared and held in a polythene bottle. Use it only under the direct supervision of laboratory staff and obey rigorously all instructions given for its handling and disposal.

Finally rinse the specimens by immersion in fresh industrial spirit and dry them quickly in warm air.

Discussion of Results

Plate D6 shows a typical result and one should note the region A where the original fine uniform grain structure exists and the central region B where single crystals of much larger size are observed.

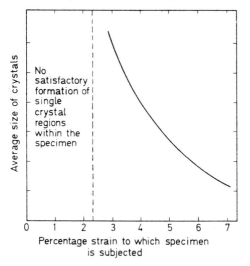

Figure D40. Sketch graph showing decrease of the size of single crystal regions within a strain-annealed specimen for strains exceeding 2·5%

The results are not of the highest order in the general laboratory experiment since the tendency is to hasten the processes of annealing in order to complete the experiment in reasonable time. If high quality crystals are desired then much longer periods than are indicated here are required.

It is found that below about 2·5% deformation in the Hounsfield tensometer, no satisfactory crystal growth takes place. The optimum strain for the production of large single crystal regions within the specimen appears to be from about 2·5 to 3% (*Plate D6*), and at high values of percentage strain the average single crystal size (*Figure D40*) falls off markedly.

One of the critical phases of the experiment is the maintenance of the temperature at 600° C, just below the melting point of aluminium (625° C). It is very easy to melt the specimen unless caution is observed.

Further Work

If time permits, variations in annealing times and temperature gradients should be investigated.

If a portion of the stressed specimen is electrolytically polished it is possible to observe the slip lines within the material. (Try 0·3 A and 10 V for 3 minutes using the methanol nitric acid electrolyte of Experiment No. D1.6.)

Reading and References

The strain anneal method of crystal growth is dealt with in *The Art and Science of Growing Crystals* by J. J. Gilman (Wiley, 1963), where a number of photographs of aluminium specimens subjected to different conditions of strain are shown. A sketch graph of the fall in grain size upon increasing percentage strain is also given (by K. T. Aust, p. 453).

Experiment No. D2.1

THE GROWTH OF SINGLE CRYSTALS OF HEXAMETHYLENE TETRAMINE FROM THE VAPOUR PHASE

Introduction

Single crystals may be grown by any of the following general methods.

(1) by fusion from the melt,
(2) by precipitation from solution,
(3) by condensation from the vapour phase.

The aim of this experiment is to illustrate the process of growing hexamethylene tetramine from the vapour.

PHYSICS OF ENGINEERING MATERIALS

This method of growth generally requires complex instrumentation for stringent temperature control but advantage is taken here of the fact that the most desirable temperature for the growth of hexamethylene tetramine is 100° C (see first reference).

Theory

Problems of temperature control are simplified here since a water bath may be used. A copper tube (*Figure D41*) conducts the heat away from the plate soldered to its base, which is in contact with the tube containing the hexamethylene tetramine.

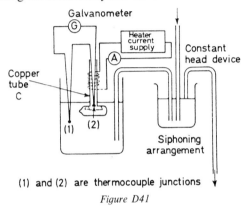

(1) and (2) are thermocouple junctions

Figure D41

The temperature difference is measured by a thermocouple which has one junction in the water bath and one soldered to the copper plate. Control of the temperature of the copper tube is made by means of a heater coil wrapped round its upper end.

A siphoning device enables adjustment of the water level to be made, and using this in conjunction with the heater control it is possible to vary the temperature difference between the water bath and the plate over a range of 0° C to about 15° C and to maintain a particular difference to within 0·5° C over fairly long periods of time.

At a temperature difference of the order of 4 or 5° C, seed crystals of the hexamethylene tetramine begin to grow beneath the plate.

Method

Place the sealed Pyrex tube (*Figure D42*) containing the finely powdered hexamethylene tetramine in the water bath and hold it in

GROWTH OF CRYSTALS FROM VAPOUR PHASE D2.1

position about 10 cm below the water surface in the bath. Place the copper plate firmly over the Pyrex tube and heat the bath using a hot plate. Pass current through the heater attached to the copper tube C until the specimen tube and the plate are at the same temperature.

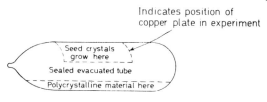

Figure D42

Using the siphon arrangement (*Figure D41*), lower the water level slowly until a difference of about 5° C is indicated by the thermocouple. After two hours or so allow the water bath to cool to room temperature and remove the Pyrex tube gently from the bath. A large number of small seed crystals will form below the plate when the conditions are correctly adjusted and those which one does not wish to grow further must be removed.

This is done by clamping the tube firmly in position on a clean dry area of the bench and then applying the pointed tip of a small electric soldering iron on the surface of the tube just above the seed to be vaporized. (Great care is necessary here to avoid possible danger of electric shock. The hands must be perfectly dry before touching the soldering iron and no metal which may be earthed must be touched.) Repeat this procedure until there remain only those seeds selected for further growth.

Replace the tube but adjust the heater current and water level this time until a temperature difference of just under 0·5° C is obtained. (The initial adjustment of equality of bath and plate temperature must be made with the plate NOT in contact with the pyrex tube or the crystals will be lost.) This temperature difference should encourage the established seed to grow large and at the same time prevent the growth of new seeds although it may be found that occasionally undesired seeds have to be removed.

Discussion of Results

It is most important that the calibration of the thermocouple is checked very carefully before use, since it is essential to have accurate knowledge of the temperature differences.

It will be found that variations in the pressure inside the sealed tube, as well as the quantity of hexamethylene tetramine within it, do affect the growth rate in the experiment.

The tube should have 15 g of hexamethylene tetramine sealed in the tube at a pressure of about 10^{-5} mm of mercury.

Currents of the order of 1 A or less should enable one to maintain the maximum required temperature difference but the necessary current is critically dependent upon the depth of the copper plate below the water surface.

The method of removing unwanted seeds is a 'tricky' one and requires considerable patience, practice and skill. Crystals selected for growth are easily lost when the remaining seeds are few in number and this is perhaps the most unsatisfactory part of the experiment from a student point of view.

Further Work

The effects of (*a*) slight variations in tube pressure and (*b*) slight changes in the established temperature differences between plate and bath should be noted.

Reading and References

The conditions of growth of hexamethylene tetramine were investigated by B. Honigmann, *Zeitschrift Electrochem*, vol. 58, p. 322, 1954 (in German).

Preparations of Single Crystals by W. D. Lawson and S. Neilsen (a Butterworth Semi-conductor Monograph, 1958) gives a survey of growing single crystals and a more detailed account of nucleation and crystal growth.

Experiment No. D2.2

TO DETERMINE THE DIFFERENCE BETWEEN THE PRINCIPAL MAGNETIC SUSCEPTIBILITIES OF AN ANISOTROPIC WEAKLY PARAMAGNETIC CRYSTAL

Introduction

In measuring the magnetic susceptibility of a powdered crystalline specimen such as (for example) copper sulphate, the 'average' susceptibility of a statistically large number of randomly orientated small crystals is in fact measured.

PRINCIPAL MAGNETIC SUSCEPTIBILITIES D2.2

If, however, a large single crystal is chosen for careful investigation it is found that in different directions through the crystal the value of the magnetic susceptibility may vary. This experiment is designed to measure the difference between the principal susceptibilities of such a crystal.

Theory

If the crystal is suspended from a fine long silk thread in the uniform field between the poles of a strong magnet then assuming that the torsion fibre is parallel to the third principal magnetic axis corresponding to susceptibility k_3 the couple Γ acting on the specimen is given by

$$\Gamma = V\mu_0 H^2 \theta (k_1 - k_2)$$

(compare the formula for the force on a Gouy specimen),

where $V =$ the volume of the crystal
$H =$ applied field
$\theta =$ angle between the direction of the susceptibility k_1 and the applied field
$\mu_0 =$ permeability of free space.

When no magnetic field is present the period of oscillation of the crystal is given by

$$T = 2\pi \left(\frac{I}{c}\right)^{\frac{1}{2}}$$

where $c =$ torsional control per radian twist
$I =$ moment of inertia of the specimen.

In the magnetic field the effective torsional control is modified to c' such that

$$c' \left(= \frac{4\pi^2 I}{T_1^2} \right) = c + \frac{\Gamma}{\theta} = c + V\mu_0 H^2 (k_2 - k_1)$$

hence we get a new time of oscillation since

$$c' = \frac{4\pi^2 I}{T_1^2}$$

i.e.

$$\frac{4\pi^2 I}{T_1^2} = \frac{4\pi^2 I}{T_0^2} + V\mu_0 H^2 (k_1 - k_2)$$

337

therefore

$$k_1 - k_2 = \frac{4\pi^2 I}{V\mu_0 H^2}\left(\frac{1}{T_1^2} - \frac{1}{T_0^2}\right)$$

or

$$\chi_1 - \chi_2 = \frac{k_1 - k_2}{\text{Density } \rho} \frac{4\pi^2 I}{m\mu_0 H^2}\left(\frac{1}{T_1^2} - \frac{1}{T_0^2}\right)$$

$$= \frac{4\pi^2 I}{T_0^2 m\mu_0 H^2}\left(\frac{T_0^2 - T_1^2}{T_1^2}\right)$$

$$= \frac{c}{m\mu_0 H^2}\left(\frac{T_0^2 - T_1^2}{T_1^2}\right).$$

Method

Note the time for 20 oscillations when the magnetic field is zero. Switch on the magnetic field and increase the current slowly to give a significant change in the period of oscillation of the crystal. Again note the time for 20 swings.

A glass shield about the crystal is advisable to obviate draughts.

Discussion of Results

The choice of crystal for the experiment has to be made very carefully since from the above it will be seen that the theory assumes the three principal magnetic axes to be at right angles. Ideally a crystal of simple geometry is to be preferred (such as those grown from the melt in Experiment No. D1.9).

It has also been assumed (*Figure D43*) that the direction of the fibre lies along the third principal magnetic axis, so that although it is possible to measure a difference in susceptibilities (even if the setting is not accurate) a true difference between principal susceptibilities will not be obtained.

A microscope can often be used to advantage to measure the period of oscillation accurately but try to avoid the condition where the motion shows considerable pendulum-type swinging rather than oscillation.

Further Work

As pointed out in the discussion of results, accurate positioning of the crystal is required if true differences between principal susceptibilities are to be obtained. This process though involving much more time and more complicated arrangement of apparatus with a suitable goniometer head makes a worthwhile project-type experiment.

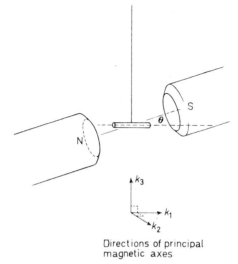

Directions of principal magnetic axes

Figure D43

Reading and References

A chapter on magnetic anisotropy and magnetic single crystals may be found in *Modern Magnetism* by L. F. Bates (Cambridge University Press, 4th edition, 1961).

A brief account of this method may be found in *Magnetochemistry* by P. W. Selwood (Interscience, 1956).

Experiment No. D2.3

TO PREPARE A FERRITE SPECIMEN AND TO INVESTIGATE THE VARIATION OF ITS MAGNETIC PROPERTIES WITH COMPOSITION

Introduction

Ferrites are materials that whilst having high electrical resistance possess strong magnetic properties. Their high resistance enables them to be used in electrical engineering where low eddy current losses are required and in this regard they are particularly valuable at high frequencies.

Applications of ferrites are in the construction of inductor and transformer cores for radio frequencies and the permeability of such ferrite cores are found to be higher than that for powdered dust cores, the magnetic core material of comparable resistivity.

Their saturation induction is however still much less than that for iron so that alloys of the latter are still chosen for transformers and machines operating at power frequencies.

Theory

The ferrites consist of mixed oxides having the general formula

$$M.O.Fe_2O_3$$

where M represents a divalent metal such as magnesium, manganese, zinc, cobalt or even iron. Note in passing that lodestone (one of the first magnetic materials, known to the Chinese before Christ) is in fact $FeO.Fe_2O_3$ i.e. ferroso-ferric oxide and therefore a ferrite).

Method

Mix the Fe_2O_3 with the appropriate oxide of magnesium, copper, manganese, nickel and zinc in the proportions of their molecular weights indicated in *Table D2* making allowance for the fact that in

Table D2

Ferrite	Added oxide		Quantities (grams)	
	Formula	Molecular weight	Fe_2O_3	Added oxide
$MgO.Fe_2O_3$	MgO	40	8	2
$CuO.Fe_2O_3$	CuO	79.5	8	4
$MnO.Fe_2O_3$	MnO_2*	87	8	4.4
$NiO.Fe_2O_3$	NiO	74.7	8	3.8
$ZnO.Fe_2O_3$	ZnO	81.4	8	4

* The oxide of manganese is added as the dioxide. It decomposes to MnO upon being heated above about 750°C.

$$\therefore \text{ Weight added is } \left(\frac{\text{Molecular weight of } MnO_2}{\text{Molecular weight of } MnO} \right)$$

times the appropriate amount indicated by the formula for the ferrite.

heating manganese dioxide it is converted above about 750° C to MnO so that the weight of the dioxide initially required is

$$\left(\frac{\text{Molecular weight of MnO}_2}{\text{Molecular weight of MnO}}\right)$$

times the amount of MO indicated by the formula (*Table D2*).

Grind and mix the weighed oxides thoroughly in a mortar. Into suitable steel moulds such as shown in *Figures D44* and *D45* pour

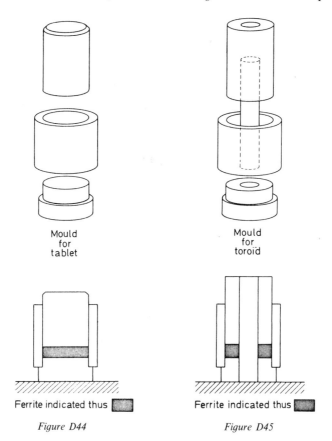

Figure D44 Figure D45

sufficient mixture to one third fill them, ensuring that the powder surface is level before inserting the compressing plunger. Compress the mixture for a few seconds at about 27·6 MN m^{-2} (i.e. 4 000 lb in^{-2}) (packing blocks not only obviate excessive movement of the press but make for easy removal of the specimen).

Slowly release the pressure and remove the moulds from the press.

Using an ejector pin apply gentle pressure to remove the ferrite from the mould, then place it in a jig (*Figure D46*) to saw it to the desired shape BEFORE 'firing' it. (For the oscillation experiment to determine the Curie point, the specimen is cut into the form of a bar.)

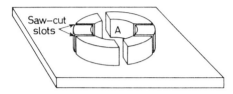

Figure D46. Jig used in cutting ferrite. The specimen is placed at A

There is a tendency for the oxides to stick to the moulds and to obviate this the surfaces should be 'painted' with stearic acid dissolved in a little ether (care should be taken to keep well away from any naked flame since ether is highly inflammable).

Place the specimen on a sheet of refractory material and put it in a suitable furnace. Raise the temperature to about 1 150° C, maintaining this for approximately 15 minutes before allowing the sample to cool as *slowly* as conveniently possible (to prevent cracking).

Discussion of Results

Produce a number of specimens of each type of ferrite and then use only those which are free from cracks and gross distortion, when investigating their physical properties.

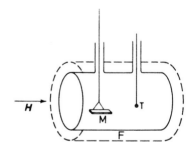

H, Weak magnetic field supplied by bar magnet
F, Furnace winding in fireclay
T, Operational thermocouple
M, Oscillating magnet

Figure D47

Further Work

Try various heat treatments and note the effect this has upon the magnetic (and other) properties of the ferrite.

Try making mixed ferrites, for example, nickel-zinc, manganese zinc and magnesium nickel ferrites in order to study their properties. Such ferrites have some commercial importance of which detail may be found in the last reference.

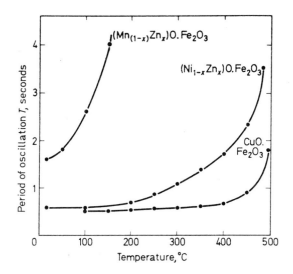

Figure D48. Graph of period of oscillation of ferrite specimens of different composition against temperature, for determination of their Curie points

Figure D47 shows a diagrammatic representation of a type of furnace which may be used to investigate the Curie temperature of ferrites by an oscillation method. Use is made of the classical expression

$$T = 2\pi \left(\frac{I}{MH}\right)^{\frac{1}{2}}$$

where H = the weak magnetic field provided by bar magnets either side of the furnace
I = the moment of inertia of the oscillating magnetic specimen.

As temperature θ rises the magnetic moment M decreases and for

high θ, $M \to 0$. Therefore the period of oscillation $T \to \infty$ (*Figure D48*).
Alternatively

$$\frac{1}{T^2} = (\text{constant})\, M$$

and $1/T^2$ plotted against θ tends to 0 (*Figure D49*). Note the graph does not actually fall to zero due to the torsional control exhibited by the 'refrasil' fibre suspension.

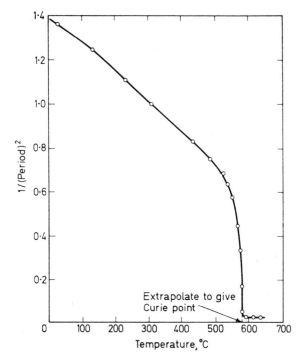

Figure D49

Figure D50 shows the result of a preliminary furnace temperature calibration using a second thermocouple in the position that the oscillating magnetic specimen will later occupy and it is notable that differences between the temperature at this point and the temperature at the point of location of the operational thermocouple do occur.

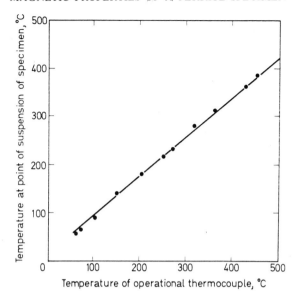

Figure D50

Reading and References

The theory of the Néel theory of ferrimagnetism is given in *Effective Field Theories of Magnetism* by J. S. Smart (Saunders, 1966).

Data concerning the important physical properties of ferrites can be found in *Ferromagnetism* by R. M. Bozorth (Van Nostrand, 1955).

A comprehensive treatment of ferrites will be found in *Ferrites* by J. Smit and H. P. J. Wijn (Phillips Technical Library, 1959).

Applications data on commercial ferrites may be found in the *Mullard Technical Handbook*, vol. 6 (1963).

Experiment No. D2.4

TO INVESTIGATE THE EFFECT OF FERROMAGNETIC IMPURITY IN A WEAKLY MAGNETIC SPECIMEN

Introduction

In measurements upon the magnetic susceptibilities of weakly magnetic materials the presence of small proportions of ferromagnetic impurity in the specimen or the container can seriously falsify the results obtained. This method shows how one may detect the presence of ferromagnetic impurities.

Theory

The Gouy method (see first reference) is used here with the specimen, in the form of a cylinder, hanging from one arm of a balance so that the base of the cylinder lies at the centre of the magnetic field (*Figure D51*).

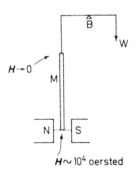

Figure D51. Principle of the Gouy method for the measurement of magnetic susceptibilities.
B, Balance arrangement
M, Specimen which experiences a 'pull' due to the magnetic field. Note (1) that the lower end lies in the maximum magnetic field region and that the value of the magnetic field is sensibly constant over the central regions of the magnet pole tips N and S, (2) that at the upper end of the specimen the magnetic field tends to zero by comparison with the field ($\sim 10^4$ oersteds) along the axis of the pole pieces
W, Added weight needed to counterbalance the specimen under various conditions of magnetic field

The force on the specimen when the field is switched on is given by

$$mg = A \int_0^H I\mu_0 \left(\frac{dH}{dy}\right) dy$$

where mg = total force on specimen
 A = cross sectional area of the specimen
 I = intensity of magnetization in the specimen
 dH/dy = magnetic field-gradient
 μ_0 = permeability of free space.

EFFECT OF FERROMAGNETIC IMPURITY D2.4

Assuming that for a weakly magnetic specimen the magnetic volume susceptibility is constant $I = \mu_0 k H$ and the force on the weak magnetic may be written

$$m_1 g = \mu_0 A k \int_0^H H \left(\frac{dH}{dy}\right) dy$$

However, any ferromagnetic impurity present reaches saturation at comparatively low values of magnetic field hence the force produced on the impurity by the application of the field is

$$m_2 g = \mu_0 A \int_0^{I_s} I_s \, dH$$

where I_s = saturation value of intensity of magnetization.

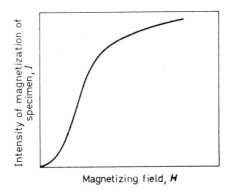

Figure D52. Diagrammatic representation of the variation of the intensity of magnetization with change in magnetizing field for a ferromagnetic substance

Figure D52 shows the I/H curve for a ferromagnetic and it is seen that $\int I \, dH$ = area under the curve.
Hence the total force on the specimen

$$mg = m_1 g + m_2 g = \mu_0 \left\{\frac{AkH^2}{2} + A(IH)\right\}$$

where (IH) represents the area under the ferromagnetic magnetization curve,
i.e.

$$mg = \frac{\mu_0 A H^2}{2}\left\{k + \frac{2I}{H}\right\} \quad \text{or} \quad \frac{2mg}{\mu_0 A H^2} = \left\{k + \frac{2I}{H}\right\}$$

347

but $(2mg/\mu_0 AH^2)$ = the effective magnetic susceptibility $k_{\text{effective}}$ of the specimen as a whole in the field H, i.e.

$$k_{\text{effective}} = k + \frac{2I}{H}$$

where k represents the magnetic susceptibility of the weakly magnetic substance.

Method

Fill a pyrex tube with finely powdered manganese oxide of technical grade and tap the base of the tube gently so that the powder is well packed.

Suspend the tube from one arm of a balance so that the bottom of the tube lies at the central axis of the pole pieces of an electromagnet capable of providing up to 10 000 oersted (i.e. 1 Wb m^{-2}).

Switch on the field current and increase slowly to about 1 A.

Balance the pull on the specimen and calculate $k_{\text{effective}}$ from the equation

$$mg = k_{\text{effective}} \mu_0 \frac{AH^2}{2}$$

Repeat the procedure for say 2, 3, 4 and 5 A and plot $k_{\text{effective}}$ against $(1/H)$ as shown in *Figure D53*. The intercept gives the volume susceptibility of the weak magnetic body and from the slope of the graph the intensity of magnetization in the ferromagnetic impurity may be found (=Slope/2).

Warning. With such large magnets, always reduce the magnet current SLOWLY to zero BEFORE switching off.

Discussion of Results

If the quantity of ferromagnetic impurity is large then it may be found that the tube tends to cling to the magnet pole tips in which case a wider field gap must be used (it may be convenient to run a permanent magnet over the powder to remove a proportion of the ferromagnetic impurity).

The particles should be well ground in a mortar. The particle size does introduce another variable factor when employing the Gouy method.

The result for the intensity of magnetization may at first seem very low when one compares it with the accepted values for ferromagnetics but reflect that the volume of impurity is small compared with the volume of the specimen as a whole. In addition the result will depend upon the shape of the particles and their distribution throughout the specimen. (In the case of spherical particles

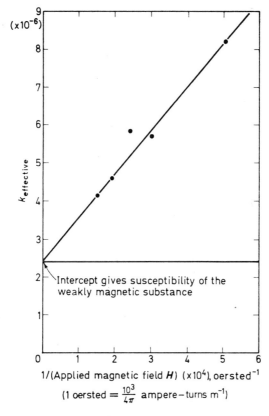

Figure D53. Determination of the magnetic susceptibility of a weakly magnetic powder and detection of the presence of ferromagnetic impurity

the demagnetization factor, which has been neglected in the simple theory above, is large.) It is this, above all, that imposes strict limitations upon the value of the method as an aid to determining the actual amount of impurity present because in general the form and distribution of impurity particles are unknown.

Further Work

Before using the electromagnet a full survey of its capabilities should be made. In particular graphs giving actual field values at the centre of the region between the pole tips (*a*) for a full range of current values, (*b*) for a large number of different pole gaps, should be obtained.

Reading and References

The Gouy method is described in *Modern Magnetism* by L. F. Bates (Cambridge University Press, 4th edition, 1961), as well as in many standard degree texts.

The theory given of this experiment together with a treatment of the demagnetization effects (omitted here) may also be found in *Modern Magnetism*.

The effect of particle size on Gouy measurements is dealt with by A. Earnshaw in a paper entitled 'Magnetochemistry' to be found in *Laboratory Practice*, vol. 10, pp. 89 to 91 and 157 to 159, Feb. and March, 1961.

The paper will also suggest further work, even where laboratory magnet facilities are limited.

Experiment No. D2.5

TO PREPARE DOPED CADMIUM SULPHIDE PHOTOSENSITIVE DEVICES AND TO INVESTIGATE THEIR PERFORMANCE

Introduction

The aim of the experiment is to determine what effect variation in copper doping has upon the characteristic response of a cadmium sulphide photosensitive surface. The variation of cell current with intensity of illumination is also investigated and the dark current of the cell measured.

Theory

Cadium sulphide photocells are basically variable resistance devices whose resistance depends upon the intensity level of the incident light, the resistance falling as the illumination increases.

PERFORMANCE OF PHOTOSENSITIVE DEVICES   D2.5

For applications involving short periods of illumination after a long period in darkness (when the resistance is very high and the 'dark' current is small) the value of the initial cell current is of particular interest.

For those applications where small changes of light intensity occur, the variation of cell current or voltage with resistance is of importance as also is the voltage/current characteristic of the cell.

Method

A slurry of cadmium sulphide should be made by placing 2·9 grams (20 millimoles) of cadmium sulphide (CdS) in a thoroughly cleaned bottle and adding 3 millimoles of cadmium chloride as an aqueous solution in about 20 ml of distilled water from a burette. Further distilled water is then added to make the solution up to

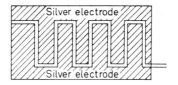

Figure D54. Suitable photocell geometry

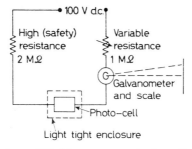

Figure D55. Measurement of photocell dark current

40 ml and the whole shaken well. 1·5 mg (approximately 0·01 millimoles) of cupric chloride ($CuCl_2$) solution are added and the liquor shaken for one hour. After allowing the resulting solution to stand for sufficient time to settle, the supernatant liquid is decanted (or preferably aspirated) off to leave the doped slurry. This is

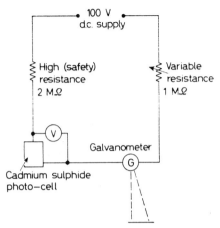

Figure D56. Circuit for the determination of the voltage-current characteristic of a doped cadmium sulphide photocell. It may be found more convenient to use a 0–50 mA meter for the more sensitive cells and to substitute a microammeter for the lower current ranges

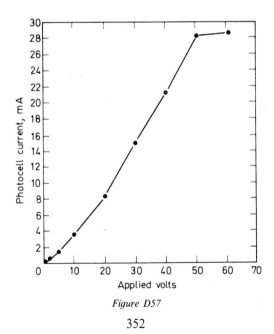

Figure D57

painted onto a piece of fireclay such as is used to make furnace boats and then baked in an oven at 550° C for from 2 to 10 minutes.

The procedure is repeated using 0·92 g (5 millimoles) of $CdCl_2$ and 1·28 g (i.e. 7 millimoles) instead of the 3 millimoles used originally. After firing in the furnace or oven, two electrodes are painted on the resulting surface using silver paint. The electrodes should be about 1 mm apart and of the order of 10 mm long. An effective way of doing this is shown in *Figure D54*.

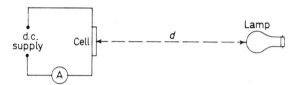

Figure D58. Circuit for the investigation of variation of photocell current with luminous intensity

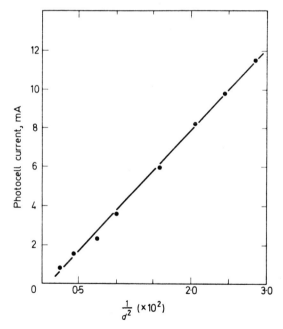

Figure D59. Variation of photocell current with intensity of illumination of cell. d = distance of source of illumination from the cell

PHYSICS OF ENGINEERING MATERIALS

The important measurements that can be made with the photo-sensitive cell are:

(1) Measurement of the dark current.
(2) Variation of cell current with intensity of illumination.
(3) The voltage-current characteristic.

Figure D55 shows the circuit used to measure the dark current with typical values of resistance in the circuit. Note the fixed resistor included in the circuit for safety. Occasionally specimens may break down and arc.

The voltage current characteristic may be measured using the circuit of *Figure D56* but the applied voltage should be increased slowly since the cell may fatigue rapidly when voltages of too high an order are applied. A typical characteristic is shown in *Figure D57*.

It is also valuable to investigate the variation of cell current with intensity of illumination (*Figure D58*). The most convenient way to do this is to plot a graph (*Figure D59*) of $1/(\text{distance})^2$ against photo-cell current.

Discussion of Results

Table D3 shows the way in which variation in doping and firing time can drastically affect the sensitivity of the cell.

The measurements can be made with fairly good accuracy as *Figures D57, D59* and *D61* indicate and the response of the cell is reasonably linear.

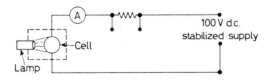

Figure D60. Measurement of photocell fatigue

In addition to the measurements quoted above the following parameters might be investigated:

(1) The fatigue associated with the photo-sensitive device (see *Figure D60*), i.e. the drop in response which occurs after prolonged exposure of the surface to light (*Figure D61*).

PERFORMANCE OF PHOTOSENSITIVE DEVICES D2.5

Table D3. Typical data for doped cadmium-sulphide photosensitive surfaces

Number of boat	Firing time (min at 550° C)	Preparation	Dark current in μA (100 V applied)	General remarks on sensitivity
1	2	(a)	<0·2	Insensitive
		(b)	,,	,,
		(c)	,,	,,
2	3	(a)	<0·2	Insensitive
		(b)	,,	,,
		(c)	Broke down	,,
3	4	(a)	<0·2	Low sensitivity
		(b)	0·5	,,
		(c)	<0·2	,,
4	5	(a)	<0·2	Average sensitivity
		(b)	,,	,,
		(c)	7	,,
5	6	(a)	<0·2	Average sensitivity
		(b)	,,	Insensitive
		(c)	,,	Good sensitivity
6	7	(a)	<0·2	Average sensitivity
		(b)	,,	Good sensitivity
		(c)	,,	Extremely sensitive

(2) Variation of resistance with temperature.
(3) Response of the cell to variation in wavelength of incident light.

Reading and References

The production of these devices is taken from a paper by W. F. Sheehan entitled 'Experiments with photoconductive cadmium sulphide' in the *American Journal of Chemical Education*, October 1962, p. 540.

This paper also gives an alternative method of preparation and lays more stress upon the chemical aspects of the topic as well as giving results of much more extensive investigations of cell performance.

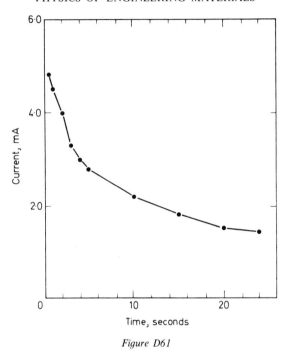

Figure D61

An excellent pamphlet on 'Cadmium sulphide photoconductive cells—their properties and applications' may be obtained from Mullard Ltd.

Experiment No. D2.6

THE ELECTRON SPIN RESONANCE METHOD USED TO DETERMINE THE LANDÉ SPLITTING FACTOR

Introduction

The classical theory of the Bohr atom predicts that the ratio of the angular momentum of a system to its magnetic moment is $(2m/e)$ but in fact the ratio is $(2m/ge)$ where g is the Landé splitting factor. This factor for the spinning electron is 2 and is shown to have a firm theoretical basis in wave mechanics.

LANDÉ SPLITTING FACTOR D2.6

Theory

If an electron of mass m is moving with angular velocity ω about a nucleus its angular momentum p may be written

$$p = (mr^2)\omega$$

where r represents the radius of the orbit. This motion gives rise to a magnetic dipole moment M which according to the magnetic shell theory $= iA$ where $i =$ the current and $A =$ the area of the orbit.

If it is assumed that we have a circular orbit $A = \pi r^2$ and since $2\pi ri = e\omega r$ therefore $i = e\omega/2\pi$ hence $M = e\omega r^2/2$. Substituting for $\omega r^2 = p/m$ from the first equation, the magnetic moment M of the electron due to its orbital motion $= ep/2m$.

In a similar manner we can attribute a magnetic moment due to the *spin* of the electron. It is found that

$$M_{(spin)} = \frac{gep}{2m}$$

where g is the Landé spectroscopic splitting factor and for the spinning electron

$$M_{(spin)} = \frac{ep}{m} \quad \text{i.e. } g = 2.$$

Quantum theory shows that if a magnetic field H is applied to the system two possible energy states can arise and the energy difference ΔE between the states is given by

$$\Delta E = \Delta M_{(spin)} B = \frac{geB}{2m} \Delta p$$

where $B =$ flux density in the specimen. The electron states correspond to electron spin quantum numbers $+\frac{1}{2}$ and $-\frac{1}{2}$ and thus Δp corresponds to $h/2\pi$
i.e.

$$\Delta E = g\left(\frac{eh}{4\pi m}\right) B.$$

If electromagnetic radiation energy is now applied to the system the electron may at one particular frequency be raised from the lower to the higher level and a resonance absorption takes place
i.e.

$$h\nu = -g\left(\frac{eh}{4\pi m}\right) B.$$

The quantity ($eh/4\pi m$) represents the fundamental unit of magnetic moment and is known as the Bohr magneton μ_B.
Hence we have

$$h\nu = g\mu_B \, \boldsymbol{B}.$$

Method

The circuit is set up as shown in *Figure D62*. The sample of diphenyl picryl hydrazyl (DPPH) is contained in a small ignition

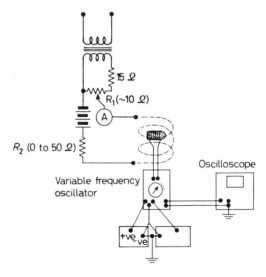

Figure D62

tube around which is wrapped a coil of a few turns of thick copper wire. Through this small coil is fed the radio frequency energy which excites the electrons.

The sample and its coil lie in the static magnetic field $\boldsymbol{H}$ provided by the solenoid and this may be adjusted by variation of resistances R_1 and R_2. Then by varying the oscillator frequency to satisfy the condition $h\nu = g\mu_B \, \boldsymbol{B}$ transitions between the two energy levels are induced.

The resulting absorption of energy may be displayed upon the oscilloscope screen as shown (*Figure D63*).

The transformer shown in *Figure D62* superimposes a small alternating field upon the static field $\boldsymbol{H}$ so that the resonance condition is caused to recur, when, for convenience of this display, the differential of the absorption pattern is often used (*Figure D63*).

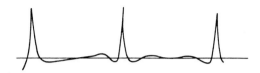

Figure D63. Typical oscilloscope trace showing absorption peaks due to electron spin resonance

Note the oscillator frequency v and calculate the value of H so that g may be found.

Discussion of Results

The answer obtained should lie very near to the theoretical value of 2 but in considering the accuracy of the result it must be noted that this value is for a single free electron. Here we have many electrons absorbing energy. Also if the formula

$$H = n_1 I \text{ ampere turns m}^{-1} \quad \left(\text{or } H = \frac{4\pi ni}{10} \text{ oersted}\right)$$

is used to give the value of the static field remember that this strictly applies only to an infinite solenoid (which of course we do not have)

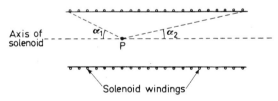

Figure D64. Field due to a solenoid of finite length. If H = magnetic field at point P, n_1 = number of turns cm^{-1} on the solenoid, α_1 and α_2 = angles made with the axis of the solenoid by the lines joining P to the ends of the solenoid. then,

$$H = \frac{n_1 I}{2}(\cos \alpha_1 - \cos \alpha_2)$$

although the necessary correction (see fourth reference) can be made (*Figure D64*).

Further Work

The type of set-up described here can be bought completely (for example Scientifica Kits, Ltd.) but it is possible to construct a piece of apparatus.

Such equipment is described in the fifth reference, although in this case it was designed as a field measuring instrument. Notice that it uses a pair of Helmholtz coils, thus avoiding the approximation involved with the solenoid.

Reading and References

Background theory concerning the Landé splitting factor and the Bohr magneton may be found in *Modern Magnetism* by L. F. Bates (Cambridge University Press, 4th edition, 1961) as well as many other standard degree physics texts.

The method is described in more detail in the *Physics of Magnetism* by S. Chikazumi (Wiley-Interscience, 1968) and in more elementary form in *Electronic Processes in Materials* by L. V. Azaroff and J. J. Brophy (McGraw-Hill, 1963).

Theory of the field due to a solenoid of finite length may be found in *Intermediate Physics* by C. J. Smith.

A magnetic field meter based upon this principle is described in *Wireless World*, vol. 69, p. 403, 1963.

A note upon electron spin resonance experiments in low fields may be found in the *American Journal of Physics*, December 1962, p. 927, which gives detail of an alternative circuit arrangement suitable for making up in an undergraduate laboratory as well as indicating its use as a magnetometer. See also a resource letter giving a guide to literature and teaching aids on electron paramagnetic resonance which may be found in the *American Journal of Physics*, vol. 33, No. 2, Feb. 1965, entitled 'Resource Letter NMR-EPR1' by R. E. Norberg.

A most detailed report upon the construction of apparatus for electron paramagnetic resonance at low fields is given in the *American Journal of Physics*, vol. 29, No. 8, August, 1961, p. 492, including a complete list of materials required and the workshop equipment needed for fabrication.

Experiment No. D2.7

DETERMINATION OF THE GYROMAGNETIC RATIO USING THE NUCLEAR MAGNETIC RESONANCE METHOD

Introduction

It was seen in Experiment No. D2.6 that the spin of the electron gave rise to a magnetic moment and that a definite value could be attributed to the ratio of the angular momentum to the magnetic moment. In general nuclei have spins and the associated magnetic moments may be measured in a manner similar to that used for the electron although the resonance frequencies involved will be lower by a factor of (1/1836) than those of the electron.

Theory

The unit of magnetic moment used here is the nuclear magneton (where m_p = the mass of the proton rather than the Bohr magneton, e = fundamental electronic charge and h = Planck's constant)

$$\mu_{(nuclear)} = \frac{eh}{4\pi m_p} \text{ in absolute units.}$$

The gyromagnetic ratio

$$\rho = \frac{1}{g}\frac{2m}{e}$$

and the nuclear g factor may have values from about 0·1 to 5·6.

Putting fundamental values into the above equation will show that the resonance frequency lies in the radio frequency range when using magnetic fields of the order of 10^4 oersted.

Absorption of radio frequency energy will produce a characteristic 'dip' on the cathode ray oscilloscope (*Plate D7*).

As in the case of electron spin resonance

$$h\nu = -g\left(\frac{eh}{4\pi m_p}\right)H \qquad (1)$$

but in this case as indicated above m_p is appropriate to the proton and not the electron.

Method

With the apparatus connected as shown (*Figure D65*) the radio receiver is tuned to 14·9 MHz and the test tube containing a sample of dilute copper sulphate solution is placed in position in the probe unit coil between the magnet pole pieces.

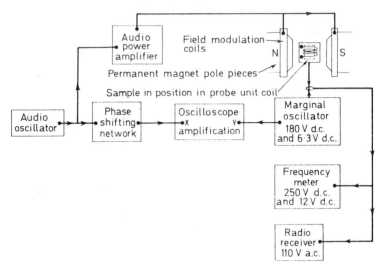

Figure D65. Block diagram of nuclear magnetic resonance apparatus

Adjust the frequency control on the radio frequency oscillator until audio frequency oscillations are heard on the radio receiver.

Set the X amplication on the cathode ray oscilloscope to give full screen deflections and adjust the reactance control on the radio frequency oscillator to cause the audio frequency oscillations to just cease.

The absorption due to nuclear magnetic resonance appears as shown in *Plate D7b*.

Adjustment of Y amplification to give suitable gain is first made followed by fine adjustment of reactance and frequency controls on the oscillator before manipulating the phase shifting network to superimpose the two peaks (*Plate D7*) on the screen.

The frequency corresponding to the condition where the absorption is observed is noted on the frequency meter, then using equation (1) the gyromagnetic ratio (or more important, the nuclear g factor) may be determined.

Discussion of Results

The accuracy of the result for g will depend chiefly upon the accuracy to which the magnetic field can be measured. If this is done by search coil it will not be as accurate as if a material of standard susceptibility is used in order to estimate the magnetic field value.

It is therefore probably more convenient with the apparatus available to accept the accurately measured value of the g factor for the proton and to use the method to make an accurate calculation of the field of the magnet.

Further Work

Depending upon the resolving power of the set-up it may be possible to measure the spin lattice relaxation time associated with the chosen specimen, i.e. the short but finite time that the nuclear magnetic resonance signal is observed to take to reach its equilibrium amplitude after the sample has been placed in the magnetic field. (It increases to its maximum value in accordance with the equation

$$a = A \{1 - \exp(-t/T)\} \quad \text{where } a \text{ is the amplitude}$$

of the observed signal at time t after placing the sample in position. A is the final maximum amplitude and T is the spin lattice relaxation time which is due to the fact that the nuclear dipole is not an isolated one but is surrounded by the atomic lattice.)

Reading and References

A simple treatment of the theory of nuclear magnetic resonance and of magnetic resonance generally may be found in *Electronic Processes in Materials* by L. V. Azaroff and J. J. Brophy (McGraw-Hill, 1963).

An article on 'A low cost nuclear magnetic resonance apparatus' by J. P. Stuart, upon which pattern the above set-up was based may be found in *Laboratory Practice* for January 1964.

A detailed text upon the subject is *Nuclear Magnetic Resonance* by E. R. Andrew (Cambridge University Press, 1958).

Experiment No. D2.8

TO OBSERVE MAGNETIC DOMAINS IN A FERRIMAGNETIC GARNET AND TO USE THE CHANGES IN THE DOMAIN PATTERN TO INTERPRET THE MAGNETIZATION HYSTERESIS CYCLE

Introduction

In investigating the rare earth garnets Dillon (see references) discovered that not only were they ferrimagnetic but that thin slices of them were transparent to visible light. This enables their domain structure to be readily observed in polarized light.

The garnets have the chemical formula $M_3Fe_2(FeO_4)_3$ where M represents any one of the rare earth atoms, samarium, yttrium, erbium or gadolinium. They have been shown by microwave and optical resonance experiments to be of the highest fundamental importance and a number of large industrial concerns are at present engaged in heavy programmes of work upon them.

Theory

Ferromagnetic and ferrimagnetic materials are found to consist of small regions in which the magnetic moment vectors are already aligned, i.e. each small region is spontaneously magnetized although the alignment in neighbouring regions differs and hence the resulting spontaneous magnetization of the whole specimen is determined by the vector sum of the moments of the discrete regions or domains and can take any value from zero to saturation.

In low magnetic field conditions a number of preferred directions of magnetization exist and as the field increases one group of domains grows at the expense of neighbouring domain areas. The changes of the domain pattern can be observed in polarized light as the specimen is taken through changing magnetizing field conditions around the hysteresis cycle.

Method

Gadolinium iron garnet specimens are chosen since comparatively low fields are required to produce magnetic saturation within the specimen. The garnets can be obtained in convenient form, about 3 mm diameter and of the order of a few hundredths of a millimetre thick already mounted upon a microscope slide (from the Laboratory Optical Co., P.O. Box 387, Plainfield, New Jersey, U.S.A.).

Place the field coil (*Figure D66*) on the microscope stage so that the mounted garnet may slide freely beneath it. (Two pieces of

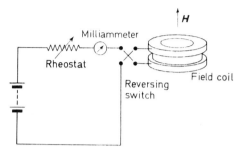

Figure D66. Field coil circuit

perspex glued either side of the coil form suitable lugs to clamp the coil and also form a bridge for the microscope slide.)

Fit up the circuit shown in *Figure D66* with a suitable battery, rheostat and reversing switch.

Illuminate the specimen by means of a high power mercury lamp such as the one shown in *Plate D8* and after first crossing the analyser and polarizer rotate the latter slightly to give optimum contrast on viewing the specimen.

Switch on the field current and increase slowly to saturation then decrease to zero and reverse the current to 'negative' saturation before returning to zero. Note significant pattern changes and tabulate the field current values at these points. Then photograph the specimen at each of these stages.

If a suitable photomultiplier cell is available a graph such as the one shown in *Figure D67* can be plotted.

Discussion of Results

Note in the photographs shown how the pattern first changes very rapidly corresponding to the steeply sloping region of the magnetization curve (*Plate D9*, 1–3) then less rapidly as saturation is approached (*Plate D9*, 4 and 5). Reduction of the field restores the rapid pattern change and this continues as the field is reversed (*Plate D9*, 7–8 and 9–10) until saturation is again approached in the opposite sense. Finally the pattern changes once more first slowly then more rapidly as the current returns to zero affording a striking demonstration (and explanation!) of hysteresis in terms of domain structure.

The photocell shown on the microscope in *Plate D8* is in fact a 'home-made' one selected from a number produced in accordance

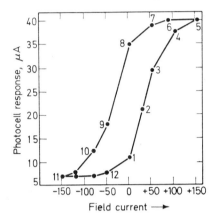

Figure D67. Typical domain hysteresis curve in polarized light

with the instructions given by W. F. Sheehan in the *Journal of Chemical Education*, October 1962, but the use of a commercial cell is preferable. Great care is necessary to eliminate the effects of stray light since the total light associated with the pattern and its changes is very low compared with the level of illumination in the surroundings. (A compensating 'dark current' cell may be used.)

Further Reading and References

Dillon, J. F. *Bulletin of the American Physical Society, Series 2*, vol. 2, p. 238, 1957.

Elementary reading on domains may be found in *Textbook of Electricity and Magnetism*, by G. R. Noakes (Macmillan, 4th edition, 1968), and a more advanced account by L. F. Bates in *Endeavour*, July 1957, p. 151, as well as in standard degree-level texts.

Dillon also supplies other references concerning the garnets in the *Journal of Applied Physics*, vol. 29, No. 3, p. 539, March 1958, as well as a number in the well-known Bell Telephone Company's Monographs.

Experiment No. D2.9

TO PREPARE A THERMISTOR AND TO INVESTIGATE THE VARIATION OF ITS RESISTIVITY AT ROOM TEMPERATURE WITH VARYING PROPORTIONS OF ITS CONSTITUENTS

Introduction

Thermistors are semiconducting devices which have a high negative temperature coefficient of resistance (see also Experiment No. A2.4).

Theory

Conductance in the thermistors is purely electronic and at *constant temperature* the current is proportional to the applied voltage.

The resistivity of the thermistor material depends upon the proportions of the compounds forming its constituents (the oxides of nickel, manganese, copper and cobalt), and may therefore be controlled by variation of these proportions.

In addition the resistivity appears to depend upon the 'firing' temperature.

Method

Weigh out the following mixtures and mix and grind them well in a mortar.

Mixture

A   To 1 g nickel oxide add 4 g manganese oxide.
B   To basic mixture A add 0·05 g cupric oxide.
C   To basic mixture A add 0·10 g cupric oxide.
D   To 1 g nickel oxide add 4 g cobalt oxide.

Place about half a gram of one of the prepared mixtures in a crucible, add a small quantity of distilled water and mix to a heavy slurry. (An organic binder such as a mixture of n-butyl acetate and pyroxylin may be found to be a useful alternative to distilled water.)

Using thin platinum wire (of the order of 0·09 mm, 0·0035 in., diameter) thread it across a nickel jig such as the one shown in

Figure D68, fastening the ends at A and B. By means of a glass rod transfer globules of the thermistor slurry across pairs of wires as

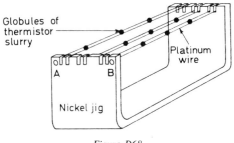

Figure D68

shown in *Figure D69*, noting the positions of the various mixtures for identification. Mount a number of each type of specimen, in case of failure to obtain good contact of the wire leads after sintering.

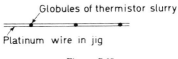

Figure D69

Allow the beads to dry in an oven at about 110° C for 30 minutes or so, then place the jig in a furnace and raise the temperature to 600° C fairly quickly. Continue heating more slowly to about 1 150° C to fuse the oxides onto the wires then switch off the furnace and allow it to cool *slowly* to room temperature. Remove the jig and cut the thermistors away leaving two wire leads in the manner indicated in *Figure D70*.

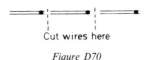

Figure D70

Mount the thermistors on a device similar to the one shown in *Figure D71*, in order to measure their resistivity at room temperature, and note how this varies with the composition of the material.

Discussion of Results

Unless grinding and mixing are very thorough, variations in thermistor properties will occur even with different samples of the same mixture, due to lack of homogeneity.

Further Work

Experiment with different firing temperatures to sinter the thermistors and note what effect, if any, this has upon the properties of the devices.

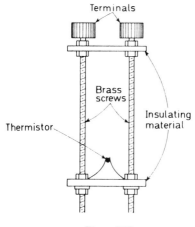

Figure D71

Experiment number A2.4 describes the way in which the characteristics of the thermistors may be investigated and comparison should certainly be made between the properties of these and the commercial types (e.g. Stantel F15 and F23 thermistors).

Reading and References

Applications data upon thermistors may be found in the *Mullard Technical Handbook*, vol. 5, 1963, under 'Non-linear resistors'.

A brief note of introductory nature on the thermistor is to be found in the *Encyclopaedic Dictionary of Physics*, edited by J. Thewlis (Pergamon, 1962). A typical thermistor characteristic is shown and a brief bibliography given.

Example No. D2.10

TO INVESTIGATE THE VARIATION OF CAPACITANCE (AND POWER FACTOR) WITH TEMPERATURE CHANGE IN A FERROELECTRIC (BARIUM TITANATE)

Introduction

When an electric field is applied to a dielectric crystal the positive charges are displaced in one direction and the negative charges in the opposite direction so that electric dipoles are produced in the material.

Certain materials possess permanent dipole moments and are said to be ferroelectric (they can be considered as analogous to ferromagnetic materials in the theory of magnetism). Barium titanate is chosen here since it displays ferroelectric properties to a marked degree.

Theory

The analogy with magnetism may be carried further and we find that the relative permittivity in ferroelectrics obeys a Curie-Weiss type law
i.e.

$$\varepsilon_r = \frac{\text{constant}}{(T - \theta)}$$

where ε_r = the relative permittivity, T = the absolute temperature of the specimen, and θ = transition temperature at which electric moments are randomly orientated and above which ferroelectric effects disappear.

Upon plotting a graph of relative permittivity against temperature an anomalous increase in the former occurs at the transition temperature (*Figure D72*).

In fact a change in the structure of the barium titanate (see second reference) takes place here and above the transition temperature it becomes piezoelectric.

Method

The specimen may be conveniently used in the form shown in *Figure D73* where the silver coatings effectively act as capacitor plates. It is placed in one arm of a Schering (or other standard) capacitance-measuring bridge set at a known frequence (say 1 kHz)

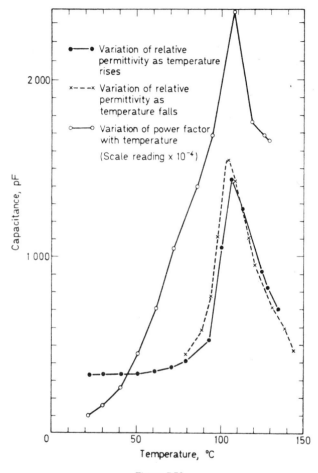

Figure D72

and immersed in an oil bath whose temperature is raised in convenient steps to about 130° C. Take care here and preferably use a fume cupboard. The temperature can be recorded by thermometer or thermocouple.

It may be assumed that the permittivity of air remains constant over the range of temperature of the experiment and hence the capacity of the arrangement

$$C = \frac{\varepsilon_0 \varepsilon_r A}{t}$$

enables the relative permittivity ε_r to be calculated since both A the area of each plate and t the separation of the plates are constant, and ε_0 the permittivity of free space is known.

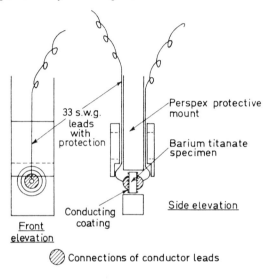

Figure D73. Diagram showing detail of barium titanate specimen in protective mount

Figure D72 shows a typical graph of variation of relative permittivity (and power factor) with temperature.

Discussion of Results

By far the least accurate of the measurements will be that of temperature and some estimate of the overall accuracy of the experiment may be made by taking readings not only as the temperature rises but as the specimen cools in the oil bath.

Reading and References

An account of ferroelectricity and ferroelectric crystals may be found in *Introduction to Solid State Physics* by C. Kittel (Wiley, 3rd edition, 1966).

A graph of the variation of the relative permittivity of barium titanate is given here. A more elementary treatment may be found in *Electronic Processes in Materials* by L. V. Azaroff and J. J. Brophy (McGraw-Hill, 1963).

Section E. Applied Heat, Mechanics of Fluids and Solids

Experiment No. E1.1
TO INVESTIGATE THE FORM OF THE LAW GOVERNING THE OPERATION OF A THERMOCOUPLE AND TO DETERMINE THE NEUTRAL TEMPERATURE OF THE COUPLE

Introduction

In this experiment a simple copper–iron thermocouple is used to illustrate the basic truths of the theory of thermoelectric effects. In practice chromel–alumel thermocouples are employed rather than the copper–iron ones, since they display the effect to a much greater extent and can be used at far higher temperatures.

Theory

The law governing the operation of the thermocouple may be assumed to be given by an equation of the form

$$E = a\theta + b\theta^2 \tag{1}$$

i.e. approximately parabolic, where E represents the thermal e.m.f., θ the temperature in °C, and a and b are constants.

The so called thermoelectric power $dE/d\theta$ (a misnomer since it is not truly power as it does not give variation of e.m.f. with time) may then be written

$$\frac{dE}{d\theta} = a + 2b\theta \tag{2}$$

which is a straight line law whose constants a and b may be determined. Having done so it should be noted that the turning point of the parabola, which gives the so called neutral temperature (*Figure E1*) occurs when $dE/d\theta = 0$. Hence using this fact in conjunction with the second equation the value of the neutral temperature may be confirmed.

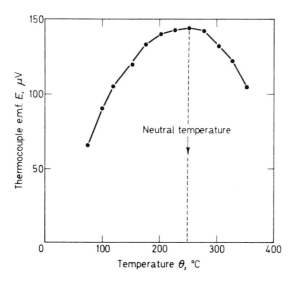

Figure E1. Graph showing that the relationship between thermocouple e.m.f. E and temperature θ is approximately parabolic, i.e.

$$E = a\theta + b\theta^2$$

A calibrated potentiometer is used in order to determine the thermoelectric e.m.f. developed at the junction. From the circuit (*Figure E2*) it is seen that

$$\frac{2}{R_1 + R_2 + R_L} = \frac{E_s}{R_1} \qquad (3)$$

where E_s represents the e.m.f. of the standard cell. Let r_1 = resistance cm^{-1} of wire of length L, i.e. $R_L = r_1 L$.
Then

$$\frac{V_L}{E_s} = \frac{r_1 L}{R_1} \qquad (4)$$

where V_L = potential drop across the wire of length L.
Hence

$$V_1 = \frac{V_L}{L} = \frac{E_s r_1}{R_1} \qquad (5)$$

so that the potentiometer is effectively calibrated.

Method

One junction of the thermocouple is placed in melting ice whilst the other is placed in a large block of metal on a suitable plate heater, the block carrying a small hole so that it may also take a thermometer reading up to $\sim 500°$ C.

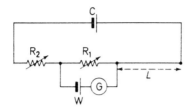

Figure E2. Circuit used in calibration of potentiometer wire.
C, 2 V cell   L, Length of potentiometer wire
W, Weston cadmium standard cell

The circuit is wired as shown in *Figure E3*. Then the potentiometer tapping key K is held at A and the key M switched to E, the Weston standard cell across R_1 being balanced by adjustment of R_2.

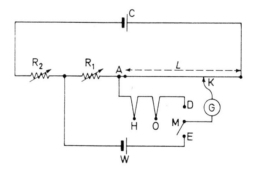

Figure E3. Investigation of the law governing the operation of a thermocouple
L, Length of potentiometer wire   C, 2 V cell
W, Weston standard cell   H, Hot junction
O, Cold junction at 0° C   G, Galvanometer
K, Potentiometer tapping key   M, Two-position key

The potential drop per centimetre of wire can then be calculated from equation (5), thus calibrating the wire potentiometer.

The key M is now switched to D and the thermocouple junction in the block heated.

Measurements of the thermoelectric e.m.f. are then made at known temperatures to obtain a graph such as *Figure E1*.

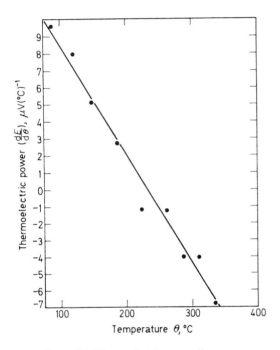

Figure E4. Thermoelectric power diagram

By taking the slope of the resulting parabola at convenient points, a graph of $dE/d\theta$ against θ may be plotted (*Figure E4*), the slope of which gives $2b$ and the intercept (at $\theta = 0°$ C!) gives a.

Discussion of Results

The most difficult portion of the experiment is the correct assessment of the slope of the graph and fairly wide 'scatter' upon the straight line graph of *Figure E4* may result. It must be remembered too that the agreement with the parabolic law is only approximate and that towards the extremes of the curve the deviations are marked. The value of the neutral temperature obtained by finding the slope and intercept of the straight line graph, putting them into equation (2) and using the fact that $dE/d\theta = 0$, gives a good indication of the accuracy of the experiment, when compared with the neutral temperature read directly from the graph of *Figure E4*. Note that the equation resulting from the knowledge of the constants *a* and *b* is effectively a calibration graph for the thermocouple.

The use of graphite or other good conductor in powder form placed in the hole carrying the thermometer, ensures good thermal contact and helps to contribute towards an accurate result.

Further Work

If a furnace is available, rather than the simple arrangement here, carry out the experiment for other thermocouples such as are actually used in practice.

Reading and References

The theory of the thermoelectric circuit is given in *Electricity and Magnetism* by J. H. Fewkes and J. Yarwood (University Tutorial Press, 1965).

A useful booklet on *Noble Metal Thermocouples* by H. E. Bennett is published by Johnson-Matthey, Ltd. (3rd edition, 1961).

The use of thermocouples in temperature measurement and control is dealt with in a book of that title by W. F. Coxon (Heywood, 1959).

Another useful publication is *Thermocouples—Their Instrumentation and Use* by B. F. Billing (Institute of Electrical Engineers, 1964).

Example No. E1.2

TO CALIBRATE A CHROMEL-ALUMEL THERMOCOUPLE BY MEANS OF A COOLING METHOD

Introduction

In this experiment the thermocouple e.m.f.s corresponding to the melting points of pure metals are found and used to plot a calibration graph for the thermocouple.

Theory

The normal cooling curve of a substance in which temperature is plotted against time may be considered to consist of three main regions as will be seen from *Figure E5a*.

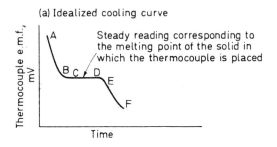

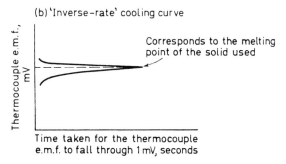

Figure E5. The calibration of a chromel-alumel thermocouple

In region AB the rate of fall of temperature is rapid. In the region CD the rate of fall is zero, i.e. temperature remains constant over a considerable period whilst the substance changes state, and in the region EF the rate of fall is once again rapid.

Hence if rate of fall is plotted against time a sharp horizontal 'peak' (*Figure E5b*) results corresponding to the temperature passing through the melting point of the substance.

It is convenient to use a dial type precision potentiometer to measure the e.m.f.s of the thermocouple and hence the graphs show (*Figure E6*) the times taken for the e.m.f. to fall in 0·1 mV steps rather than giving the actual rate of fall of e.m.f.

Method

Connect up the circuit as shown in *Figure E7* so that the e.m.f. generated by the arrangement of thermojunctions may be measured by a suitable potentiometer.

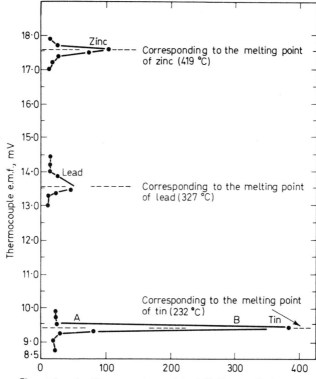

Figure E6. Graphs of thermocouple e.m.f. against time taken for the e.m.f. to fall 1 mV in the case of three pure metals

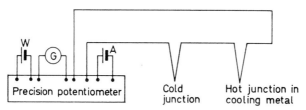

Figure E7. Apparatus used in the calibration of a thermocouple
W, Weston standard cell   G, Galvanometer   A, 2 V cell

Fill a small foundry type crucible with a quantity of pure tin and heat the crucible until all the metal has melted. Place the hot junction of the chromel–alumel thermocouple into the molten tin and continue heating for some minutes so that the temperature is a little above the melting point of the metal. Remove the source of heat and surround the crucible with asbestos plates cut to shape so that the sides and the top of the vessel are insulated, thus slowing down the cooling rate.

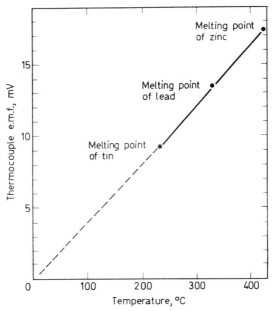

Figure E8. Calibration graph for chromel-alumel thermocouple

If a 'decade dial' type potentiometer is used find the approximate position of balance then reduce the dial by a 1 mV setting and as the specimen cools begin timing as the galvanometer deflection passes through the zero position. Reduce the setting once again and note the time taken for the specimen to cool through a temperature corresponding to 1 mV drop on the potentiometer. Repeat this procedure in order to determine the melting point of the tin (the time intervals will first increase and then decrease).

Plot the graph of the e.m.f. against the time taken for the e.m.f. to fall through 1 mV as in *Figure E6*.

Finally reheat the crucible, remove and clean the thermocouple and carry out the same procedure with first lead and then zinc.

Knowing the melting points of the three metals it is then possible to construct a calibration graph (*Figure E8*) for the thermocouple.

Discussion of Results

The decade type of potentiometer has the disadvantage that it is not possible to measure to subdivisions below 1 mV, hence as indicated by the graphs, it may not be possible to obtain confirmatory readings in such regions as AB (*Figure E6*)

A good calibration graph should preferably have more points than the one actually shown (aluminium and sodium chloride may be suggested as providing these). Note that the resulting graph displays slight curvature but over the restricted range shown here may be assumed to approximate to a straight line.

Further Work

The cooling method can also be utilized in differential thermal analysis. This requires two substances as similar in characteristics as possible but such that the one under test suffers a phase change (or other type of transformation, e.g. an order–disorder transformation), whilst the second (the reference medium) shows no such change over the chosen experimental range. This is illustrated in *Figures E9* and *E10* where naphthalene and anthracene are used as illustration.

The naphthalene melts at approximately 80° C, giving a discontinuity at B as shown in *Figure E9*, but the anthracene does not melt over the temperature range from 20° C to 150° C, the graph being approximately linear.

If the junctions of a chromel–alumel thermocouple are now placed in opposition, one in each substance, and the temperature of the bath is recorded by a second thermocouple, then the temperature difference between the two specimens may be plotted against the reference temperature of the bath. Over the range A to B (*Figure E9*) the two substances are at the same temperature hence the temperature difference is zero and the 'differential' graph is a horizontal straight line (*Figure E10*).

In the graph shown, the temperature difference is not zero— this can be due to a number of factors, e.g. different cooling rates of the samples, incorrect matching of the thermocouple junctions or

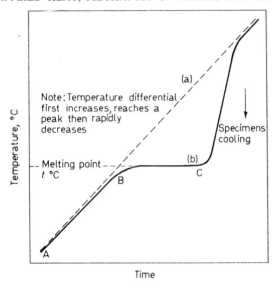

Figure E9. The method of differential thermal analysis showing the cooling curves of (a) the reference substance with no change in state, (b) the test sample which melts at $t°$ C and provides the temperature differential actually measured

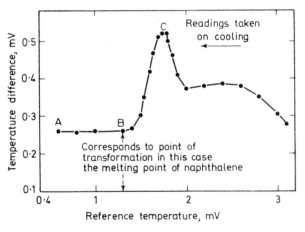

Figure E10. The method of differential thermal analysis. Graph showing the sharp temperature differential occurring at the melting point of naphthalene

simply variations in temperature within the bath and between the specimens in the initial phase of the experiment.

At B a difference in temperature is seen and this increases, hence the graph of *Figure E10* rises sharply until the point corresponding to C on *Figure E9* is reached. From this point the temperature of the

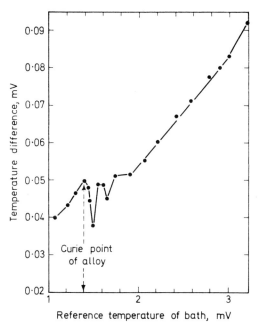

Figure E11. Detection of order-disorder transformation by differential thermal analysis method at the Curie point of a strongly magnetic Monel-metal alloy

(now liquid) naphthalene rises rapidly towards that of the reference substance, i.e. the temperature differential decreases, hence a 'hump' appears in the graph of *Figure E10*.

It was noted earlier that the method can be used for transformations other than those associated with melting, e.g. order–disorder transformations, and as a case in point strongly magnetic alloys such as Monel metal suffer a re-orientation of structure and lose their strong magnetic characteristics as the temperature rises above the Curie point. *Figure E11* shows the differential temperature curve of Monel metal and a 'dip' is seen here corresponding to the transformation at the Curie point. Note that the graph slopes on either side of the 'dip' but it must be remembered that the y axis is highly magnified and the few degrees difference represented by the slope is typical of that obtained in an experiment of only two or three hours' duration since the cooling rate of the bulk material of the bath is seldom as slow as one would wish it to be. The vestigial second dip is probably due to lack of homogeneity in the alloy. Thus differential thermal analysis forms a convenient way of detecting the small thermal changes that occur with such order–disorder transformations which would be difficult to measure in any other way. The method must be used with caution, however, if reproducible results are to be obtained.

Reading and References

Thermocouples and temperature measuring instruments are dealt with in *Metallurgy for Engineers* by E. C. Rollason (Arnold, 1949).

Practical Physical Metallurgy by Rawlings (Butterworths, 1961), besides giving a brief account of the method of differential thermal analysis will provide many ideas for further work.

Differential Thermal Analysis—Theory and Practice by W. J. Smothers and Y. Chiang (Chemical Publishing Company, 1958) provides a deeper treatment of the subject.

Experiment No. E1.3

TO INVESTIGATE THE VARIATION IN THE SPECIFIC HEAT OF GRAPHITE WITH CHANGE IN TEMPERATURE

Introduction

We so often make use of an average value of the specific heat of a substance in elementary calorimetric calculations that it is sometimes overlooked that in general the specific heat of a substance varies with temperature.

In this experiment graphite is chosen as the experimental material since its specific heat varies considerably over the range 100 to 400°C.

SPECIFIC HEAT OF GRAPHITE   E1.3

Theory

The quantity of heat Q given to the system is given by

$$Q = ms\theta$$

where m = the mass of the block
s = specific heat of graphite
and θ = change in temperature,

hence the rate of loss of heat as the block cools

$$\frac{dQ}{dt} = ms\frac{d\theta}{dt}$$

but

$$\frac{d\theta}{dt} = \text{the slope of the cooling curve (\textit{Figure E13})}$$

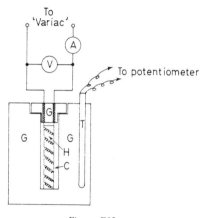

Figure E12
A, Ammeter (up to 10 A)   C, Cavity to contain heater
G, Graphite   T, Thermocouple
V, Voltmeter (0–250 V)   H, Heater

at the chosen temperature and at any particular equilibrium temperature

$$\frac{dQ}{dt} = \frac{I^2 R}{J}$$

where I = current
R = resistance,
and J = the electrical equivalent of heat.

i.e. is proportional to the power supplied to the heater.

Thus if the mass m of the block is known, the specific heat s at any temperature may be found

$$s = \frac{60P}{m \times \text{Slope} \times 4\cdot 2}.$$

(Note the conversion factors. The factor J ($=4\cdot 2$) appears only if the quantity of heat is expressed in calories—not if expressed in S.I. units, i.e. joules.)

From the calculation a graph such as *Figure E15* may then be plotted.

Method

The apparatus consists of a large block of graphite weighing some 500 g with a hole drilled in it (*Figure E12*) to take a heating element of nickel–chrome wire down the centre and also a chromel–alumel thermocouple within the main body of the block.

The heater is supplied with its maximum recommended power and heating is continued until an equilibrium temperature is achieved, after which time the block is allowed to cool, thermocouple readings being taken every three minutes. A graph is plotted such as shown in *Figure E13*.

A lower value of power is now supplied to the heater until (a new) equilibrium temperature is obtained and this is repeated for a series of different values of input power. The equilibrium temperature is then plotted against the input power (*Figure E14*).

Discussion of Results

The most important source of error is the estimation of the slope of the cooling curve.

It should be remembered too, that the system cools as a whole and no correction has been made for the heater about whose 'specific heat' (and its variation) little or nothing is usually known.

Further Work

Some materials, of which mild steel (or iron) is an example, display discontinuities in the variation of their specific heats corresponding to order–disorder transformations within the material itself.

Plate D6. Specimen after strain annealing showing the growth of single crystal regions over the central part of the aluminium

 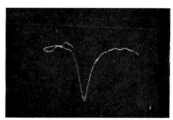

Plate D7. Nuclear magnetic resonance. Traces obtained (left) in the absence of resonance and (right) with resonance absorption taking place

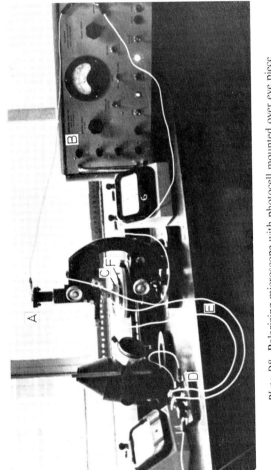

Plate D8. Polarizing microscope with photocell mounted over eye piece to give hysteresis curve readings of light transmission through specimen
A, Photocell   B, Photocell supply
C, Field coil   D, Field coil current reversing switch
E, Field coil supply   F, Specimen on microscope slide

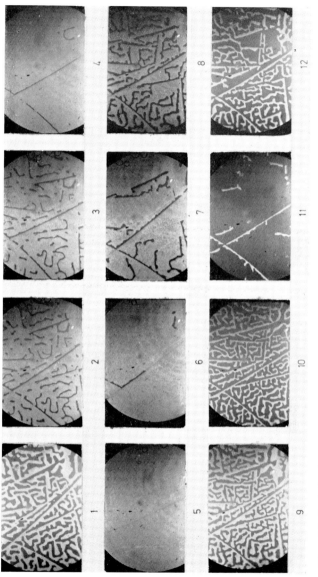

Plate D9. Variation in the domain pattern of a ferrimagnetic garnet, with variation of applied magnetic field to the specimen. Numbers correspond to those giving the approximate positions on the hysteresis curve (*Figure D67*)

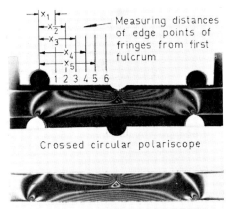

Plate E1. Photographic negatives of fringe patterns obtained in the experiment with a three point loaded beam

SPECIFIC HEAT OF GRAPHITE

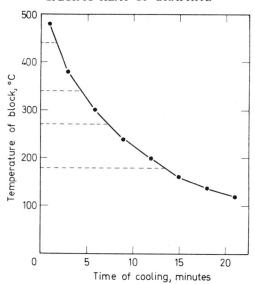

Figure E13. Cooling curve for graphite

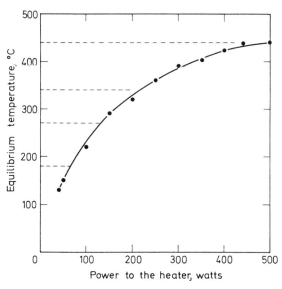

Figure E14. Graph showing the equilibrium temperatures obtained corresponding to the powers supplied to the heater

It is worthwhile to investigate these using the differential thermal analysis method (see Experiment No. E1.2). Copper forms a convenient reference material and the specimens should be allowed to

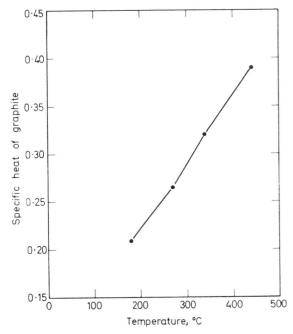

Figure E15. Graph showing the variation in the specific heat of graphite with change in temperature

cool slowly from about 1 000° C. (Don't forget that a Curie point transformation occurs as well as two phase changes within the iron.)

Reading and References

The simple treatment of this experiment should be compared with that given in Apparatus for the Measurement of Specific Heat of Metals by W. H. Martin and J. Rixon in *Journal of Scientific Instruments*, vol. 36, pp. 179–182. Whilst this latter method is one where temperature increase is employed rather than cooling, many parallels may be drawn, and the theory given there will give a deeper insight into the errors involved in the simplified method used above. Detail of a suitable heater is also shown.

Experiment No. E1.4

TO DETERMINE THE THERMAL CONDUCTIVITY OF A BAD CONDUCTOR IN THE FORM OF A SOLID CYLINDRICAL SHELL

Introduction

The method usually encountered by students for the determination of the thermal conductivity of a bad conductor is that of Lees. The aim of this experiment is to measure this quantity by a method depending upon the radial flow of heat through a cylindrical shell, a method used by many workers but not as familiar to students in practice as Lees disc.

Theory

If the flow of heat across a cylindrical shell (*Figure E16*) is considered we have

$$Q_1 = 2\pi r l k \frac{d\theta}{dr}$$

where Q_1 = heat flowing second^{-1}
k = thermal conductivity of the shell
r = the radius of the shell
$\frac{d\theta}{dr}$ = temperature gradient across the shell.

Rearranging

$$\int_{\theta_1}^{\theta_2} d\theta = \frac{Q_1}{2\pi l k} \int_{r_1}^{r_2} \frac{dr}{r} \quad \text{and} \quad k = \frac{Q_1}{2\pi l} \frac{\log(r_2/r_1)}{(\theta_2 - \theta_1)}$$

where θ_1 represents the temperature on one side of the shell (radius r_1) and θ_2 that on the other (radius r_2).

For convenience of experimental arrangement steam passes on the outside of the glass tube chosen as specimen here, (i.e. $\theta_1 \simeq 100°$ C), and water flows along the inner side of the tube. As it does so its temperature is noted at the inlet (θ_3) and at the outlet point of the tube (θ_4) so that the average temperature on the inner surface

APPLIED HEAT, MECHANICS OF FLUIDS AND SOLIDS

may be taken as $(\theta_4 - \theta_3)/2$ corresponding to θ_2 in the above theory. Hence the equation becomes

$$k = \frac{Q_1}{2\pi l} \frac{\log(r_2/r_1)}{\left[\theta_1 - \left(\frac{\theta_4 - \theta_3}{2}\right)\right]}.$$

It now remains to evaluate Q_1 and this may be done from calorimetric considerations for $Q_1 = ms(\theta_4 - \theta_3)$

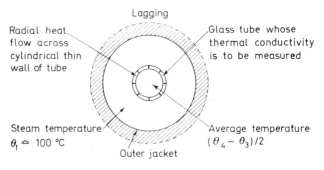

Figure E16. Determination of thermal conductivity by radial heat flow through a cylindrical shell

where m represents the mass of water flowing per second, and s = specific heat of water. Hence if the total mass of water M collected in time t is noted, $m = M/t$

and

$$k = \frac{Ms(\theta_4 - \theta_3)\log(r_2/r_1)}{t2\pi l\left\{\theta_1 - \dfrac{\theta_4 - \theta_3}{2}\right\}}.$$

Method

The bad conductor chosen in this case is a thin walled glass tube which is surrounded by a lagged steam jacket. It is tilted slightly upwards so that water from a constant head arrangement can flow safely through without air bubbles being entrapped.

Steam enters the outer jacket in such a way that it first encounters the warmest water (*Figure E17*).

The flow of water is adjusted so that the inlet and outlet temperatures show a difference of some 15–20° C when a steady state has been achieved.

THERMAL CONDUCTIVITY OF A BAD CONDUCTOR E1.4

The flow of water through the inner tube is collected in a measuring cylinder over a given time so that the mass of water flowing per second is known.

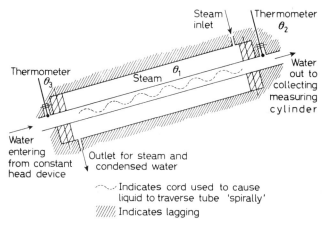

Figure E17. Determination of thermal conductivity of glass by radial flow of heat across a cylindrical shell

The inner and outer radii of the tube are measured by vernier microscope and also the length of the tube inside the steam jacket. Then using the formula previously quoted k is determined.

Discussion of Results

The largest error here is due to the fact that the thermometers lie outside the jacket and the distance between the points at which temperatures θ_4 and θ_3 are taken cannot be easily determined with accuracy. The thermometers themselves should read to $\frac{1}{5}°$ C (or better). The temperature of the steam corresponding to the atmospheric pressure at the time of the experiment should be noted from tables rather than assuming $\theta_1 = 100°$ C. Generally the temperature in the outer jacket differs markedly from this value and is the source of serious error in the final result.

The theory assumes that the temperature is taken immediately at the inner wall and that across the section of the inner tube it remains constant. To help effect this a spiral of rubber or cord may be pushed inside the tube so that water traverses the tube in a spiral motion.

APPLIED HEAT, MECHANICS OF FLUIDS AND SOLIDS

Further Work

The experiment should be repeated using tubes of different insulating materials and the effect with and without the cord spiral in the inner tube should be noted.

Reading and References

Details of radial flow methods of measuring thermal conductivity are to be found in *Dictionary of Applied Physics* by R. Glazebrook (McMillan, 1923).

General methods of thermal conductivity measurement are treated in standard degree texts and in particular in *Text-Book of Heat* by H. S. Allen and R. S. Maxwell (McMillan, 1962) as well as in the next reference.

A brief description of this experiment suggesting a graphical treatment is given in *Advanced Practical Physics* by B. L. Worsnop and H. T. Flint (Methuen, 1962).

Experiment No. E1.5

THE RADIAL HEAT-FLOW METHOD USED TO DETERMINE THE THERMAL CONDUCTIVITY OF A LIQUID

Introduction

The method depends essentially upon the radial transfer of heat by conduction across the liquid, contained in a suitable cell (*Figure E18*), from an electrically heated filament. The thermal conductivity is a function of the variation of the power supplied to the filament, with the resistance of the filament, as well as of the physical parameters of the cell. The cell constants are eliminated from the calculations by first using (in turn) two liquids, water and toluene, of known thermal conductivity. The outer wall temperature of the cell is kept constant by a suitable thermostatic bath.

There are thus two distinct parts to the experiment. (1) The determination of the cell parameters by the use of two liquids of known thermal conductivity. (2) The determination of the unknown thermal conductivities of other liquids.

THERMAL CONDUCTIVITY OF A LIQUID

Theory

The power P supplied by the wire when steady state conditions have been achieved, is

$$P = \frac{k_L 2\pi l (\theta_f - \theta_I)}{\log (r_I/r_f)} \quad \text{or} \quad \frac{\theta_f - \theta_I}{P} = \frac{\log (r_I/r_f)}{k_L 2\pi l} \quad (1)$$

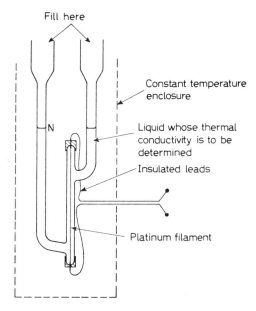

Figure E18. Detail of cell for determination of thermal conductivity of liquids. N.B. The cell should be filled with liquid to a point well above the limb carrying the filament

where k_L = thermal conductivity of the liquid in the cell
 l = length of the heated wire
 r_I = internal radius of the tube of the cell
 r_f = radius of the filament
 θ_f = temperature of the filament
 θ_I = temperature at internal wall of glass tube,

Across the glass wall of the cell the heat dissipated

$$P = \frac{2\pi l k_g (\theta_I - \theta_0)}{\log_e (r_0/r_I)} \quad \text{or again} \quad \frac{\theta_I - \theta_0}{P} = \frac{\log (r_0/r_I)}{2\pi l k_g} \quad (2)$$

where θ_0 = temperature at outer wall of containing tube (i.e. temperature of thermostatic bath)
r_0 = radius of outer wall of tube
and k_g = thermal conductivity of glass.

It may be assumed that the resistance is proportional to the (absolute) temperature and hence since we are concerned only with *changes* in temperature and resistance

$$(R_f - R_B) = b(\theta_f - \theta_B)$$

where R_f = resistance of heated filament
R_B = resistance of unheated filament
θ_f = temperature of heated filament
θ_B ($= \theta_0$ above) = temperature of unheated filament
and b = constant.

Adding equations (1) and (2)

$$\left[\frac{\theta_f - \theta_1}{P} + \frac{\theta_1 - \theta_0}{P}\right] = \frac{\log_e (r_1/r_f)}{2\pi l k_L} + \frac{\log (r_0/r_1)}{2\pi l k_g} = \frac{\theta_f - \theta_0}{P}$$

but

$$(\theta_f - \theta_0) = \frac{R_f - R_B}{b}$$

hence

$$\frac{R_f - R_B}{bP} = \frac{\log_e (r_1/r_f)}{2\pi l k_L} + \frac{\log_e (r_0/r_1)}{2\pi l k_g}$$

i.e.

$$\frac{(R_f - R_B)}{P} = \frac{A}{k_L} + \frac{B}{k_g} = \frac{A}{k_L} + B'$$

where

$$A = \frac{b \log_e (r_1/r_f)}{2\pi l} \quad \text{and} \quad B' = \frac{B}{k_g} = \frac{b \log (r_0/r_1)}{2\pi l k_g}$$

which are essentially constant parameters of the cell used.

To determine A and B' first water [$k_w = 6\cdot1 \times 10^{-3}$ watts cm^{-1} (°C)$^{-1}$] and then toluene [$k_T = 1\cdot33 \times 10^{-3}$ watts cm^{-1} (°C)$^{-1}$] are used in the cell and graphs of (dR/dP) are plotted in each case. Knowing the thermal conductivity of glass (k_g), A and B' can be found.

The apparatus may then be used to determine the thermal conductivity of other liquids.

Method

The cell is filled to the point N with distilled water. The bridge circuit is fitted up as shown in *Figure E19*. The decade resistance box D must be carefully chosen (a suitable box is made by H. W. Sullivan, Ltd.) since very small changes in resistance of 0·001 Ω (or

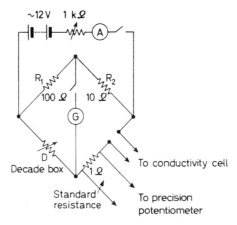

Figure E19. Bridge circuit for thermal conductivity cell

better) have to be measured, see *Figure E20*), and care must be taken not to exceed the current rating appropriate to such a box (it will be stated on the box).

The current is increased in suitable steps to the recommended maximum (which will vary with each individual cell) and balance is found by adjustment of the variable resistance D. The current through the cell may be determined by placing a standard resistance (1 Ω in the case quoted) in series with the cell so that the total resistance in that arm of the bridge is $(R_c + 1)$, where R_c = resistance of the cell itself.

The voltage drop across the cell is measured by a suitable potentiometer, hence the current i through the standard resistance (and thus through the cell) is known. (The potentiometer is not shown on the diagram. A Cambridge potentiometer was used to obtain the results quoted here.)

The power supplied to the cell is then $i^2 R_c$ and a graph of the resistance of the cell against the corresponding power reading is drawn (*Figure E20*). The process is repeated using toluene in the cell and the slopes of these graphs, for water and toluene of known

APPLIED HEAT, MECHANICS OF FLUIDS AND SOLIDS

thermal conductivities, enable the parameters of the cell A and B' to be determined since

$$m_w = \frac{A}{k_w} + B', \qquad m_T = \frac{A}{k_T} + B'$$

where m_w = the slope of the graph drawn in the case of water
m_T = the slope of the graph drawn in the case of toluene.

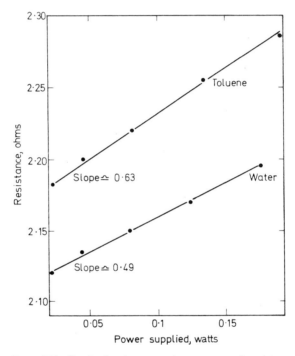

Figure E20. Graph of resistance against power used to determine the operational parameters of a cell employed in the determination of the thermal conductivity of a liquid

Finally the cell is filled with the liquid whose thermal conductivity is to be found (i.e. methanol) and the procedure is repeated.

Discussion of Results

Clean and secure contacts throughout the electrical circuit are essential for an accurate result to be obtained.

The cell itself must be carefully cleaned before use. (First with dilute hydrochloric acid, then with distilled water and finally rinsed with the liquid to be used and thoroughly dried.) Care should be taken to fill well above the top of the limb containing the filament and to avoid entrapped air bubbles in the cell.

The temperature must be maintained constant by a suitable thermostatic bath. For use here it has been found sufficient to place the cell in an empty vacuum flask. (The use of a liquid bath introduces difficulty with regard to effective insulation of the cell and its electrical leads.)

Note too that the leads of the cell are arranged to leave the cell symmetrically (they may be bound to the limb of the cell with electrical tape). This is important only if the cell is to be immersed in a liquid bath, when a temperature difference may arise between the two junctions of the platinum wire, giving rise to a small thermal e.m.f.

Convection currents within the liquid can create serious error in the results obtained. The effect varies with the liquid under test but is clearly indicated by departure from linearity in the resistance-power graphs at high current ratings.

For optimum accuracy several minutes should elapse at each power setting in order that equilibrium conditions in the cell may be achieved.

Further Work

A whole range of liquids may be investigated as well as the variation of thermal conductivity at different temperatures.

An extensive project would be provided by the design of a direct-reading instrument such that the changes in filament resistance (resulting in lack of bridge balance) are shown as a calibrated deflection on a suitable galvanometer. Thus the meter would give a direct reading of thermal conductivity.

Reading and References

The method and construction of the cell is given in detail in a paper by D. T. Jamieson and J. S. Tudhope entitled, 'A Simple Device for Measuring the Thermal Conductivity of Liquids with Moderate Accuracy' published as the *National Engineering Laboratory's Report 81*, April 1963.

The results obtained with a range of twelve liquids are given, together with an extensive bibliography.

APPLIED HEAT, MECHANICS OF FLUIDS AND SOLIDS

Experiment No. E1.6

ÅNGSTRÖM'S METHOD USED TO MEASURE THE THERMAL DIFFUSIVITY OF A COPPER BAR

Introduction

The thermal diffusivity of a substance is a numerical measure of the rate of diffusion of heat through the substance, and the mathematical treatment of the topic shows that the diffusion takes place in accordance with a law whose expression involves not only the thermal conductivity but also the specific heat s and the density ρ of the substance, so that the diffusivity may be written as $(k/\rho s)$.

Theory

If we consider the temperature θ at a point P distance x along a metal bar then the heat flowing into the section (*Figure E21*) of area A is given by

$$Q_x = -kA \frac{\delta\theta}{\delta x} \sec^{-1}.$$

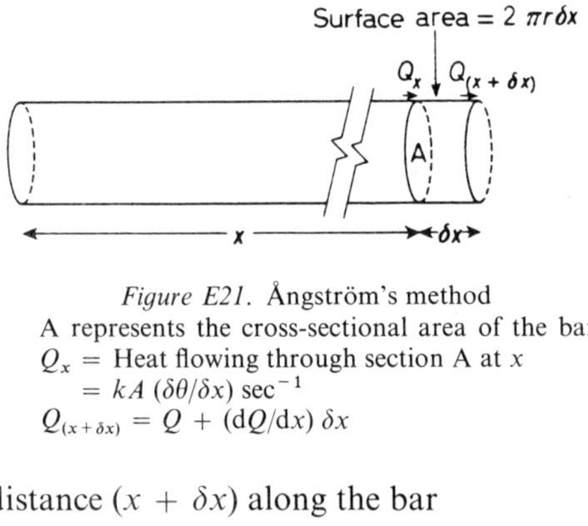

Surface area = $2\pi r \delta x$

Figure E21. Ångström's method
A represents the cross-sectional area of the bar
Q_x = Heat flowing through section A at x
   = $kA\,(\delta\theta/\delta x)\sec^{-1}$
$Q_{(x+\delta x)} = Q + (\mathrm{d}Q/\mathrm{d}x)\,\delta x$

At a point distance $(x + \delta x)$ along the bar

$$Q_{(x+\delta x)} = Q + \left(\frac{\mathrm{d}Q}{\mathrm{d}x}\right)\delta x.$$

Therefore the resultant flow of heat

$$= -\left(\frac{\mathrm{d}Q}{\mathrm{d}x}\right)\delta x = \frac{\mathrm{d}}{\mathrm{d}x}\left[-kA\left(\frac{\mathrm{d}\theta}{\mathrm{d}x}\right)\right] = -kA\frac{\mathrm{d}^2\theta}{\mathrm{d}x^2}.$$

This heat is expended in two ways:
(1) in heating up the bar,
(2) in heat loss from the surface of the bar.

Now consider the rate of change of temperature ($d\theta/dt$) given by the equation

$$\frac{dQ}{dt} = A\rho s(\delta x)\frac{d\theta}{dt}$$

since $A\rho(\delta x)$ represents the mass of the portion of the bar under consideration. Also by Newton's law the rate of loss of heat is proportional to the excess temperature, i.e.

$$\frac{dQ}{dt} = \beta 2\pi r\, \delta x(\theta - \theta_0)$$

where β is a constant
hence

$$kA\,\delta x\left(\frac{d^2\theta}{dx^2}\right) = a''(\theta - \theta_0)\,\delta x + A\rho s(\delta x)\frac{d\theta}{dt} \quad \text{(where } a'' = \beta 2\pi r\text{)}$$

and

$$\left(\frac{k}{\rho s}\right)\left(\frac{d^2\theta}{dx^2}\right) - a'(\theta - \theta_c) = \frac{d\theta}{dt} \quad \left(a' = \frac{a''}{A\rho s}\right)$$

Now if

$$\Theta = (\theta - \theta_0)$$

then

$$\frac{d\Theta}{dt} = \frac{d\theta}{dt} \quad \text{and} \quad \frac{d\Theta}{dx} = \frac{d\theta}{dx} \text{ etc.}$$

since θ_0 the temperature of the surroundings may be assumed to be constant, therefore

$$\left(\frac{k}{\rho s}\right)\frac{d^2\Theta}{dx^2} - a'\Theta = \frac{d\Theta}{dt} \quad \text{or} \quad \frac{d^2\Theta}{dx^2} - \frac{a'\Theta}{D} = \frac{1}{D}\left(\frac{d\Theta}{dt}\right)$$

$D = (k/\rho s)$ represents the thermal diffusivity of the material of the bar.

If one end of the bar is now heated and cooled alternately for fixed periods of time, then the solution of the function may be written

$$\Theta = \sum_{n=0}^{n\to\infty} C_n \exp(-a_n x)(\sin n\omega t - b_n x).$$

To solve the equation

$$\frac{d^2\Theta}{dx^2} - \frac{a'}{D}\Theta = \frac{1}{D}\left(\frac{d\Theta}{dt}\right)$$

we assume

$$\theta(xt) = f(x)f_1(t)$$

then we can write

$$f_1 t \frac{\partial^2}{\partial x^2}(fx) - \left(\frac{a'}{D}\right) f(x)f_1(t) = \frac{1}{D} f(x) \frac{d}{dt}(f_1 t).$$

Dividing by $f(x)f_1 t$ we obtain

$$\frac{1}{f(x)} \frac{\partial^2}{\partial x^2}[f(x)] - \left(\frac{a'}{D}\right) = \frac{1}{D} \frac{1}{f_1(t)} \frac{d}{dt}(f_1 t)$$

and it is important to note that the left-hand side of the equation is time independent whilst the right-hand side is independent of x. The two sides can thus only balance if they are equal to a constant and thus

$$\frac{1}{f(x)} \frac{d^2}{dx^2}(fx) - \alpha = p \tag{1}$$

$$(\alpha = a'/D)$$

and

$$\left(\frac{1}{D}\right) \frac{1}{f_1 t} \frac{d}{dt}(f_1 t) = p. \tag{2}$$

From (2) we obtain, by separating the variables

$$\frac{df_1 t}{f_1 t} = Dp \, dt$$

and integrating both sides

$$\log_e f_1(t) = Dpt + \text{constant } C$$

taking antilogs

$$f_1(t) = \exp(Dpt + C) = \exp(Dpt) + \exp(C)$$

but $\exp(C)$ is a constant since c is constant,

i.e.

$$f_1 t = A \exp(Dpt). \tag{3}$$

Again

$$\frac{1}{f(x)}\frac{d^2}{dx^2}fx - \alpha = p \quad \text{i.e.} \quad \frac{d^2}{dx^2}fx = (\alpha + p)f(x)$$

The solution is of the form

$$f(x) = B_1 \exp\{(\alpha + p)^{\frac{1}{2}}x\} + B_2 \exp\{-(\alpha + p)^{\frac{1}{2}}x\}$$

as can be confirmed by simply differentiating twice.

Also by the physical nature of the experiment, as x increases the temperature decreases, hence only the second term is appropriate and we have

$$f(x) = B \exp\{-(\alpha + p)^{\frac{1}{2}}x\} \tag{4}$$

Since the temperature variation is a function of both x and time t we must combine the two solutions obtained in equations (3) and (4) and we get

$$\Theta = C \exp\{-(\alpha + p)^{\frac{1}{2}}x\} \exp(Dpt) \tag{5}$$

In the experiment we in fact apply a periodically varying source of heat and may write

$$Dp_n = in\omega$$

Note that p has now been replaced by p_n since we are considering the possibility of n cycles and equation (5) becomes

$$\Theta = C_n \exp\left\{-\left(\alpha + \frac{in\omega}{D}\right)^{\frac{1}{2}}x\right\} \exp(in\omega t)$$

Let

$$\left(\alpha + \frac{in\omega}{D}\right)^{\frac{1}{2}} = (a_n + ib_n)$$

then the equation becomes

$$\Theta = C_n \exp\{(-a_n - ib_n)x + in\omega t\}$$
$$= C_n \exp(-a_n x) \exp\{i(n\omega t - b_n x)\}$$
$$= C_n \exp(-a_n x)\{\cos(n\omega t - b_n x) + i \sin n\omega t - b_n x)\}.$$

Since we have a linear differential equation *both* real and imaginary parts of this equation satisfy it and in particular the imaginary part gives

$$\Theta = \sum_{n=1}^{n\to\infty} C_n \exp(-a_n x) \sin(n\omega t - b_n x).$$

The subscript n is introduced since a different value of C is appropriate for each value of n.

We had

$$\left(\alpha + \frac{in\omega}{D}\right)^{\frac{1}{2}} = (a_n - ib_n) \quad \text{above,}$$

therefore

$$\alpha + \frac{in\omega}{D} = a_n^2 - b_n^2 + 2ia_nb_n$$

and equating real and imaginary terms

$$\alpha = a_n^2 - b_n^2 \quad \text{and} \quad \frac{n\omega}{D} = 2a_nb_n.$$

From the last equation $a_n = n\omega/2b_nD$ and substituting in the equation for α one obtains

$$\alpha = \frac{n^2\omega^2}{4b_n^2 D^2} - b_n^2$$

hence

$$4b_n^4 D^2 + 4D^2\alpha b_n^2 - n^2\omega^2 = 0$$

and

$$b_n^2 = \frac{-4D^2\alpha \pm (16D^4\alpha^2 + 4n^2\omega^2 4D^2)^{\frac{1}{2}}}{8D^2}$$

$$= \frac{-\alpha \pm (\alpha^2 + n^2\omega^2/D^2)^{\frac{1}{2}}}{2}$$

whence taking the positive (physically real) solution

$$b_n = \left\{\left(\alpha^2 + \frac{n^2\omega^2}{D^2}\right)^{\frac{1}{2}} - \alpha\right\}^{\frac{1}{2}} \bigg/ 2$$

and

$$a_n = \frac{n\omega}{2D} \frac{2^{\frac{1}{2}}}{\{(\alpha^2 + n^2\omega^2/D^2)^{\frac{1}{2}} - \alpha\}^{\frac{1}{2}}}$$

or

$$a_n = \frac{n\omega}{2^{\frac{1}{2}}D} \frac{\{(\alpha^2 + n^2\omega^2/D^2)^{\frac{1}{2}} + \alpha\}^{\frac{1}{2}}}{\{\alpha^2 + n^2\omega^2/D^2 - \alpha^2\}^{\frac{1}{2}}} = \left\{\frac{(\alpha^2 + n^2\omega^2/D^2)^{\frac{1}{2}} + \alpha}{2}\right\}^{\frac{1}{2}}.$$

Method

The metal rod (brass or copper are suitable) should be about 1 metre long and about 25 mm diameter. It carries a number of thermocouples at measured distances along it and is lagged in asbestos for the majority of its length. At one end a 750 W heater is screened from a small brass cylinder C, which surrounds the rod, by an asbestos screen (*Figure E22*).

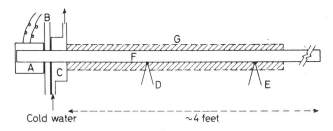

Figure E22. Apparatus used in the determination of the diffusivity of a metal bar by Ångström's method
A, Tubular heater (750 watts)   B, Insulating screen
C, Metal chamber through which cold water flows
D and E, Thermocouples attached to bar
F, Brass or copper bar ($\sim$ 1 metre long)
G, Insulation

The rod is heated for 10 minutes and then cold water is passed through the cylinder for 10 minutes to cool the end of the rod.

Cold water flowing over the end of the rod maintains it at constant temperature throughout the experiment.

The process of heating and cooling for 10 minute intervals is repeated and temperature readings are noted every half minute. The 'cycling' of heating and cooling is continued until the bar is observed to have reached a steady state. Typical results are shown on the graph of *Figure E23*, and from this basic data of temperature against time, further graphs are plotted, viz.:

(1) Of the logarithm of *twice* the temperature amplitude of the first graph against the distance of the temperature recording points from the chosen origin of the bar (*Figure E24*).

(2) The times at which wave crests occur on the first graph are plotted against x and on the same scale the times also that (*a*) troughs, and (*b*) median points occur on the original graph against x (*Figure E25*).

APPLIED HEAT, MECHANICS OF FLUIDS AND SOLIDS

From the slopes of these graphs the diffusion coefficient can be calculated as follows.

For the fundamental mode we have

$$\Theta = C_1 \exp(-a_1 x) \sin(\omega t - b_1 x)$$

and $\sin(\omega t - b_1 x)$ may be written

$$\sin 2\pi \left(\frac{t}{T} - \frac{x}{\lambda} \right)$$

whence

$$b_1 = \frac{2\pi}{\lambda}.$$

But

$$\lambda = v_1 T = v_1 \left(\frac{2\pi}{\omega} \right)$$

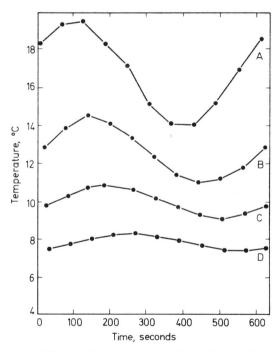

Figure E23. Graph of temperature fluctuations at four points along a bar used to determine diffusivity

where T represents period, and v_1 the velocity of the fundamental, i.e.

$$b_1 = \frac{\omega}{v_1}$$

Then

$$D = \frac{\omega}{2a_1 b_1} = \frac{v_1}{2a_1}.$$

The slope of the log $\Delta\Theta$ against x graph (*Figure E24*) gives the value of a, and the average of the slopes of the graphs recording crests and

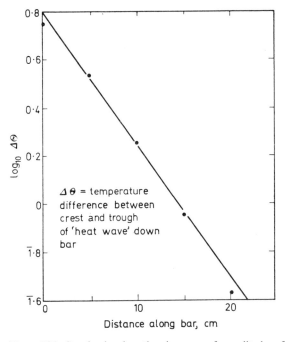

Figure E24. Graph showing the decrease of amplitude of maximum temperature fluctuations with distance along the bar, in Ångström's method

trough movements (*Figure E25*) gives the velocity of the fundamental, hence thermal diffusivity D is given by

$D = $ Slope of graph of *Figure E25*/2 $\times$ Slope of graph of *Figure E24*.

Discussion of Results

The graph of temperature against time at the recording point nearest the source will vary significantly from a true sine wave due to the presence of harmonics (of higher frequency) if the thermocouple is very near the source of heat.

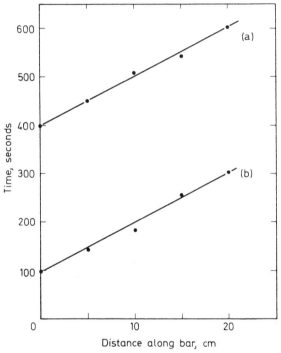

Figure E25. Times of occurrence of (a) crests and (b) troughs plotted against distance of points of observation along the bar

The curves corresponding to the more distant recording points follow the simple sine curve more closely, due to the fact that the harmonics are rapidly attenuated.

One should ensure that a steady state has been achieved along the bar BEFORE temperatures are tabulated.

Further Work

D. J. McNeil, in a note in the *American Journal of Physics*, vol. 32, No. 8, p. 642, August 1964, gives an alternative treatment of this experiment. The specimen is in the form of a copper wire only 1 mm in diameter and 16 cm long. The heating and cooling cycles are reduced to 120 seconds and a large number of results can be obtained in a short space of time.

In the main experiment described above, the points at which temperatures are taken are sufficiently remote from $x = 0$ for the harmonic to be drastically attenuated but in McNeil's treatment the harmonics are allowed to be significant, and using a digital computer it provides a valuable exercise in Fourier analysis.

Reading and References

This method is described in a slightly different form, to determine thermal conductivity in *Advanced Practical Physics* by B. L. Worsnop and H. T. Flint (Methuen, 1962) where a useful reference to the mathematical methods of Fourier analysis is also given.

Without the detail of the theory given here the method is described by H. C. Bryant in the *American Journal of Physics*, 1963, vol. 31, p. 325.

Experiment No. E1.7

TO FIND THE TEMPERATURE DISTRIBUTION ACROSS THE CLADDING OF A NUCLEAR REACTOR FUEL ELEMENT BY AN ANALOGUE METHOD, USING THE SERVOMEX FIELD PLOTTER

Introduction

By the inherent nature of the heat transfer problem in a nuclear reactor an analogue testing method recommends itself. Such an analogy exists between heat flow and the flow of electricity since both are governed by laws of the same form. Hence to solve such a steady state problem a 'scale model' of the reactor system is cut from 'Teledeltos' conducting paper and wired into a bridge circuit in just the same way as used in Experiment Nos. A1.11 and E2.6. The potential can then be measured at every point on the model, to produce equipotential lines corresponding to isothermals in the system under investigation.

Theory

When a balance point is obtained for values R_1 and R_2 of the ratio arms of the Wheatstone bridge, the resistance (R_3) between the first electrode and the probe and (R_4) between the second electrode and the probe are such that

$$V_1 = i_1 R_1 = i_2 R_3$$

and $$V_2 = i_1 R_2 = i_2 R_4 \quad (Figure\ E26)$$

or $$\frac{V_1 + V_2}{V_1} = \frac{R_1 + R_2}{R_1}$$

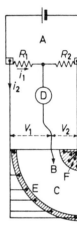

Figure E26. Simulation of heat transfer problem
A, Wheatstone bridge arrangement or 'Servomex' field plotter
B, Probe   C, 'Teledeltos' analogue   D, Balance detector
E and F, Electrodes painted in silver conducting paint

but potential difference and temperature difference are analogous hence

$$\frac{\theta_2 - \theta_1}{\theta} = \frac{R_1 + R_2}{R_1}$$

where θ_2 and θ_1 represent the boundary temperatures and θ the temperature difference between the isothermal upon which the probe lies and the first boundary. Therefore knowing the resistance values and the specified boundary temperatures the isothermal temperature may be calculated.

Two factors are important here:

(1) The rate of flow of heat Q across the bulk material given by the equation

$$Q = kA \left(\frac{\delta \theta}{\delta x} \right)$$

TEMPERATURE DISTRIBUTION IN NUCLEAR REACTOR E1.7

where k = thermal conductivity of the material of the cladding
A = the effective cross sectional area across which the heat is transferred
$(\delta\theta/\delta x)$ = the temperature gradient across the cladding.

(2) The transfer of heat at the surface of the material which depends upon the heat transfer coefficient h such that

$$Q = hA(\delta\theta)$$

and the thermal resistance

$$= \left(\frac{\delta\theta}{Q}\right) = \frac{1}{hA}.$$

By the analogy with Ohm's law one may write $Q = \delta\theta/R_T$ where R_T represents thermal resistance—hence across the bulk of the medium the thermal resistance

$$R_T = \left(\frac{\delta\theta}{Q}\right) = \frac{(\delta x)}{kA}.$$

Thus if the thermal resistance is to be represented on the 'Teledeltos' paper then the boundary must effectively be extended in the same ratio to take account of the thermal resistance upon heat transfer at the surface, i.e. a border of thickness k/h is drawn at the geometrical boundaries representing the slab of the medium across which heat transfer takes place.

In the problem chosen here, it is assumed that the bulk temperature of the long cylindrical fuel element of radius R is constant (not strictly true of course) and hence the central electrode on the conducting paper may be painted in as a circle of silver paint, in scale to represent the actual cross section of the fuel element. The element is assumed to be clad in a 'Magnox' cylinder of thickness t and the heat transfer coefficient from fuel to can is h_1. Hence the boundary is extended to scale, a distance k/h_1—it may be shown for convenience on the paper in NON-conducting ink and now represents the inner boundary of the cladding. A further increase of radius k to represent the effect of the cladding is then marked and finally another increase k/h_2 (where h_2 is the coefficient appropriate to heat transfer from 'Magnox' to coolant). The circle is now used to form the inner periphery of the second electrode (*Figure E27*).

Method

Fit up the circuit as shown in *Figure E26* (compare Experiment Nos. A1.11 and E2.6) with a potential across the electrodes to represent the known temperature drop from fuel to coolant and by using the probe, find points on the equipotential of the system (*Figure E28*). Using the analogy equations of the theory calculate the heat transfer rate from the isothermals.

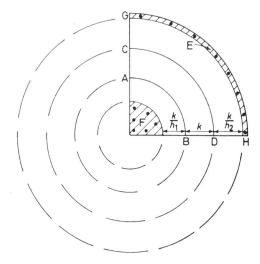

Figure E27. Simulation of heat flow across the cladding of a reactor fuel element. N.B. Drawn out of scale for clarity
AB, Periphery of extended boundary (can be marked in non-conducting ink) to simulate heat transfer at surface. k = thermal conductivity coefficient; h = surface heat transfer coefficient
BD, Represents cladding through which heat passes
DH, Represents second boundary extension to simulate heat transfer at the second surface
E and F, Electrodes painted in silver paint on conducting paper

Discussion of Results

In the example chosen, note that the average bulk temperature of the coolant (another over-simplification) has been taken as 270° C and that this corresponds to a fall in temperature of some 100° C across the cladding. If the coolant were more efficient such that the external temperature was only say 70° C then the corresponding drop across the cladding would be greater and the temperature at the inner face of the can higher—it would in fact enable a prediction to be made as to whether the more efficient coolant resulted in the Magnox canning being brought to a dangerously high temperature where failure would ensue.

It should be noted that the method in this form is only suitable for the solution of steady state problems of heat transfer. Networks of resistors (and capacitors, which are analogous to thermal capacity) are required for more complicated problems. An inherent

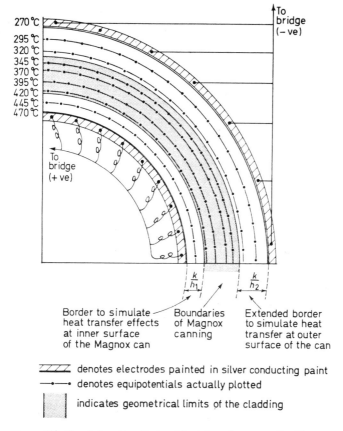

Figure E28. Simulating the effects of heat transfer across the Magnox cladding of a nuclear reactor fuel element. Temperatures corresponding to the equipotentials are shown on the left

error in the method is due to the slight anisotropy of the 'Teledeltos' paper which has a different resistance measured in one direction, than in the direction at right angles. Check that this, as it should be, is small. For purposes of illustration the full circular section of the analogue is discussed here but because of the symmetry it is only necessary to use a small sector in the experiment itself (*Figure E28*).

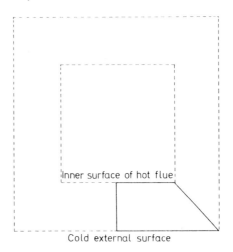

Figure E29. Simulation of the problem of heat transfer across a chimney of square cross-section. The solid line shows the one-eighth section which owing to the symmetry of the system is all that need be considered

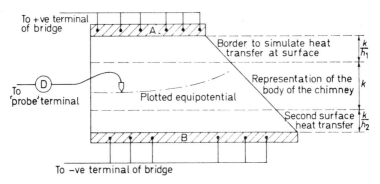

Figure E30. Chimney analogue cut from 'Teledeltos' paper
A and B, Electrodes in silver paint
k = thermal conductivity of body of chimney
h_1 = heat transfer coefficient at first (inner) surface
h_2 = heat transfer coefficient at external surface

Note that *Figure E28* shows a number of electrode points of electrical connection within the periphery of each electrode in order to ensure that good conduction at each electrode is achieved. It is of interest to investigate this facet of the method as suggested below.

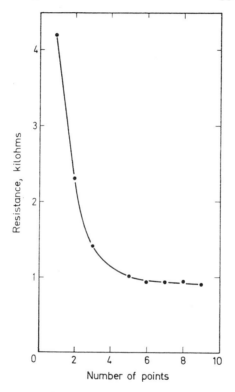

Figure E31. Graph showing the variation of the resistance of a fixed area of 'Teledeltos' paper with the number of electrode point attachments

Further Work

A number of problems of heat transfer will present themselves; for example *Figure E30* shows the way in which the flow of heat across a chimney of square cross section may be simulated.

Note again that because of the symmetry of the system, only a one-eighth section has been drawn (*Figure E29*).

It is worthwhile to take a number of rectangular sections of the conducting paper and to investigate the effects of increasing the number of electrode points of attachment on either side of the test pieces. As the graph of *Figure E31* indicates, not until six or more points of connection have been made do conditions approach those suggested as the 'theoretical' resistance of the conducting paper for any chosen area.

Reading and References

The method is described in much greater detail and a number of problems given in *Principles of Heat Transfer* by F. Kreith (International Textbook Co., 2nd edition, 1965).

Experiment No. E1.8

TO DETERMINE THE LATENT HEAT OF VAPORIZATION OF LIQUID NITROGEN FROM VAPOUR PRESSURE MEASUREMENT

Introduction

Liquid nitrogen in a Dewar vessel rapidly assumes a condition of equilibrium at a temperature which is determined by its vapour pressure. This in turn may be adjusted by sealing the Dewar and exhausting the space over the liquid nitrogen at a controlled rate. The speed of pumping itself is simply controlled by a manually operated 'throttle' valve in the pumping line. Since extensive tables of variation of vapour pressure with temperature are available it is possible to use the vapour pressure readings as measurements of temperature and from these to determine the latent heat of vaporization of the liquid nitrogen.

Theory

The variation of boiling point with pressure is given by the Clausius-Clapeyron equation

$$\frac{dP}{dT} = \frac{L}{T(V_2 - V_1)}$$

where P = pressure
T = absolute temperature
V_2 = volume of 1 gram-molecule of vapour
V_1 = volume of 1 gram-molecule of liquid

LATENT HEAT OF VAPORIZATION OF NITROGEN E1.8

i.e. $V_2 \gg V_1$. Hence $(\mathrm{d}P/\mathrm{d}T) \simeq (L/TV)$ and assuming the ideal gas law to hold $PV = RT$ (R = universal gas constant) then

$$\frac{\mathrm{d}P}{\mathrm{d}T} = \frac{LP}{RT^2}$$

and integrating

$$\int \frac{\mathrm{d}P}{P} = \frac{L}{R}\int \frac{1}{T^2}\mathrm{d}T \quad \text{or} \quad \log_e P = \frac{-L}{R}\left\{\frac{1}{T}\right\} + B$$

where B is a constant of integration.

Hence plotting $\log P$ against $1/T$ results in a straight line of slope $(-L/R)$ (*Figure E32*).

Method

First note the atmospheric pressure. Carefully pour the liquid nitrogen into the Dewar which has a rubber collar A at the top (see *Figure E33*). Place the top cap B firmly over the rubber, then fold back the rubber over the flange of the cap to form an effective seal. Connect one lead C to the rotary vacuum pump with a 'throttling' valve T in the pumping line. Connect the second lead D to a mercury U-tube manometer. With the throttle valve open switch on the pump and when a steady state has been achieved note the readings l_1 and l_2 in each limb of the manometer.

The pressure head is then $(l_1 - l_2)$ and by subtracting this head from the atmospheric pressure on the open side of the U-tube the vapour pressure in the Dewar is obtained.

Adjust the flow of gas from the Dewar to the pump, by means of the valve, until a significant change in the steady state condition on the manometer is observed and again calculate the vapour pressure. Repeat this procedure at convenient pressure intervals and from standard tables tabulate the temperatures corresponding to the calculated vapour pressures.

Plot the natural logarithm of pressure against the reciprocal of temperature as suggested in the theory and from the slope of the graph obtain a value for the latent heat of vaporization of liquid nitrogen.

Discussion of Results

Care should be taken to observe any precautions given for the transfer of the liquid nitrogen appropriate to the particular form of apparatus used. Damage to the apparatus and loss of liquid nitrogen may otherwise result. The operation of the experiment is critically dependent upon good sealing of the system and time spent on this is worthwhile.

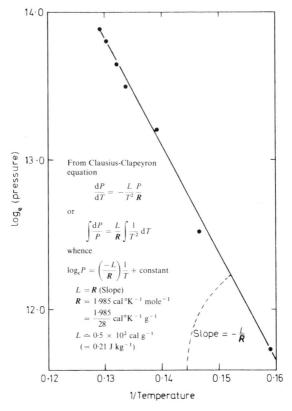

Figure E32. Graph of $\log_e$ (pressure) against 1/(temperature) to determine the latent heat of vaporization of nitrogen T_p = temperature (°K) found from vapour pressure measurements

A comparison of the result obtained from the experimental graph given with the accepted value shows that a fortuitously good one has been obtained here. The accuracy indicated by the deviations of graphical points about the 'best fit' line is not so high as the numerical result itself at first suggests.

In carrying out the calculation remember L is required in kg^{-1} or calories gram^{-1} whereas the universal gas constant is generally found to be expressed in cals $(°K)^{-1}$ mole^{-1} or alternatively in ergs $(°K^{-1})$ mole^{-1} so that the gram molecular weight of nitrogen

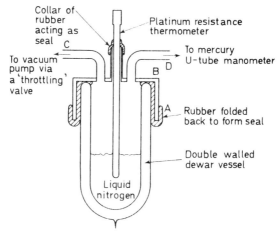

Figure E33. To determine the latent heat of vaporization of liquid nitrogen

must be introduced. (The pressure too must be in Nm^{-2} or dynes cm^{-2} and not left in terms of centimetres of mercury!)

Further Work

A considerable improvement to the experiment is affected by using an independent means of measuring temperature, viz. a

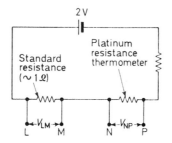

Figure E34. Circuit for independent measurement of temperature of liquid nitrogen at various pressures by means of a platinum resistance thermometer. The voltages across the standard resistance (V_{LM}) and the platinum resistance thermometer (V_{NP}) may be measured by a suitable potentiometer

platinum resistance thermometer (in fact this is shown in the diagram of the apparatus, *Figure E33*).

A suitable circuit for including such a thermometer is shown in *Figure E34*.

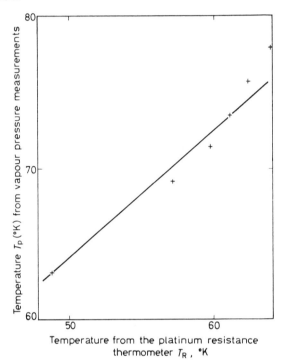

Figure E35. Graphical comparison of temperatures given on the platinum resistance and vapour pressure scales. Note that although a large discrepancy ($\sim 14°$ K) occurs in the readings here, it does not affect the calculation of the latent heat of the liquid nitrogen significantly since it is the slope of the log P against $1/T_p$ graph that is involved

An experimental graph comparing platinum resistance thermometer temperatures and temperatures estimated from vapour pressure measurements gives a good indication of the order of magnitude of errors which may occur in such an experiment as this. The graph of *Figure E35*, whilst displaying a serious constant discrepancy in the estimate of temperature, does not result in a widely differing value of the latent heat since only the slopes are involved in the calculation.

Reading and References

The background theory of the thermodynamic reasoning behind the Clausius-Clapeyron equation is given in *Thermal Physics* by P. M. Morse (Benjamin, 1965), as well as in the many standard degree texts upon heat and thermodynamics.

For students having no knowledge of cryogenics *Low Temperature Techniques* by A. C. Rose-Innes (English Universities Press, 1964) may be recommended.

Vapour pressure/temperature tables may be found in *Experimental Techniques in Low Temperature Physics* by G. K. White (Oxford University Press, 1959).

Experiment No. E2.1

CORNU'S METHOD TO DETERMINE THE ELASTIC MODULI OF A PERSPEX SPECIMEN USING AN INTERFERENCE TECHNIQUE

Introduction

The specimen, in the form of a beam, is bent by the application of a force at either end (*Figure E36*), and the interference fringes

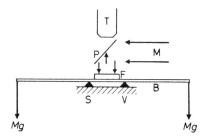

Figure E36. Cornu's method—experimental arrangement
F, Small flat plate of glass
B, Bar undergoing bending
T, Telescope
M, Monochromatic beam of parallel light from a sodium lamp ($\lambda = 5893$ Å)
P, Reflector of Newton's rings viewing arrangement
S and V, Fulcra supporting specimen
Mg, Force applied to each end of bar

APPLIED HEAT, MECHANICS OF FLUIDS AND SOLIDS

produced between the upper surface of the specimen and the lower surface of a flat glass plate, enable the elastic constants of the material of the beam to be calculated.

Theory

The fringes produced in the air gap between the perspex and the glass appear as shown in *Figure E37a* and are viewed using an

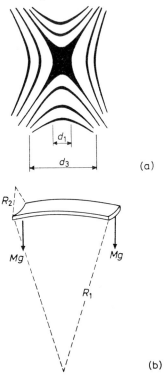

Figure E37. Diagram showing the two curvatures produced in the specimen (transverse to each other) by the applied load and an enlarged view of the resulting fringes

arrangement analogous to that employed in the observation of Newton's rings. In the latter case we know

$$\frac{d^2}{4R} = n\lambda \qquad (1)$$

422

where d = diameter of the circular fringes
 R = radius of curvature of the surface at which interference occurs
 n = order of the fringe considered, and
 λ = wavelength of the light used.

Thus
$$\frac{1}{R} = \frac{4n\lambda}{d^2} \qquad (2)$$

The analogy holds for the hyperbolic fringes produced here, if d is now reckoned as the distance between corresponding fringes on either side of the symmetrical fringe system (*Figure E37a*).

The curvature R is created by the application of force at either end of the specimen and from 'bending' beam theory the couple C is related to the curvature by the formula

$$C = \frac{YAk^2}{R} \qquad (3)$$

where Ak^2 = geometrical moment of inertia, and
 Y = Young's modulus of elasticity for the bar.

The couple $C = Mgl$ where l is the length of the beam, hence

$$\frac{1}{R} = \frac{Mgl}{YAk^2} \qquad (4)$$

and if $(1/R)$ is plotted on the y axis against M on the x axis the slope of the resulting straight line is gl/YAk^2.

The geometrical moment of inertia (Ak^2) may be found from the dimensions of the bar since for a bar of rectangular cross section $k^2 = t^2/12$ where t is the thickness of the bar and k the radius of gyration. The cross sectional area is (bt) hence

$$(Ak^2) = \frac{bt^3}{12} \qquad (5)$$

so that the

$$\text{Slope of the graph} = \frac{12gl}{Ybt^3}$$

or

$$Y = \frac{12gl}{bt^3 \, (\text{Slope})} \qquad (6)$$

Method

Set up the bar as shown in *Figure E36* and increase the load in convenient steps such as are indicated by the graph of *Figure E38*.

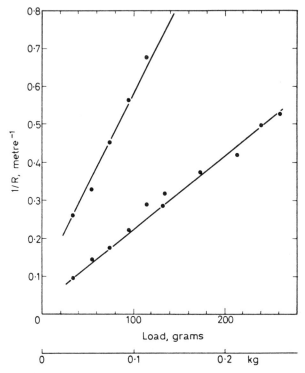

Figure E38. Graph of the reciprocal of the radius of curvature of a bent beam against the applied load, used to determine the elastic constants by Cornu's method

Using a travelling microscope measure the distance d between the first corresponding fringes out from the centre of the system and then of the sixth fringe for each load in turn. Hence from equation (2) determine $(1/R)$. Plot this value against the value of the applied load (*Figure E38*) and from the slope of the graph calculate Young's modulus for the specimen using equation (6).

Discussion of Results

Adjustment of the microscope is aided by placing a small blob of ink on the upper surface of the beam and focusing first of all upon this. It should be possible to see the location of the system of fringes by eye before aligning the telescope.

A small clip and screw adjustment to hold the reflecting plate in the 45° position allows adjustment to be made to obtain optimum conditions of illumination.

The application of the load by placing the weights in the centre of a platform suspended below the bar by cords from the ends, helps in producing a uniform bending of the beam.

Further Work

It should be noted that the method may be applied to any optically transparent isotropic specimen (e.g. glass).

Note too that the graph of *Figure E38* shows two straight lines corresponding to the two systems of hyperbolic fringes (*Figure E37a*). The gradients of these lines are in the ratio of the curvatures which occur in perpendicular planes in the specimen (*Figure E37b*) and thus

$$\frac{\text{Slope 2}}{\text{Slope 1}} = \sigma$$

Poisson's ratio for the material of the bar.

For non-transparent specimens a strain gauge method may be used (see Experiment No. E2.2).

Reading and References

The theory of the method of observing Newton's rings may be found in any degree textbook, e.g. *Fundamentals of Optics* by F. A. Jenkins and H. E. White (McGraw-Hill, 3rd edition, 1957), and that of the bent beam in *Properties of Matter* by F. H. Newman and V. H. L. Searle (Arnold, 5th edition, 1957).

APPLIED HEAT, MECHANICS OF FLUIDS AND SOLIDS

Experiment No. E2.2

TO INVESTIGATE THE LAW GOVERNING THE OPERATION OF A STRAIN GAUGE AND TO USE THE GAUGE TO DETERMINE YOUNG'S MODULUS FOR A METAL BEAM

Introduction

The strain gauge consists essentially of a length of wire in the form of a flat 'concertina' (*Figure E39*). It is cemented to the surface of

Plan 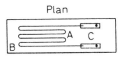   Elevation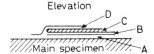

A, Resistance wire
B, Paper mounting
C, Electrodes and bridge lead connections

A, Lower film of cement
B, Mounting paper of the gauge
C, Wire of strain gauge
D, Film of cement over the gauge

Figure E39. Use of the strain gauge—detail of the gauge

the specimen in which strain is to be investigated so that changes in length of the latter, result in changes in resistance in the wire which may be measured using a suitable Wheatstone bridge arrangement as shown (*Figure E40*).

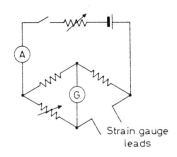

Figure E40. Wheatstone bridge arrangement

Theory

The resistance of a wire is given by

$$R = \frac{SL}{a}$$

where S = specific resistance of the wire
  l = length of the wire
  a = cross sectional area of the wire,

i.e.

$$R = \frac{4Sl}{\pi D^2}$$

where D is the diameter of the wire, thus

$$\log R = \log\left(\frac{4S}{\pi}\right) + \log l - 2 \log D$$

and

$$\delta(\log R) = \frac{\delta}{\delta l}(\log l)(\delta l) - 2\frac{\delta}{\delta D}(\log D)(\delta D)$$

i.e.

$$\frac{\delta R}{R} = \frac{\delta l}{l} - 2\frac{\delta D}{D}$$

but the ratio of fractional change in diameter to the fractional change in length represents Poisson's ratio σ, i.e.

$$\sigma = -\frac{(\delta D/D)}{(\delta l/l)} \qquad (1)$$

hence

$$\frac{\delta R}{R} = \frac{\delta l}{l} + 2\sigma \frac{\delta l}{l} = (1 + 2\sigma)\frac{\delta l}{l}.$$

Again the ratio

$$\frac{(\delta R/R)}{(\delta l/l)}$$

represents the sensitivity constant of the strain gauge (also known as the gauge factor λ).

Since theoretically $\sigma < 0.5$ the gauge factor is therefore found to have values up to about 1·8.

Again Young's modulus Y is given by

$$Y = \frac{(\delta F/A)}{(\delta l/l)} \quad \text{or} \quad \left(\frac{\delta F}{A}\right) = Y\left(\frac{\delta l}{l}\right)$$

where $(\delta F/A)$ = the increase in applied stress (A being the cross sectional area of the beam)
hence

$$\delta F = \frac{AY}{\lambda}\left(\frac{\delta R}{R}\right) \qquad (2)$$

and taking A, Y, λ and R constant the graph of δF, the change in

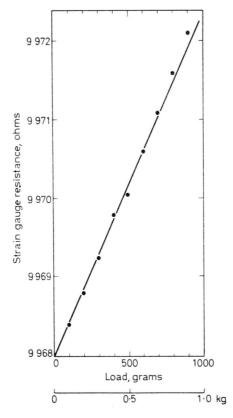

Figure E41. Graph showing the change in strain gauge resistance with loading of the specimen

OPERATION OF A STRAIN GAUGE

applied load against change in the gauge resistance, should be a straight line (*Figure E41*).

From the theory of the bending beam, the couple *C* acting on the beam is given by

$$C = \frac{Y(Ak^2)}{r}$$

where (Ak^2) = geometrical moment of inertia of the beam, and
r = radius of curvature of the beam.

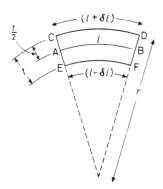

Figure E42. Diagram showing extension and contraction in the surfaces of a bent beam. AB is the neutral filament which remains unchanged in length l. N.B. $\delta l/(t/2) = l/r$

From *Figure E42* it is seen that

$$\frac{\delta l}{(t/2)} = \frac{l}{r}$$

where t = thickness of the beam,

i.e.

$$\frac{1}{r} = \left(\frac{\delta l}{l}\right)\frac{2}{t}.$$

Again

$$\frac{\delta l}{l} = \frac{1}{\lambda}\left(\frac{\delta R}{R}\right).$$

Therefore

$$C = \frac{Y(Ak^2)}{r} = \frac{YAk^2}{\lambda}\left(\frac{\delta R}{R}\right)\frac{2}{t}$$

and

$$Y = \left(\frac{\delta F}{\delta R}\right)\frac{lt\lambda R}{2} = \left\{\frac{lt\lambda R}{2\,(\text{Slope})}\right\}$$

Method

The beam will be supplied with the strain gauge cemented on as shown in *Figure E39*. The leads from the gauge are connected into one arm of a simple Wheatstone bridge arrangement, a dummy 'compensation' gauge (not shown on the diagram) being incorporated with the proper gauge.

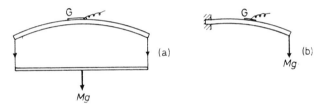

Figure E43. Strain gauge in position (a) on beam under uniform bending, (b) on cantilever beam. G is the gauge (not to scale)

The resistance R_3 is adjusted so that with the beam unloaded the bridge is balanced and the initial resistance R of the gauge is noted.

The beam may be set up as shown in *Figure E43a* under conditions of uniform bending or fixed at one end as a cantilever (*Figure E43b*).

A suitable load (of say 100 g) is applied to the beam when the bridge will go out of balance due to change in the gauge resistance. Resistance R_3 is then readjusted to restore the balance and the new resistance of the gauge is computed.

The procedure is repeated for suitable step increases of the load and a graph of strain gauge resistance against the load is plotted to verify the linear law predicted by the theory (*Figure E41*).

Finally, knowing the gauge constant (which will be given with it) and the cross sectional area of the beam, Young's modulus for the material of the bar may be calculated using equation (3).

Discussion of Results

Resistance readings should be taken not only as the beam is loaded but also as the load is removed. The discrepancies that arise between the two sets of readings give good indication of the accuracy of the method.

Further Work

It is of interest to investigate bars of other types of cross section (and therefore other geometrical moments of inertia), e.g. 'I' section such as a curtain rail under different conditions of loading, e.g. direct linear extension when

$$Y = \frac{\lambda R}{(\text{Slope of graph}) \times A}$$

directly from equation (2).

Reading and References

Background theory upon the bending of beams may be found in standard texts such as *Properties of Matter* by F. H. Newman and V. H. L. Searle (Arnold, 5th edition, 1957).

An elementary introduction to strain gauges and strain measuring equipment is given in a pamphlet of that title by Phillips Ltd.

Experiment No. E2.3

TO ANALYSE THE PRINCIPAL STRESSES IN A LOADED 'ARALDITE' SPECIMEN USING THE PHOTOELASTIC TECHNIQUE

Introduction

Glass and many transparent plastics although isotropic when in a strain free condition become anisotropic when subjected to stress.

If then viewed in polarized light the application of stress produces coloured patterns corresponding to the strains produced in the material.

The development of new synthetic resins such as araldite (CT200) and of polaroid have enabled photoelastic stress analysis to become a powerful laboratory tool.

Theory

A plane polarized beam of light entering a stressed araldite (CT200) specimen is split into two components each travelling with different velocity. The components therefore emerge with a rela-

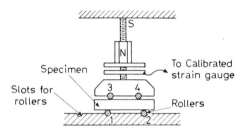

Figure E44. Typical loading rig arrangement

tive path retardation which is found to be proportional not only to the thickness of the specimen but also to the difference between the principal stresses within it, i.e.

$$\text{Retardation } R = C(\sigma_1 - \sigma_2)t \tag{1}$$

where σ_1 and σ_2 = principal stresses
         t = thickness of the specimen
and C = stress optical coefficient.

Perhaps the simplest example to take is that of a rectangular beam viewed in monochromatic light under simple uniform bending. The principal stress parallel to the long axis of the specimen is given by

$$\sigma_1 = \frac{My}{I} \tag{2}$$

where M = the bending moment
         I = the appropriate geometrical moment of inertia
and y = the distance of the point under consideration from the neutral filament of the specimen.

The second principal stress at right angles to the long axis may be assumed to be zero at the upper and lower edges of the beam.

Method

To produce bending of the type to which the above formula may be applied, a four point loading frame may be used. *Figure E44* shows such a system where rollers are fitted in slots 1 and 2, the specimen placed on them and the pressure head gently lowered so that rollers in slots 3 and 4 make contact with the beam.

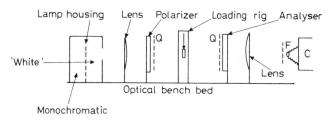

Figure E45. Polariscope for observation of strain patterns. Polarizer and analyser consist of polaroid sheet mounted in graduated circular scales
Q, Quarter-wave plates   F, Optical wavelength filters
C, Camera

Variations in the form of the frame are common but the one shown has a coarse adjusting screw S and a fine adjustment nut N. The applied stress may be conveniently read off using a strain gauge and transducer.

The optical arrangement is shown in *Figure E45*, a 'monochromatic' source and a suitable filter being used. The polarizer and

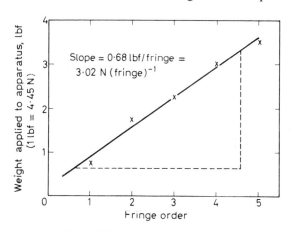

Figure E46. Calibration of a specimen

analyser are 'crossed' 90° to one another and the quarter-wave plates adjusted to be at +45° and −45° to the polarizer and analyser.

Adjustment of the lens system is now made to give a sharp geometrical image on the screen.

The whole field of view now appears dark and a small applied load tends to concentrate the zero order fringe along the central axis of the specimen.

The beam should then be loaded until the first order fringe appears along the edges of the specimen and this load then tabulated. Further loading will then render fringes of higher order visible (*Table E1*) and *Plate E1* shows a typical pattern produced.

Table E1. Results for strain analysis experiment on 'Araldite' specimen

Fringe number	For Upper Face Distance along beam		Stress lb in^{-2}*	Fringe number	For Lower Face Distance along beam		Stress lb in^{-2}*
	mm	in			mm	in	
1	8·636	0·34	417	1	8·128	0·32	417
2	16·256	0·64	834	2	17·018	0·67	834
3	23·368	0·92	1 251	3	25·146	0·99	1 251
4	33·782	1·33	1 668	4	33·274	1·31	1 668
4	43·180	1·70	1 668	4	43·942	1·73	1 668
3	52·070	2·05	1 251	3	50·546	1·99	1 251
2	60·452	2·38	834	2	59·182	2·33	834
1	68·072	2·68	417	1	67·056	2·64	417

* 1 lb in^{-2} = 6·894 kN m^{-2}

A calibration graph may now be plotted of applied load against fringe order (*Figure E46*) the slope of which gives the load per fringe. Then by using equation (2) and knowing the dimensions of the specimen, the maximum bending stress per fringe may be calculated (*Figure E47*).

The beam should now be arranged for three point loading with rollers in position as shown in *Figure E47b* and the process repeated.

Photograph the fringe pattern and measure the distance of the points where each fringe reaches the edge of the specimen from the first fulcrum. Tabulate these values for both upper and lower surfaces and also the stress indicated by the appropriate fringe (*Table E1*).

Now plot a graph of the stress against distance from the applied load (*Figure E48*). For comparison one can calculate the maximum theoretical stress and indicate this upon the same graph.

Discussion of Results

Take into account the previous history of the specimen and the distortion that occurs at the points of application of the forces in assessing the accuracy of the experiment.

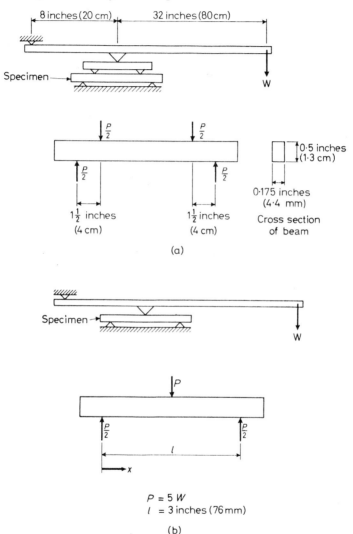

Figure E47. (a) Four-point loading system. (b) Three-point loading system

Further Work

This might include investigation of more complicated geometrical forms, e.g. from ring specimens to those possessing abrupt changes in section.

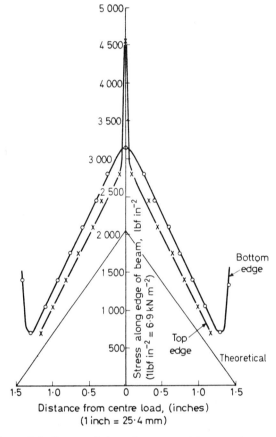

Figure E48. Stress variations along top and bottom edges of a three-point loaded beam

Reading and References

The theory concerning the optical system may be found in standard physics texts, e.g. *A Text-book of Light* by G. R. Noakes (Macmillan, 1937) or *Fundamentals of Optics* by F. A. Jenkins and H. E. White (McGraw-Hill, 3rd edition, 1957).

The theory of the bent beam may be found in, e.g. *Physics* by S. G. Starling and A. J. Woodall (Longmans Green, 1963) and an elementary treatment of photoelasticity in *Introduction to Photo-elastic Analysis* by A. W. Hendry (Blackie, 1948).
Further useful references are *Strength of Materials* by G. H. Ryder (Cleaver-Hulme, 1961), *Photoelasticity*, vol. 1, by M. M. Frocht (Wiley, 1941) or *Introduction to the Theoretical and Experimental Analysis of Stress and Strain* by A. J. Durelli, E. A. Phillips and C. H. Tsao (McGraw-Hill, 1958).

Experiment No. E2.4

TO DETERMINE THE POROSITY OF REFRACTORY MATERIALS

Introduction

In a refractory sample the ratio of the volume of all the pores to that of the entire body is known as the porosity of the material of the refractory.

In a wide variety of commercial commodities from paper of all types to bricks for different uses a range of specified porosities may be required. In this experiment the method of measuring the porosity of refractory firebricks of various grades is outlined.

Theory

In the steel making industry the standard method of measurement of porosity is of great importance and is regularly applied, when bricks fail in use, to provide background data as an essential step to solving the problem.

In order to use the standard method of testing it is necessary to familiarize oneself with the following definitions:

(*a*) The apparent porosity P_A, i.e. the ratio of the volume of the open pores within the refractory material to the total volume (the bulk volume) of solid material together with the volume of all the pores within it, both open and sealed.

(b) The true porosity P_T, the ratio of the volume of all the pores (open and sealed) to the bulk volume of the material.

(c) The sealed porosity P_S, the ratio of the volume of the sealed pores to the volume of the material.

The bulk volume

$$V_B = \frac{W_3 - W_2}{\rho_L}$$

where W_3 = the weight of the soaked sample in air
W_2 = the weight of the soaked sample immersed in water
ρ_L = the density of the water at the temperature of the experiment ($= 1$ for all practical purposes in terms of c.g.s. units but 10^3 kg m^{-3} in S.I. units).

The apparent solid volume

$$V_{AS} = \frac{W_1 - W_2}{\rho_L}$$

where W_1 = weight of the dry refractory sample. Hence by definition the bulk density

$$\rho_B = \frac{W_1 \times \rho_L}{W_3 - W_2}$$

the apparent solid density

$$\rho_{AS} = \frac{W_1}{W_1 - W_2} \rho_L$$

the apparent porosity

$$P_A = \left(\frac{W_3 - W_1}{W_3 - W_2}\right) 100 = 100 \left(1 - \frac{\rho_B}{\rho_{AS}}\right)$$

and the true porosity

$$P_T = 100 \left(1 - \frac{\rho_B}{\rho_T}\right).$$

Then

$$P_T = P_A + P_S.$$

The porosities are usually expressed in terms of the bulk density ρ_B the apparent solid density ρ_{AS} and the true density ρ_T which in turn may be defined as the ratio of the mass of the bulk volume to its apparent solid volume and to its true volume respectively.

Thus

$$P_A = 100\left(1 - \frac{\rho_B}{\rho_{AS}}\right)$$

$$P_T = 100\left(1 - \frac{\rho_B}{\rho_T}\right)$$

$$P_S = 100\,\rho_B\left(\frac{1}{\rho_{AS}} - \frac{1}{\rho_T}\right)$$

The method involves immersing the refractory specimens in boiling water and weighing them, when thoroughly soaked, in air and in water as well as taking the weight when dry.

Method

The specimens to be tested should be about 50 cm³ in volume and prior to the experiment should be suspended from a glass rod by cotton threads and heated at 110° C in a suitable oven to ensure that initially they are thoroughly dry. They are then transferred directly to a desiccator to cool slowly and are then weighed when clean and dry (W_1).

The samples are slowly raised to 110° C again and maintained at that temperature for several hours if possible. They are then rapidly transferred so that they are suspended in an adjacent tank of boiling water and boiling is continued for at least ten minutes, after which time heating is stopped and cold water is slowly introduced at the *bottom* of the tank by means of a glass tube down the inside of the tank, the hot water being displaced and allowed to flow out at the top.

After the cold water temperature has been reached the specimens may be placed in turn in a large beaker of cold water standing on a balance bridge and carefully weighed when fully immersed in the water (W_2). Finally the beaker and sample are removed from the bridge, the test piece drawn out of the water and surplus liquid removed on filter paper. (The paper should not however be allowed to actually touch the surface of the specimen.) The soaked sample is then reweighed in air (W_3) from which, using the formulae given above, the porosity may be determined when the true density has been found using the usual density bottle method.

The material of the brick is crushed in an agate or other suitable non-ceramic mortar until it passes through a 120 mesh sieve (if an iron mortar has to be used, pass a strong magnet over the powder to remove any iron particles before proceeding). The powder is then dried at 110° C for two hours or so and introduced into a carefully weighed density bottle (W_A) a second weighing then being made (of the bottle and the powder W_B). The bottle is then transferred to a vacuum desiccator and allowed to stand under a pressure of 20 torr for several hours, the desiccator being gently tapped, at intervals, to help remove any air entrapped in the powder.

The bottle is then filled with water and the stopper inserted before placing the whole in a constant temperature bath of 25° C for 20 minutes, after which surplus liquid is removed, the bottle is dried and reweighed (W_C).

Finally, if an immersion liquid other than water has been used the bottle is cleaned out and reweighed when completely filled with the liquid.

Discussion of Results

The results obtained in the ordinary laboratory experiment (where drying and other stabilization procedures are usually much shortened in order to carry out the measurement in reasonable time) may differ markedly from the results quoted for the material under test.

One way in which the experiment may be shortened is for previously dried specimens of the refractory brick to be immersed at room temperature rather than at boiling water temperature and for the vacuum desiccator procedure to be omitted in the density measurement. A comparison of results taken in this way, with those made at 110° C, gives further insight into the orders of accuracy involved in the experiment.

At least three pieces of each type of test material should be cut from the refractory brick under investigation.

Further Work

An alternative evacuation method may be used to determine the porosity of refractory materials if suitable evacuation equipment is available.

Reading and References

Standard procedures for the determination of the porosity and density are dealt with in detail in *Steelplant Refractories* by J. H. Chesters (United Steel Cos. Sheffield, 1963).

Experiment No. E2.5

THE REDWOOD NO. 1. VISCOMETER USED TO INVESTIGATE THE KINEMATIC VISCOSITY OF AN OIL WITH TEMPERATURE CHANGE

Introduction

The viscosity of an oil varies considerably with change in temperature and a number of empirical formulae have been used to describe the relationship. Whilst this has been done with reasonable success for pure liquids, no satisfactory one has been found for commercial lubricating oils, primarily because they are such complex mixtures.

Since changes in temperature affect the density ρ as well as the absolute viscosity η of the liquid, it is convenient to use the kinematic viscosity η/ρ in investigating variation of viscosity with temperature.

Of the many types of viscometer used the one chosen here is the Redwood No. 1, which is the one that is often employed to determine the viscosity of lubricating oils over their normal operational temperatures.

Theory

The Redwood viscometer is based upon the theory of the flow of liquid through a capillary tube and hence upon the well known Poiseuille's law, viz.

$$V_1 = \frac{\pi r^4 p}{8\eta l} \quad (1)$$

where V_1 = volume of liquid flowing per second
  r = radius of the capillary tube
  p = pressure = $(h \cdot \rho g)$
  h = head of liquid
  ρ = density of the liquid
  g — acceleration due to gravity
  l = length of tube
  η = the absolute viscosity.

APPLIED HEAT, MECHANICS OF FLUIDS AND SOLIDS

Remembering that the kinematic viscosity $v = \eta/\rho$ equation (1) may be rewritten

$$V_1 = \frac{\pi r^4 h \rho}{8l}\left(\frac{1}{v}\right)$$

and since the mean velocity of flow

$$v_m = \frac{V_1}{\text{Cross sectional area of the tube}}$$

$$v_m = \left(\frac{hgr^2}{8l}\right)\frac{1}{v}.$$

The time of discharge from the tube is inversely proportional to the velocity hence

$$v = \left(\frac{8l}{hgr^2}\right)t$$

i.e. $\qquad v = (\text{constant } A)t$

since l, h, g and r are all constant.

A graph of v against t is thus a straight line and the slope of the line gives the value of the apparatus constant A (*Figure E50*).

The kinematic viscosity of a liquid however varies rapidly with temperature and the aim in this experiment is to plot a graph show-

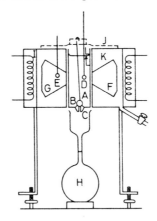

Figure E49. The Redwood Viscometer
A, Oil bath B, Ball valve C, Jet
D and E, Thermometers G and F, Stirrers
H, Measuring flask I, Heater J, Insulating cover
K, Water bath L, Fiducial oil level

ing this variation (*Figures E51* and *E53*) and to investigate the agreement that exists between the experimental results obtained and the empirical law which forms the basis of the viscosity temperature charts of the American Society for Testing Materials (equivalent charts in England are the Refutas charts by Baird and Tatlock Ltd.), namely

$$\log_{10}\{\log_{10}(v + 0{\cdot}8)\} = n\log_{10} T + c$$

where n and c are constant and T is absolute temperature (usually expressed by the engineer in degrees FAHRENHEIT!).

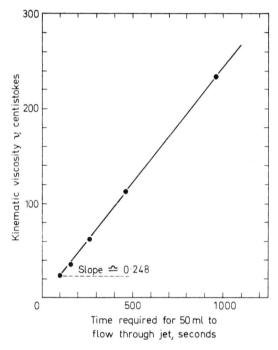

Figure E50. To verify that kinematic viscosity is proportional to the time of flow of the liquid through the orifice of a Redwood Viscometer

The charts are such that if two points for any one liquid are known the straight line drawn through the points represents approximately the variation of viscosity with temperature.

Method

At the outset check that the jet and the orifice C at the base of the oil cup of the viscometer are clear and clean (*Figure E49*). Place the ball valve in position to close the jet and fill the cup with the oil. For verifying the viscosity–time relationship, it is convenient to make up a series of aqueous solutions of glycerol (say 70, 75, 80, 85, 90% glycerol content—whose viscosities at 20° C are known and may be found in standard tables).

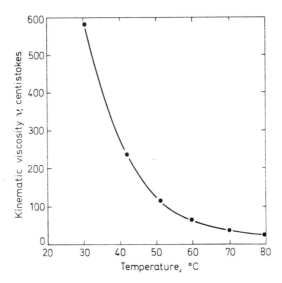

Figure E51. Graph showing the variation of the kinematic viscosity of glycerol with temperature

The water bath surrounding the oil cup is heated slightly if necessary to maintain the oil at 20° C, and as the valve is opened, a stop clock is set in motion to take the time for 50 ml of the oil to flow into a measuring flask.

The process is repeated for each of the glycerol solutions in turn and a graph drawn (*Figure E50*) which should result in a straight line whose slope gives the instrument constant A of the theory.

An oil of unknown viscosity may now be investigated and measurements of the time of flow for the oil at various temperatures are taken.

Using the constant A obtained from the first graph the kinematic viscosity = (Constant A × Time of discharge) is plotted against temperature (*Figure E51*).

Values of $\log_{10} \{\log_{10}(v + 0.8)\}$ are computed and plotted against $\log_{10} T$ as indicated in the theory.

Discussion of Results

In obtaining the straight line law governing the variation of kinematic viscosity with time of flow several grave assumptions have

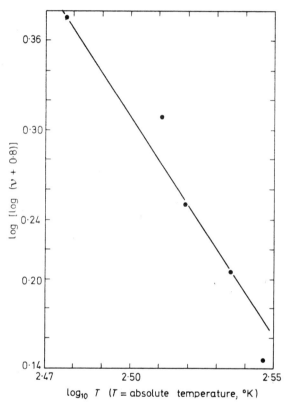

Figure E52. Verification of formula
$$\log \{\log (v + 0.8)\} = nT + c$$
upon which viscosity–temperature charts for oils are based

been made, e.g. the end correction for finite kinetic energy upon emergence has been neglected with regard to the Poiseuille equation and the theory assumes that the oil should emerge from the jet with no appreciable velocity—seldom possible in practice.

Provided with the instrument will be data giving two parameters (A and B) for its operation corresponding to an empirical formula

$$v = At - \frac{B}{t}$$

where t represents time of flow.

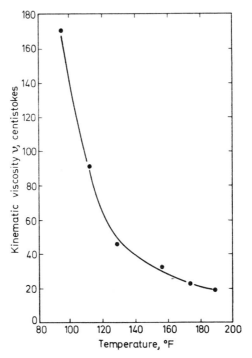

Figure E53. Graph showing variation of kinematic viscosity with temperature for a typical lubricating oil

These parameters (A and B) may vary for the instrument over different ranges and comparison of values of kinematic viscosity obtained in this way (rather than from the simple proportionality relation) gives deeper insight into the accuracy attainable with the instrument.

Note that the graph of $\log_{10} \{\log_{10} (v + 0.8)\}$ against $\log_{10} T$ can show some quite widely scattered points (*Figure E52*) unless

KINEMATIC VISCOSITY OF OIL

great care in temperature stabilization is taken. (Is the 0·8 significant for a No. 1 viscometer?) One should note the temperatures of the water bath and the oil bath at the commencement and end of flow to check what errors the variations in temperature observed can produce in the final graphs (see *Figure E53*).

Further Work

When experience has been gained with the instrument it is instructive to plot v/t against $1/t^2$ which should result in a straight line

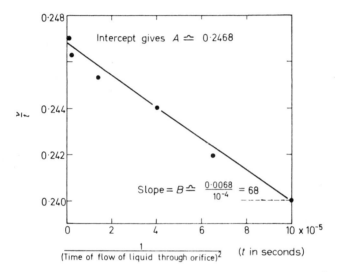

Figure E54. Graph of (kinematic viscosity v/time t) against $1/(\text{time})^2$ used to verify the operational formula of the Redwood Viscometer

$$v = At - \frac{B}{t} \quad \text{or} \quad \left(\frac{v}{t}\right) = A - B\left(\frac{1}{t^2}\right)$$

A and B are the operational constants of the Redwood Viscometer over the chosen range, i.e.

$$v = 0.247t - \frac{68}{t}$$

the slope of which gives A as the intercept (A is already known to some accuracy from the kinematic viscosity-time graph) (*Figure E54*) and B as the slope of the graph. The 'scatter' of points about the 'best fit' straight line may be very considerable in this case unless elaborate care is taken since very little variation of v/t

occurs and the y axis of the graph is on a highly magnified scale. The result obtained for B in this way however should fall within plus or minus 5% of the figure given by the instrument manufacturers and this is consistent with the overall accuracy of the instrument in practical use.

Reading and References

Details of the Redwood Viscometer may be found in *Lubrication and Friction* by P. Freeman (Pitman, 1962). This also shows graphs of viscosity variation with temperature and treats of both A.S.T.M. viscosity-temperature charts and of the empirical logarithmic formula forming their basis.

The Viscometer formula as well as the instrument itself is dealt with in *Dictionary of Applied Physics* by R. Glazebrook (McMillan, 1923).

Experiment No. E2.6

THE USE OF THE SERVOMEX FIELD PLOTTER TO INVESTIGATE THE VARIATIONS OF PRESSURE AND VELOCITY OVER A SIMULATED AEROFOIL

Introduction

The aim of this experiment is to plot, by means of the Servomex analogue device, the streamlines of flow which occur over the simulated profile of an aerofoil and from the resulting picture to investigate the variations of velocity of flow (and of pressure) which take place over the aerofoil.

Theory

At the outset it is necessary to define what is meant by a streamline.

A streamline is a theoretical curve the tangent to which at any point lies in the direction of the fluid velocity considered at that point. Lines of flow and streamlines will coincide in the case of steady (laminar) flow.

PRESSURE AND VELOCITY OVER AN AEROFOIL   E2.6

If we consider the flow of fluid as indicated by the flow lines of *Figure E55*, since the net rate of flow into the volume represented by ABCD is equal to the rate of flow from that volume then

$$y_0 v_0 = y_1 v_1 \tag{1}$$

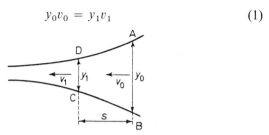

Figure E55. The simulated aerofoil
v_0 = velocity of fluid flow at chosen reference point
v_1 = velocity of flow at point to be measured
y_0 = separation of flow lines at reference point
y_1 = separation of flow lines at point at which measurement is to be made

where y_1 = separation of the lines of flow, in the region of disturbance
y_0 = separation of the lines of flow prior to disturbance
v = represents velocity.

(N.B. Here a two-dimensional picture is treated; strictly speaking areas A_0 and A_1 rather than y_0 and y_1 should be considered. The work done at the face AB = $p_0 y_0 \delta s_0$, the work done at the face CD = $p_1 y_1 \delta s_1$, where p = the pressure. Note that

$$\text{Actual work done} = (p_0 - p_1) V$$

where V = volume since $y_0 \delta s_0 = y_1 \delta s_1$ represents a measure of the volume, only assuming that no variation in profile occurs in the plane at right angles to the paper.)

Hence

$$\text{Work done} = (p_0 - p_1) \frac{m}{\rho}.$$

If we consider only the pressure difference associated with velocity changes in which we are interested here we obtain

$$(p_0 - p_1) \frac{m}{\rho} = \tfrac{1}{2} m (v_0^2 - v_1^2)$$

or

$$\left(\frac{p_0 - p_1}{\tfrac{1}{2} \rho v_0^2} \right) = 1 - \left(\frac{v_1^2}{v_0^2} \right)$$

449

and from equation (1)

$$\left(\frac{p_0 - p_1}{\frac{1}{2}\rho v_0{}^2}\right) = 1 - \left(\frac{y_0}{y_1}\right)^2$$

$(p_0 - p_1)/(\frac{1}{2}\rho v^2)$ is sometimes known as the pressure coefficient and at any point distance s from the chosen origin of observation it can be calculated from the streamline flow diagram (*Figure E56*).

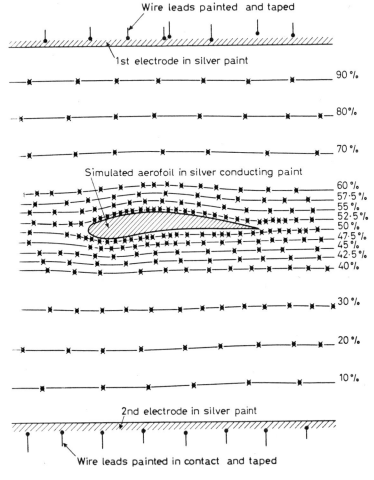

Figure E56. Simulated aerofoil on 'Teledeltos' paper for use with Servomex field plotter

A graph of v_1/v_0 against s/y_0 is shown in *Figure E57* and a graph of the pressure coefficient against s/y_0 is shown in *Figures E58* and *E59*. (s/y_0 is a measure of the angle of the flow line with the undisturbed flow direction.)

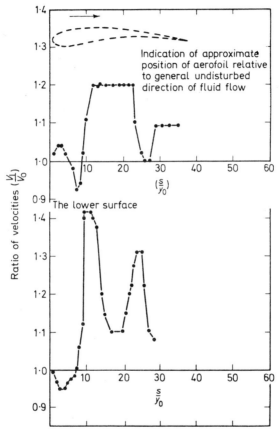

Figure E57. Graph showing variation of velocity of fluid flow over the upper surface of a simulated aerofoil

Method

Teledeltos conducting paper is firmly taped to a hardboard base and two straight line reference electrodes are painted on to it in silver conducting paint about 1 cm in from the edge as shown (*Figure E60*). The silhouette of the aerofoil is painted in the conducting silver, midway between the two electrodes.

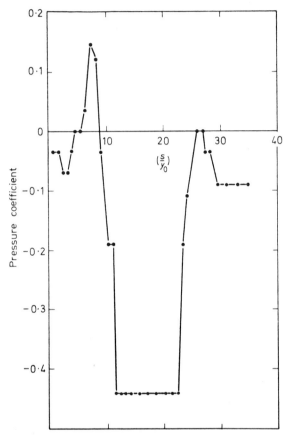

Figure E58. Graph showing the variation of pressure along the upper surface of a simulated aerofoil

The pressure coefficient $= \dfrac{p_1 - p_0}{\frac{1}{2}\rho v_0^2} = 1 - \left(\dfrac{y_0}{y_1}\right)^2$

where p_1 = pressure at first point
p_0 = pressure at reference point
ρ = density of fluid
v_0 = original velocity at reference point
y_0 = separation of flow lines at reference position
y_1 = separation of flow lines at first chosen position
s = distance along the axis between reference position and position under investigation

When the paint is dry, six (or more) wires are attached to each of the electrodes by a blob of silver paint and connected as indicated in *Figure E60* to either side of the potentiometer arrangement of the field plotter.

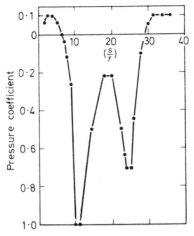

Figure E59. Graph showing the variation of pressure along the lower surface of a simulated aerofoil

s = distance along axis from reference position
y = separation of stream lines at the point under investigation

$$\text{Pressure coefficient} = \left(\frac{p_1 - p_0}{\frac{1}{2}\rho v_0}\right)$$

where p_1 = pressure at first point
p_0 = pressure at reference point
ρ = density of the fluid
v_0 = velocity of flow at reference position

Several different settings (R_1/R_2) of the 'ratio' resistances are made by the control R on the instrument and for each given ratio, the probe is adjusted to give zero deflection on the galvanometer G.

The resulting plot of streamline flow is shown in *Figure E56*. Note that well away from the 'aerofoil' the flow lines are taken in ratio steps of 0·1 but at much smaller intervals near to it.

First choose an initial reference point well away from the disturbance of flow lines, due to the obstruction and note the separation of lines y_0.

Then at convenient distance s from this reference point measure the separation of the flow lines y_1. Tabulate s and y_1.

Repeat for many different values of s.

APPLIED HEAT, MECHANICS OF FLUIDS AND SOLIDS

Calculate the ratio y_0/y_1 for each case $(=v_1/v_0)$ and also the pressure coefficient $= \{1 - (y_0/y_1)\}^2$. Plot v_1/v_0 against s/y_0 and also the pressure coefficient against s/y_0.

Repeat for both the upper and lower edges of the aerofoil.

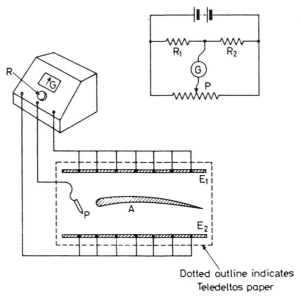

Dotted outline indicates Teledeltos paper

Figure E60. Basic Servomex field plotter circuit
G, Galvanometer with centre zero   P, Probe
R_1 and R_2, Ratio arm resistances of potentiometer
E_1 and E_2, Reference electrodes in silver paint on 'Teledeltos' paper
A, 'Aerofoil' in silver conducting paint

Discussion of Results

Note that the graph of *Figure E58* shows a large 'negative' pressure on the upper surface of the aerofoil which gives the 'lift' associated with the structure.

The distribution of pressure will of course vary with the shape of the chosen aerofoil and its orientation in relation to the original lines of flow.

By summing algebraically the areas under the curves associated with the upper and lower surfaces of the aerofoil, it is possible to estimate the resultant 'lift'.

In the example given, notice the rapid increase of velocity of flow after passing over the leading edge and after remaining fairly constant over the middle region, it drops rapidly on approaching the trailing edge (due to diffusion). In an actual experiment fluid 'eddies' would tend to form (*Figure E61*) and it should be noted that this is confirmed by the graph of pressure variation along the upper surface.

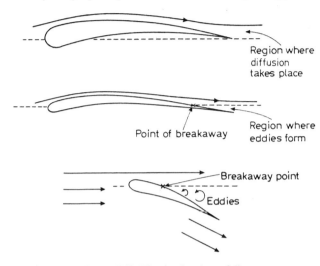

Figure E61. The simulated aerofoil

The 'breakaway' of course tends to destroy the lift and this is why flaps are used on the aircraft's trailing edge—the air being diverted through a large angle to maintain the lift (*Figure E62*). Notice

Figure E62. The effect of a flap

that to obtain more accurate graphs it is simply necessary to have a larger distance between the electrodes and a larger aerofoil.

Further Work

If the areas of the pressure variation graphs given as examples are added it will be seen that the resulting lift tends to zero (or is even slightly negative) due to the slight negative angle of incidence of the lines of flow.

APPLIED HEAT, MECHANICS OF FLUIDS AND SOLIDS

The experiment should be repeated with the aerofoil at other orientations.

It is also possible to investigate (*a*) the streamlines that occur about a cylinder and (*b*) the adverse effects associated with an abrupt constriction.

One should compare and contrast similarities of this experiment with that of plotting equipotential lines (Experiment No. A1.11).

Reading and References

The aerofoil and its applications are dealt with in *Hydraulics* by E. H. Lewitt (Pitman, 10th edition, 1958).

Detailed treatment of different types of aerofoils may be found in *Applied Aerodynamics* by L. Bairstow (Longmans Green, 1961).

Experiment No. E2.7

TO DETERMINE THE SPEED OF A VACUUM PUMP AND TO VERIFY GAEDE'S EQUATION

Introduction

In the production of low pressure in the laboratory it is important to know the 'speed' of the vacuum pump itself and of the pumping system as a whole, i.e. what volume of gas is extracted in unit time from the apparatus which is to be exhausted. The speed of the complete pumping system depends not only upon that of the pump but also upon the geometry of the connecting leads, etc. since the narrowness of the tubes and the presence of sharp bends will seriously reduce the rate at which gas may be removed.

Theory

Gaede's equation provides a relationship between the speed of the pump and the pressure in the vacuum system.

The speed may be defined as the rate at which gas is removed from the vessel to be evacuated.

Let V = volume of the vessel
        P = pressure in the vessel at time t,

SPEED OF A VACUUM PUMP

then assuming that Boyle's law applies we may write

$$P(V - \delta V) = (P - \delta P)V \quad \text{or} \quad P\left(V - \frac{\partial V}{\partial t}\delta t\right) = \left(P - \frac{\partial P}{\partial t}\delta t\right)V$$

whence

$$PV - P\frac{\partial V}{\partial t}\delta t = PV - V\frac{\partial P}{\partial t}\delta t \quad \text{and} \quad \frac{\partial P}{\partial t} = \frac{P}{V}\frac{\partial V}{\partial t}$$

but $\partial V/\partial t =$ speed S of the pump

i.e. $\quad \int_{P_1}^{P_2} \frac{\partial P}{P} = \int_{t_1}^{t_2} \frac{S}{V} \partial t \quad \text{and} \quad \log_e\left(\frac{P_2}{P_1}\right) = \frac{S}{V}(t_2 - t_1).$

Assuming P_1 is the initial pressure in the vessel we can take it as constant and a graph of log P_2 against time t should give a straight line of slope S/V so that knowing the volume V of the vessel being evacuated, S may be determined.

Method

The pump is connected to a large vessel of three or more litres capacity (*Figure E63*) and the manometer levels noted every 15

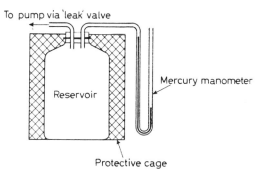

Figure E63. Measuring the 'speed' of a rotary vacuum pump

seconds or so as the pump removes air from the system (*Figure E64*). If a glass bottle is used as the reservoir then it should be protected to prevent accident in case of implosion. Tabulate the results as shown in *Table E2* and plot a graph of log P_2 against time (*Figure E65*). Note the atmospheric pressure on the Fortin barometer at the time of the experiment.

APPLIED HEAT, MECHANICS OF FLUIDS AND SOLIDS

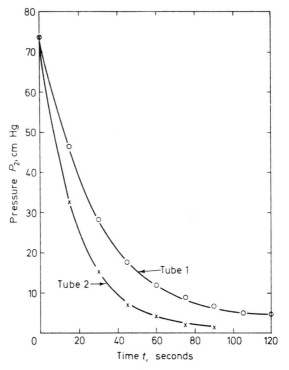

Figure E64. Graph of pressure against time

Table E2

Time t sec	Manometer limb readings cm		Pressure cm of mercury	$\log_e P$
	A	B		
0	39·5	39·5	73·9	4·30
15	53·2	25·8	46·5	3·84
30	62·4	16·6	28·1	3·34
45	67·7	11·3	17·5	2·86
60	70·6	8·4	11·7	2·46
75	72·3	6·7	8·3	2·12
90	73·4	5·6	6·1	1·81
105	74·1	4·9	4·7	1·55
120	74·5	4·5	3·9	1·36

A 'leak' tap should be provided so that the pump is not left at the end of the experiment in a condition where oil may be 'blown back' through it due to atmospheric pressure on one side only.

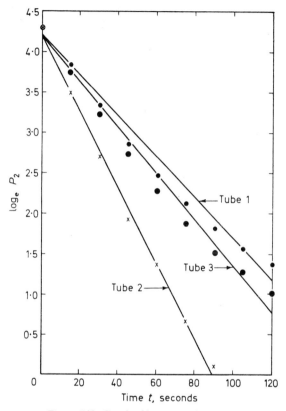

Figure E65. Graph of $\log_e P_2$ against time

Make sure at the end of the experiment that the pressure is *slowly* equalized on the input and outlet sides of the pump.

Discussion of Results

The accuracy of the result is seriously impaired by the speed with which the levels of water in the manometer change. The result will be more accurate when the speed is reduced by the inclusion of narrow bore connecting tubing.

APPLIED HEAT, MECHANICS OF FLUIDS AND SOLIDS

Further Work

In contrast to the above, the aim of workers using vacuum systems is to make all tubes as wide and short as possible in order to approach the optimum speed of which the pump is capable. Investigate the effect of (a) using connecting tubes of various bores in the system, (b) tubes of various lengths and (c) try also the effect of including sharp angular bends or tubes containing other abruptly changing geometry.

Reading and References

Information concerning apparatus for the production and measurement of vacua may be found in *Principles of Vacuum Engineering* by M. Pirani and J. Yarwood (Chapman and Hall, 1961).

Experiment No. E2.8

TO DETERMINE THE PUMPING SPEED OF A VACUUM SYSTEM BY THE STEADY STATE METHOD

Introduction

This method of determining the speed of a pump may be found preferable to the one depending upon Gaede's equation where the volume of the reservoir, or system to be evacuated, must be known.

Theory

Assuming the time taken for the steady state condition to be achieved is t seconds, the head h of liquid in the manometer (*Figure E66*) represents a volume of gas entering the manometer limb of (hA) cm^3 where A represents the cross sectional area of the manometer tubing.

We may define the throughput of the system as the volume of gas removed from the system per unit time *at an ultimate* pressure p_u of mercury.

PUMPING SPEED OF VACUUM SYSTEM E2.8

If P represents the atmospheric pressure we have

$$\text{Throughput} = \frac{PAh}{10^3 t} \text{ litre-torr-second}^{-1}$$

where 1 torr is equivalent to a pressure of 1 mm of mercury.
Again if S represents the pumping speed at pressure p, then

$$S(p - p_u) = \frac{PAh}{1000t}$$

so that plotting h values as ordinates and pressure on the x axis

$$h = \left(\frac{10^3 tS}{PA}\right)p - \left(\frac{10^3 tS}{PA}\right)p_u$$

The slope of the graph gives the speed

$$S = (\text{slope})\left\{\frac{PA}{10^3 t}\right\}$$

and the negative intercept enables a check to be made on the value of S since p_u is known.

If $p \gg p_u$ so that p_u can be neglected

$$Sp = \frac{PAh}{1000t}.$$

Method

The apparatus should be connected up as shown in *Figure E66*, and both the tap T and the needle valve closed for the start of the experiment. Note the atmospheric pressure on the Fortin barometer.

Switch on the rotary pump and continue pumping until the head of liquid in the manometer indicates that a steady state has been achieved. Then read the head of liquid corresponding to the steady state pressure.

Tabulate the pressure of the system in millimetres of mercury as registered on the vacustat (or other suitable gauge).

Open first the tap and then the needle valve so that after a time a new steady state pressure p is achieved. When this is so, close the tap, note the time for which pumping has taken place and the value of the new head of liquid.

Repeat the procedure for various values of p. Find the cross sectional area of the manometer tubing. Plot h against the pressure p in the system, as shown in *Figure E67*. A linear graph should result, the slope of which enables the speed of the pump to be found.

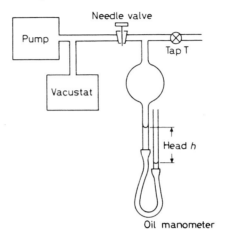

Figure E66. Apparatus to determine pumping speed using the steady state method

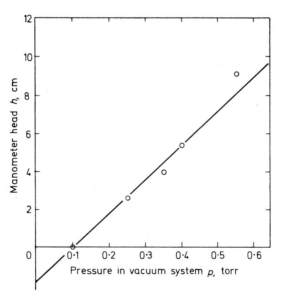

Figure E67. Determination of the pumping speed in a vacuum system using the steady state method

Discussion of Results

Note that this method of presentation enables a check to be made of the value for the speed obtained from the slope of the graph by means of that derived from considering the intercept. The scatter of the points obtained about the 'best fit' line should also help in drawing conclusions about the accuracy of the method.

Further Work

The experiment should be extended so that measurements are made using tubes of uniform bore but different lengths (see *Figure E68*) to confirm that the conductance

$$C = \left(\frac{\text{Throughput}}{\text{Pressure difference across the tube}} \right)$$

is inversely proportional to the length of the tube.

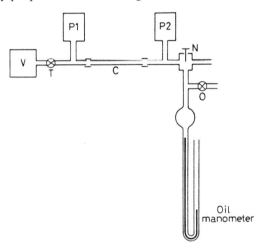

Figure E68. Apparatus used to measure the conductance along pipes of different bores in a vacuum system
P1 and P2, Pirani gauges
V, Rotary vacuum pump   O and T, Taps
C, Tube inserted for conductance investigation
N, Needle valve

Reading and References

Background reading on pumping speed and throughput may be found in *High Vacuum Engineering* by A. E. Barrington (Prentice-Hall, 1963).

The methods of measuring pumping speed may be found in *Vacuum Technique* by A. L. Reimann (Chapman and Hall, 1952).

Experiment No. E2.9

TO INVESTIGATE THE VARIATION OF PUMPING SPEED OF A MERCURY-VAPOUR DIFFUSION PUMP WITH THE PRESSURE IN A VACUUM SYSTEM

Introduction

The speed of a vapour pump depends not only upon the pressure at the jet of the diffusion pump itself but also upon the effectiveness of the 'backing' rotary pump.

The aim of this experiment is to plot the characteristic graph showing how the pumping speed in a vacuum system varies with the pressure within it.

Theory

Assuming that the backing pump is working efficiently, the diffusion pump reduces the pressure in the system and thus decreases the number of gas molecules entering the stream of vapour in the diffusion pump itself. This reduces the back-diffusion of gas molecules and thus the pumping speed rises with the decreasing pressure (*Figure E69*).

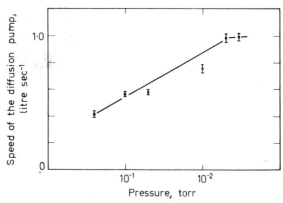

Figure E69. Variation of speed of a diffusion pump with pressure

MERCURY-VAPOUR DIFFUSION PUMP E2.9

Then theoretically the ultimate pressure is determined by the vapour pressure of the working fluid in the pump but in practice it is also a function of the design of the pump.

In Experiment No. E2.8 the theory showed that the speed S was given by

$$S = \frac{Ah}{1000t} \frac{P}{p}$$

i.e.

$$S \propto h/p$$

where A = cross sectional area of manometer tube
P = atmospheric pressure
p = the pressure in the system
t = time
h = head of liquid in the manometer.

Since the constants A, P and t are known in this experiment, measurement of h and p enables the pumping speed at any chosen pressure p to be calculated.

Method

The system is connected up as shown in *Figure E70* using a rotary pump to 'back' the mercury diffusion pump and a Pirani gauge to measure the pressure in the reservoir of the system.

The needle valve N and the tap T are closed and the vacuum pump switched on to 'rough out' the system to about 10^{-2} torr. Then (and only then) the diffusion pump heaters may be switched on and also the water supply for cooling the pump. By means of a suitable 'Variac', the power to the pump is increased to its recommended maximum and pumping continued until the Pirani gauge reading is constant. The heater current and voltage applied to the pump are then tabulated.

The tap T on the oil manometer is then opened fully and the needle valve just sufficiently to give a significant change in pressure (see the graph of *Figure E69*) when 'steady' conditions are again obtained. When stability has been achieved the tap T is closed and after the necessary time required for any slight fluctuations in the meniscus to cease, the head of oil in the manometer noted, as well as the new pressure reading on the Pirani gauge.

Using the equation given in the theory the speed at this particular pressure is calculated (atmospheric pressure is taken from the Fortin barometer).

The procedure is repeated for a number of different pressures corresponding to different rates of leakage through the needle valve and a graph plotted of pumping speed against pressure (*Figure E69*).

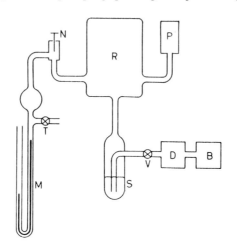

Figure E70. Diagram of the experimental arrangement
B, Rotary backing pump   T and V, Valves
D, Diffusion pump   N, Needle valve
R, Reservoir of system   M, Oil manometer
P, Pirani gauge   S, Vapour trap

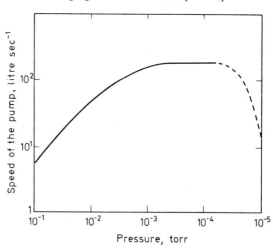

Figure E71. Graph showing the general form of the variation of pumping speed with pressure

Discussion of Results

The graph shows clearly the value of the optimum speed obtainable in the system. At the higher pressure region of the graph the gas leaking into the system is more than the pump can efficiently handle and the speed therefore falls fairly rapidly (*Figure E69*).

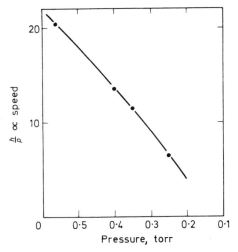

Figure E72. Variation of the speed of a rotary pump with change in pressure
h = movement of head in oil manometer in 1 minute
p = pressure in system

The graph obtained should be compared with those in standard texts (e.g. the first reference given below). Whether the fall in speed that occurs at the low pressure end of the graph (*Figure E71*) is obtained depends upon the length of time one can afford to allow the system to pump (1) before taking observation, and (2) between different sets of readings, as well as upon (3) the accuracy to which the non-linear scale of the Pirani gauge can be read in the lower ranges.

Note that the speed of the pump increases approximately linearly with decrease in pressure until about 5×10^{-3} torr, after which, owing to the vapour pressure of the mercury in the pump, the value becomes almost constant in the region of 10^{-3} torr. To obtain lower pressure a liquid nitrogen trap would have to be used.

In the pressure region above 0·1 torr the overall pumping speed is probably due to the rotary pump which is pumping weakly at its lowest pressure limit and also to the diffusion pump which at this pressure is only beginning to operate.

Further Work

Using this same steady state method it is possible also to plot the characteristic speed/pressure curve for a rotary backing pump alone (*Figure E72*). Details for particular pumps will be provided in the maker's literature (or in the references given below). An investigation of the effect of change of backing pressure upon the efficiency of a diffusion pump is also worthwhile.

Reading and References

Background reading upon the basic concepts of vacuum technology and upon pumps and vacuum systems may be found in *Introduction to the Theory and Practice of High Vacuum Technology* by L. Ward and J. P. Bunn (Butterworths, 1967).

A useful introductory text for those using vacuum equipment is *Vacuum Technology* by W. S. Spinks (Chapman and Hall, 1966).

Experiment No. E2.10

TO INVESTIGATE THE EFFECT UPON THE PUMPING SPEED OF A MERCURY-VAPOUR DIFFUSION PUMP, OF CHANGES IN THE SUPPLY OF POWER TO THE HEATER OF THE DIFFUSION PUMP

Introduction

The 'speed' of a diffusion pump is a function of the rate of diffusion of the molecules of the vapour of the working fluid and thus is critically dependent upon the power in watts supplied to the heater of the pump. The aim of this experiment is to plot a graph showing the way in which the speed varies as the heater power increases.

Theory

As the power to the vapour pump is increased the rate of production of vapour molecules increases and one would expect the speed of the pump to rise (*Figure E73*).

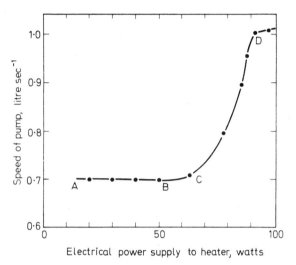

Figure E73. Graph showing the variation of pumping speed of a mercury diffusion pump with change in power supply to the pump heater

There comes a point, however, when the inlet jet of the pump can allow no greater passage of molecules and therefore further increase in the power to the heater cannot increase the speed so that the graph 'levels off' to a constant value corresponding to the maximum speed of the pump.

Method

The apparatus is set up as in the previous experiment. The needle valve and the tap T are closed and the whole system pumped down for some time (as before) using the backing pump alone.

The cooling water for the diffusion pump and the heater power supply are then switched on, the latter being set to the recommended maximum with the particular system being used.

When a steady state has been achieved in the system the ultimate pressure reading on the Pirani gauge is noted. The needle valve is then slightly opened and the tap T fully opened, pumping being continued until the new steady state condition is reached. The tap is closed and when the meniscus is again steady the manometer head and the gauge reading are taken. The speed is then calculated (h/p) as in the previous experiment and at the same time heater voltage and current should be tabulated. Using the Variac the power is reduced by say 20 W and the calculation of the speed repeated after the new steady state conditions have been achieved.

In this way a number of points are obtained to plot a graph such as shown in *Figure E73*.

Discussion of Results

The lower measurable limit on the graph (*Figure E73*, region A B) corresponds to the speed of which the backing pump alone is capable. It is seen that, as the theory predicted, the speed rises rapidly (region CD) as the power to the heater increases, finally levelling off at a maximum value.

Further Work

If possible, dismantle and examine carefully both a typical diffusion pump and a rotary pump. (Not the ones in the system being used—others will be provided.)

It is also worthwhile to carry out Experiment Nos. E2.8 and E2.9 with gases other than air to note the effect upon the pumping speed.

Reading and References

High Vacuum Technology by L. Ward and J. P. Bunn (Butterworths, 1967).

Scientific Foundations of Vacuum Technique by S. Dushman and S. M. Lafferty (Wiley, 2nd edition, 1964).

Experiment No. E2.11

TO MEASURE A LOW PRESSURE USING THE McLEOD GAUGE

Introduction

The McLeod gauge depends for its operation on isolating a known volume of gas at the pressure of the evacuated system and compressing it into a small known volume when the pressure may be

MEASURING A LOW PRESSURE

directly measured. It therefore relies upon the basic truth of Boyle's law.

Theory

The mercury reservoir shown as R in *Figure E74* is slowly raised until a volume of gas V at pressure P of the gas in the evacuated system is entrapped in the bulb of the instrument.

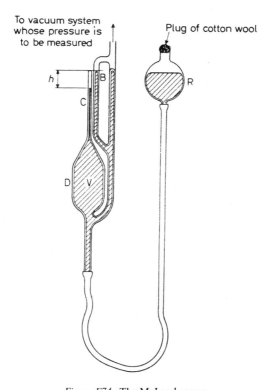

Figure E74. The McLeod gauge
V = volume of bulb and capillary
Ah = volume of entrapped gas when mercury in the limb B is level with the top of the closed capillary, i.e. head of mercury = h
and A = cross-sectional area of the capillary

On further raising the reservoir this gas is compressed into the capillary tube above the bulb and adjustment is made until the head of mercury stands level with the top of the capillary containing the gas. The head of mercury h is then noted and by Boyle's law

$$PV = (p + h)Ah = pAh + Ah^2$$

i.e. $\qquad p(V - Ah) = Ah^2$

and $\qquad P = \dfrac{Ah^2}{(V - Ah)}$

whence since Ah may be neglected compared with V

$$P \simeq \frac{Ah^2}{V}$$

(A = cross sectional area of capillary into which gas is compressed).

Method

Pump air from the system [a large (20 litre) bottle with suitable protection will do] by means of a rotary pump as indicated in *Figure E75*. Close the tap T and open the tap S thus connecting the

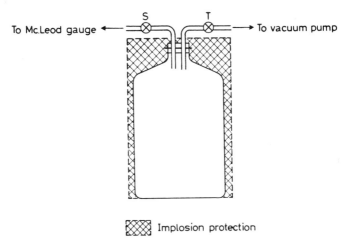

Figure E75. Calibration of the McLeod gauge

bottle directly with the McLeod gauge. Make sure first of all that the reservoir attached to the gauge is as low as is conveniently possible, and raise it slowly to entrap the air in the bulb D. (If the bulb is raised too rapidly air may be sucked back into the bottle, so take care.) Bring up the reservoir further until the mercury meniscus is level with the top of the capillary B and note the head of mercury h as indicated in *Figure E74*. Knowing the volume V of the bulb, the pressure P of the air in the bottle may now be found. Repeat the procedure for several different pressures as the pump exhausts the reservoir in small stages.

Discussion of Results

These should include comment upon the effective range of the instrument and in estimating its accuracy do not forget that a correction is necessary for the rounded head of the capillary and the limiting factor of accuracy of alignment in the two limbs. Cleanliness is also of great importance in measurements of low pressure, and dirt in the gauge can give a result which is far from true. Careful thought on the basic theory of the instrument should enable one to understand why vapour condensing traps should be used if the gauge is to be accurately calibrated.

Important Note. Before readmitting air to the bottle which has been evacuated, the reservoir must once again be LOWERED to its original position.

Further Work

With regard to the inclusion of condensing traps in the vacuum system (referred to above) it is worthwhile to operate the gauge (*a*) with the trap filled with liquid air (or nitrogen) and then (*b*) with the trap empty, noting the effect upon the readings. When in general operation, however, a liquid air trap should be used to obviate the danger of contamination by mercury vapour.

If facilities permit it is also advantageous to make measurements using systems containing different gases. Thorough cleaning and flushing out of the system is required if the gas in the one system is to be changed.

A modification of the McLeod gauge is produced commercially and is known as the 'Vacustat' (*Figure E76*). By tilting the whole arrangement on the swivel joint behind the plate of the instrument, mercury can be caused to flow into the U-tube system until, on being turned upright again, the level just touches the fixed mark M on the middle stem. The scale is then calibrated to read the pressure directly in millimetres of mercury.

The device is of use for pressures from about 10 torr down to

10^{-3} torr and it is worthwhile to become familiar with its use for rapid measurements in this range.

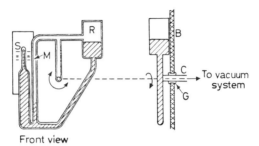

Figure E76. The Vacustat
S, Calibrated scale
R, Adjustable mercury reservoir
B, Rotatable back support to Vacustat
C, Tube for connection to the vacuum system
G, Swivelling grommet (see arrows)
M, Fixed mark on stem

Reading and References

Further reading upon the McLeod gauge and its use as well as upon the Vacustat may be found in *Introduction to the Theory and Practice of High Vacuum Technology* by L. Ward and J. P. Bunn (Butterworths, 1967). The limitations of both instruments are clearly dealt with in this book.

The cleaning and calibration of a new gauge are dealt with in detail in *Vacuum Technology* by A. Guthrie (Wiley, 1963).

Calculations on gauge parameters are given and 'square' scale and linear scale methods of pressure measurement are considered.

Experiment No. E2.12

TO INVESTIGATE THE PRINCIPLES OF OPERATION OF THE PIRANI GAUGE

Introduction

The rate of heat transfer through a gas is obviously a function of the number of molecules of gas present at the time of measurement and hence of the pressure of the gas itself.

OPERATION OF PIRANI GAUGE

This fact is used in the measurement of low pressures using the Pirani gauge.

In this experiment the principles of the gauge are investigated using a tungsten filament enclosed in a glass envelope which may be evacuated.

Theory

The gauge thus consists of a fine wire of high temperature coefficient of resistance which is enclosed in the vacuum system to be measured.

Let T_1 = initial temperature of gas molecules on striking the wire
T_2 = final temperature of gas molecules leaving the proximity of the wire. Then energy E from the wire per second is given by

$$E = a(T_1 - T_2)n$$

where a is a constant, and n represents the number of molecules of gas striking the wire per second.

Now the number of molecules present must be directly proportional to the pressure of the gas at given temperature, and hence

$$E = k(T_1 - T_2)p$$

where k is a constant.

The electrical heat input to the wire is proportional to V^2/R where V = voltage, R = resistance, and is consumed in three ways: (1) by radiation, (2) by conduction, (3) by loss to filament supports.

At very low pressures the conduction is negligible compared with the radiation, hence

$$\frac{V^2}{R} = b(T_1^4 - T_2^4) + \text{'end losses'}$$

At pressure p the conduction cannot be neglected, and hence

$$\frac{V_1^2}{R} = b(T_1^4 - T_2^4) + kp(T_1 - T_2) + \text{end losses.}$$

V_1 = the new value of voltage required to maintain temperature conditions as before.

Assuming the temperature of the wire is constant

$$\frac{V_1^2 - V_0^2}{R} = k(T_1 - T_2)p$$

or p = constant $(V_1^2 - V_0^2)$.

This is strictly true only over limited range, as is demonstrated by the graph of voltage squared against pressure (*Figure E77*).

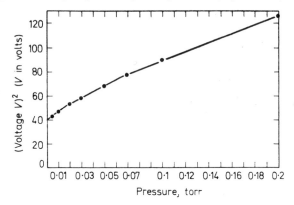

Figure E77. Graph of (voltage)² against pressure for a Pirani gauge

There are thus two ways in which changes in pressure may be measured.

(1) By observing the increase in resistance of the filament which takes place as the pressure decreases.

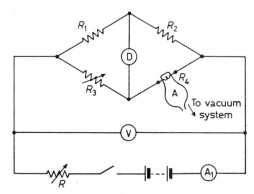

Figure E78. Wheatstone bridge circuit arrangement for a Pirani gauge

R_1 and R_2, Bridge ratio arms A_1, Ammeter
R_3, Variable resistance arm   V, Voltmeter
R_4, Gauge resistance   R, Variable resistance
D, Centre zero galvanometer
A, Tungsten filament in glass envelope connected to the vacuum system

OPERATION OF PIRANI GAUGE   E2.12

(2) By measuring the power required to keep the filament temperature (and hence its resistance) constant. This may be done by plotting the square of the voltage across the filament against the pressure as shown above.

Method

Place the tungsten filament, enclosed in its glass envelope, in one arm of a Wheatstone bridge (*Figure E78*) so that the envelope may be evacuated by a suitable vacuum system arrangement (*Figure E79*).

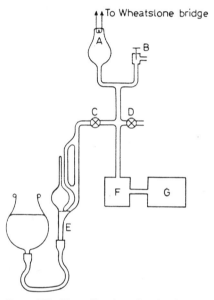

Figure E79. The calibration of a Pirani gauge
A, Tungsten filament in evacuated envelope
B, Needle valve   C and D, Taps
E, McLeod gauge   F, Diffusion pump
G, Rotary backing pump

The needle valve B (*Figure E79*) is closed and the system pumped down by rotary pump. When a steady state (due to the one pump) is achieved, first the water supply to cool the mercury vapour pump is switched on and then the recommended power to the diffusion pump heater. Pumping is continued until the ultimate pressure remains steady.

Voltage is applied across the Wheatstone bridge arrangement (*Figure E78*) and a balance point found. The values of the variable resistances R_1 and R_2 are then noted.

The resistance of the tungsten filament is calculated

$$\left[\left\{ \frac{R_3 \times R_2}{R_1} \right\} \text{ in } Figure\ E78 \right].$$

The needle valve is then opened just sufficiently to give a significant change in the ultimate steady pressure attainable and the power to the heater, the pressure and the filament resistance noted, as

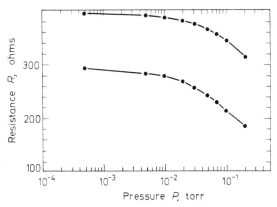

Figure E80. Graph showing variation of filament resistance with pressure for a Pirani gauge

before. Repeating the process for a number of different pressures (i.e. needle valve settings) a graph of $\log p$ against filament resistance for the selected value of bridge voltage is plotted. A second operational value of voltage is selected and the procedure repeated (*Figure E80*).

The alternative method of procedure is to maintain the filament resistance (and therefore the filament temperature) constant by varying the voltage across the bridge. When the ultimate pressure is reached the bridge may again be balanced (at 6 V) and the filament resistance noted.

Pressure is then increased as before and bridge balance restored (when steady state conditions are achieved) by raising the voltage across it. Graphs of (voltage)^2 against p, and (voltage)^2 against $\log p$ may then be plotted (*Figures E77* and *E81*).

Discussion of Results

From the resistance/pressure graphs it should be possible to assess the correct voltage to be applied in order to obtain optimum accuracy (e.g. in the 6 V curve a greater change in resistance is required to achieve balance over the range of pressures used, than is the case with the 12 V curve, thus yielding greater accuracy in measurement).

Note that the (voltage)^2 against pressure graph obtained in the second alternative way shows clearly the ranges over which the

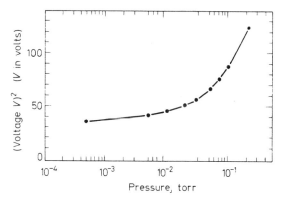

Figure E81. Graph of (voltage)^2 against pressure (on a log scale) for a Pirani gauge

theory (predicting linearity) may be strictly applied. The advantage of the (voltage)^2 plot against log p is that a greater range may be conveniently covered upon one graph without the disadvantage of the 'clustering' of points towards one end of the scale.

Note the useful (linear) range of the gauge. The lower limit of operation is imposed when radiation effects become of the same order of magnitude as those due to conduction, since the radiation, as is seen from the theory, is then not dependent upon pressure.

The upper limit marks the point beyond which, as noted in the theory, the conduction effects are no longer linearly dependent upon pressure. The typical useful range of an arrangement of this type is from 10^{-1} to about 6×10^{-3} torr.

The method employing the measurement of change of resistance is found to be the more accurate one since its accuracy can be improved to any desired degree, by suitable adjustment of the ratio arms of the bridge. It is also found fairly easy to calibrate.

Further Work

Another convenient method of calibration, which has the advantage of enabling continuous readings to be made, is to balance the bridge with the needle valve closed, then to note the 'out of balance' current through the galvanometer when the pressure has been increased by opening the leak valve for a short interval. (Allow time for stabilization before reading.) A typical graph obtained in this way is shown in *Figure E82*.

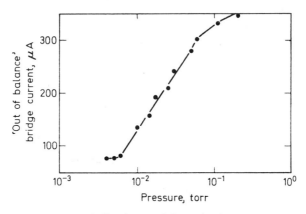

Figure E82. Calibration graph for a Pirani type gauge

If facilities permit, the calibration of a commercial Pirani gauge should also be made to compare its performance with the crude arrangement used here.

Reading and References

The theory of the Pirani gauge may be found in standard texts, such as *A Degree Physics*, Part I: *Properties of Matter* by C. J. Smith (Arnold, 2nd edition, 1960). Or in *Pressure Measurement in Vacuum Systems* by J. H. Leck (Chapman and Hall, 2nd edition, 1957).

Experiment No. E2.13

TO PREPARE A THIN FILM OF SILVER BY EVAPORATION IN VACUUM

Introduction

Of recent years research into the optical properties of thin films, particularly for interferometric and electron microscope studies, has been extensive.

This has been made possible by the refinements in vacuum technique.

Theory

No method of thin film deposition allows such complete measure of control as that of deposition by thermal evaporation in vacuum.

Distribution of film thickness can be controlled by the use of shutters, diaphragms and movement of the target during deposition.

The factors to be considered are:

(1) The nature and pressure of the residual gas in the chamber. The gas in the chamber must be compatible with the material to be evaporated onto the target surface. (For instance aluminium films break down if a mercury diffusion pump is used in the evacuation of the chamber due to interaction with the residual mercury atoms.)

The pressure must be such that the mean free path of the volatilized atom must be much greater than the source to target distance. *Table E3* gives the orders of magnitude of mean free path corresponding to various pressures and it may be seen that a pressure below 10^{-3} torr is generally required. (In more accurate work pressures of the order of 10^{-5} torr or less are used.)

Table E3. Variation in the orders of magnitude of the mean free path with change in pressure (for air)

Mercury pressure torr	Mean free path cm
10^{-2}	5×10^{-1}
10^{-3}	5
10^{-4}	5×10^{1}
10^{-5}	5×10^{2}
10^{-6}	5×10^{3}

1 torr = 1 mm of mercury pressure.

(2) The intensity of the stream of atoms which condenses on the target surface. Careful experiments show that a critical beam density exists below which no condensation of atoms occurs on the target.

(3) The nature and condition of the target surface. The surface must be clean and free from grease. Glass forms a suitable smooth target. The temperature of the substrate is also critical. Above certain temperatures films are only produced with great difficulty and with very high beam intensities.

(4) The velocity of the impinging atoms, i.e. the temperature of evaporation of the atoms. At low rates of evaporation, films displaying little or no crystalline structure are obtained, but at high rates of evaporation the films may be found to consist of large crystals.

(5) Contamination—the source material should be chosen so as not to form a solid solution with the material of the 'boat' or container at the high temperatures involved, otherwise target contamination results and the resulting thin film is not pure.

Method

The glass dome (*Figure E83*) is fitted with a neoprene sealing ring and the chamber can be evacuated by means of a conventional system of rotary and mercury diffusion pumps.

The glass microscope slide which is to form the substrate for the thin silver film is first thoroughly cleaned in strong nitric acid, washed in trichloroethylene and later industrial methylated spirits then after drying, placed in the clamp H (*Figure E83*). A small piece of tungsten foil is pressed into the form of a boat and screwed between the current-carrying leads of the apparatus, then a piece of the silver wire to be evaporated is placed in the boat.

The dome is placed firmly in position and the protective cover put over it.

The rotary pump is switched on and allowed to pump alone for 10 minutes or so, after which time the cooling water supply to the mercury diffusion pump is turned on and the recommended power supplied to the pump heaters.

When steady state conditions of 10^{-3} torr or less are achieved, the current to the 'boat' leads is switched on and slowly increased until the boat and its contents glow brightly (up to 30 A may be required!).

The silver is seen to condense on the slide, although this can be difficult because silver also deposits itself on the inside of the domed chamber.

The pumping system is then isolated from the evaporation chamber and the needle valve connected to the chamber is opened so that air slowly re-enters.

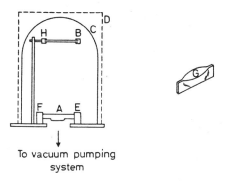

To vacuum pumping system

Figure E83. Apparatus for vacuum deposition of thin films
A, Boat containing silver wire
B, Target (microscope slide)
C, Domed evacuated chamber
D, Protective cover against implosion
E and F, Heavy current-carrying leads
G, Diagram showing construction of boat
H, Clamp holding the target slide

The protective cover and dome are then lifted off and the microscope slide which is coated with silver on one side may be removed.

Using other slides and varying (1) the length of silver wire used, (2) the time of evaporation, (3) the boat heater current density, a number of films of varying thickness may be obtained.

Discussion of Results

It will probably be found that the most difficult part of the experiment is to produce a *clean* slide, and time spent in preparation here is well worth while, for otherwise discontinuities in the film and non-uniformity of thickness may result.

Do not be impatient over pumping time—a good vacuum is essential to the production of a good film.

Make sure that no vibration of the target occurs due to that of the pumping system, as this too can hinder the production of a good result.

Finally, the vacuum pumping system will be provided with a facility for equalizing pressures on both sides of the pump. Ensure that this is done at the end of the experiment, otherwise oil from the rotary pump may be blown back through the system (*Figure E84*).

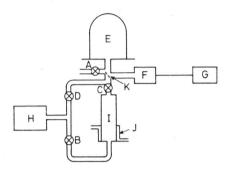

Figure E84. Block diagram showing arrangement of pumping system to vacuum deposition chamber

A, Valve to atmosphere
D, Bypass valve
F, Penning gauge
H, Rotary backing pump
J, Water cooling jacket

B and C, Isolation valves
E, Chamber to be evacuated
G, Recording meter for pressure
I, Diffusion pump
K, Baffle valve

Care should be taken not to use too high a current density or the boat may collapse.

Further Work

The thickness of the thin films made should be measured using 'Newton's wedge' method and it is instructive to determine the reflection coefficient of the films using a suitable photometer.

Reading and References

The preparation of vacuum evaporated films is dealt with in *Techniques for Electron Microscopy* edited by D. H. Kay (Blackwell, 2nd edition, 1965). This includes a table of evaporation data for a large selection of materials and discusses the cleaning of evaporation filaments.

A deeper treatise is *Vacuum Deposition of Thin Films* by L. Holland (Chapman and Hall, 1956).